Advances in Experimental Medicine and Biology

Innovations in Cancer Research and Regenerative Medicine

Volume 1432

Series Editor
Phuc Van Pham, Laboratory of Stem Cell Research and Application, University of Science, Vietnam National University, Ho Chi Minh City, Vietnam

Innovations in Cancer Research and Regenerative Medicine is based on a bi-annual conference and its topics, and represents a major contribution to the emerging science of cancer research and regenerative medicine. The series publishes review and original research contributions, short reports, conference proceedings, and guest-edited thematic volumes. Each volume brings together some of the most preeminent scientists working on cancer biology, cancer treatment, cancer diagnosis, cancer prevention and regenerative medicine to share information on currently ongoing work which will help shape future therapies.

These volumes are invaluable resources for active researchers or clinicians, those entering related fields, and professionals in industry. All contributions will be published online first and collected in book volumes. There are no publication costs.

Innovations in Cancer Research and Regenerative Medicine is a subseries of *Advances in Experimental Medicine and Biology*, which has been publishing significant contributions in the field for over 30 years and is indexed in Medline, Scopus, EMBASE, BIOSIS, Biological Abstracts, CSA, Biological Sciences and Living Resources (ASFA-1), and Biological Sciences.

Phuc Van Pham
Editor

Advances in Mesenchymal Stem Cells and Tissue Engineering

Volume 4

Springer

Editor
Phuc Van Pham
Stem Cells Institute
University of Science
Viet Nam National University Ho Chi Minh City
Ho Chi Minh City, Vietnam

ISSN 0065-2598 ISSN 2214-8019 (electronic)
Advances in Experimental Medicine and Biology
ISSN 2662-3285 ISSN 2662-3293 (electronic)
Innovations in Cancer Research and Regenerative Medicine
ISBN 978-3-031-38612-1 ISBN 978-3-031-38613-8 (eBook)
https://doi.org/10.1007/978-3-031-38613-8

This Springer imprint is published by the registered company Springer Nature Switzerland AG
The registered company address is: Gewerbestrasse 11, 6330 Cham, Switzerland

Contents

Part I

Advances in Mesenchymal Stem Cell Biology and Therapy

Adv Exp Med Biol - Innovations in Cancer Research and Regenerative Medicine (2023) 4: 3–18
https://doi.org/10.1007/5584_2021_672

Published online: 6 November 2021

Hypoxia, Serum Starvation, and TNF-α Can Modify the Immunomodulation Potency of Human Adipose-Derived Stem Cells

Binh Thanh Vu, Hanh Thi Le, Khanh Nha Nguyen, and Phuc Van Pham

Abstract

Introduction Adipose-derived stem cells (ADSCs) are mesenchymal stem cells (MSCs) that are found in adipose tissues, which are easily obtained from liposuction procedures using an enzyme mixture. The adhering cells are then selectively cultivated. ADSCs have great potential in regenerative medicine because they are plentiful, easily accessible, and less invasive. They also have an impressive proliferation ability and can be differentiated into mesenchymal lineages and trans-differentiating into many other cell types. In particular, they have extraordinary abilities in immunomodulation. This study aimed to investigate the effects of culture conditions (hypoxia, starvation, and TNF-α treatment) on the immunomodulation of human ADSCs.

Methods Human ADSCs were expanded in vitro in the standard condition before they were cultured in different stress conditions. ADSCs from passages fifth were confirmed as MSCs by some standard assays suggested by the International Society for Cell and Gene Therapy. These MSCs were used to culture in four different stress conditions: hypoxia, serum starvation, and TNF-α treatment in 48 h. After treatments, MSCs were used to evaluate their immunomodulation capacity using MSCs mixed lymphocyte reaction assay, and the concentrations of IDO, PGE2, IL-6, and IL-10 were secreted in the culture medium.

Results In different stress conditions, ADSCs exhibited different responses related to their immunomodulation. In serum starvation, ADSCs exerted a strong secretion of IDO

B. T. Vu
Laboratory of Stem Cell Research and Application, University of Science, Vietnam National University Ho Chi Minh City, Ho Chi Minh City, Vietnam

Vietnam National University Ho Chi Minh City, Ho Chi Minh City, Vietnam

H. T. Le and K. N. Nguyen
Vietnam National University Ho Chi Minh City, Ho Chi Minh City, Vietnam

Stem Cell Institute, University of Science, Vietnam National University Ho Chi Minh City, Ho Chi Minh City, Vietnam

P. Van Pham (✉)
Laboratory of Stem Cell Research and Application, University of Science, Vietnam National University Ho Chi Minh City, Ho Chi Minh City, Vietnam

Vietnam National University Ho Chi Minh City, Ho Chi Minh City, Vietnam

Stem Cell Institute, University of Science, Vietnam National University Ho Chi Minh City, Ho Chi Minh City, Vietnam
e-mail: phucpham@sci.edu.vn

and PGE2, whereas they showed strong IL-6 secretion in the TNF-α-supplemented medium. When exposed to lymphocytes, ADSCs caused an increase in the ratio of regulatory T cells (Tregs), and co-culture lymphocytes with ADSCs induced in hypoxic malnutrition conditions increased the IL-10 level the most. In addition, when exposed to dendritic cells (DCs), ADSCs inhibited the mature marker expressions of the DCs.

Conclusion The current research showed that ADSCs change their immunomodulation properties to survive in in vitro culture environments. Treatment of ADSCs in the starvation medium for 48 h can increase the immunomodulation of ADSCs.

Keywords

Adipose-derived stem cells · Hypoxia · Immunomodulation · Starvation medium · Stressed mesenchymal stem cells

1 Introduction

Adipose-derived stem cells (ADSCs) are a type of mesenchymal stem cells (MSCs) that are known to have great potential. Firstly, they are a type of adult stem cell that should be a source of autologous stem cells with a high degree of individualization. In addition, they are plentiful, easily acquired, and minimally invasive as an available source of tissue. Secondly, they can continually divide and maintain their long-term "stem" characteristics. Moreover, they are differentiated into mesenchymal cell lineages and non-mesenchymal cell lineages such as liver cells (endodermal) and nerve cells (ectodermal) (Bacakova et al., 2018; Dai et al., 2016; Si et al., 2019). Finally, they can modulate immunity and enhance angiogenesis. Their immunomodulation is related to both their surface proteins and secretomes, such as indoleamine 2,3-dioxygenase (IDO), prostaglandin E2 (PGE2), interleukin-6 (IL-6), and interleukin-10 (IL-10) (Ceccarelli et al., 2020; Al-Ghadban & Bunnell, 2020).

Despite the vast capabilities of ADSCs, the application results for their treatment are not as high as expected. This is because the microenvironment of damaged tissues is harsh to grafted cells; the damage to blood vessels temporarily stops the supply of oxygen and nutrients and causes the stagnation of inflammatory factors (Elsässer et al., 2001). This results in a huge amount of grafted ADSCs dying very quickly due to failure to keep up with the change from the in vitro culture to the very harsh microenvironment (Gyöngyösi et al., 2008; Toma et al., 2002; McGinley et al., 2013). Therefore, more effective strategies are needed to allow the number of transplanted ADSCs to survive so that they can fulfill their therapeutic function.

One of the appropriate approaches is to familiarize ADSCs with specific environmental changes, from which they can adapt to new conditions through changes in cell characteristics, including secretory mechanisms (Hu et al., 2008; Wang et al., 2014; Herrmann et al., 2010). This process is called preconditioning or licensing MSCs. By using some stress conditions, the biological characteristics of ADSCs can be enhanced or reduced. Some recent publications showed that hypoxia, serum starvation, and treatment with some inflammatory factors could modify the immunomodulation of ADSCs (Roemeling-Van Rhijn et al., 2013; Abdolmohammadi et al., 2020; Lee et al., 2010).

In 2013, Roemeling-Van Rhijn et al. showed that the immunomodulation of ADSCs was maintained when they were cultured in 1% or 20% oxygen (Roemeling-Van Rhijn et al., 2013). In other studies, hypoxia could trigger the expression of anti-apoptotic proteins that enhanced their survival (Bader et al., 2015; Hsiao et al., 2014; Rosová et al., 2008; Beegle et al., 2015). In addition, hypoxia can trigger the proliferation of BM-MSCs and enhance the production of IL-6 and IL-8 (Chen et al., 2014) or maintain stemness (Saller et al., 2012).

Starvation is another stress that can modify stem cell characteristics. In 2016, Wells et al. showed that starvation is an initiator for MSC differentiation (Wells et al., 2016). Anderson et al. (2016) primed MSCs with serum starvation and hypoxia (1% O_2) and showed that these cells produced EVs enriched in PDGF, EGF, and Nk-kB (Anderson et al., 2016). In another way,

Espagnolle et al. (2014) primed the MSCs (from bone marrow and adipose tissue) with M1 or M2 macrophage and investigated their characteristics. The results showed that in the same prime condition, the responses of BM-MSC and ADSCs are different. Indeed, BM-MSCs displayed more immunosuppressive characteristics when primed to M1, while ADSCs stimulated the T lymphocyte proliferation when primed to M1 (Espagnolle et al., 2014). Moreover, the contact of MSCs to M1 or M2 can modify the secretomes of MSCs. M1-primed BM-MSCs expressed more pro-inflammatory cytokines (IL-6, IL-8, CCL2, and Cox2), while M1-primed ADSCs produced more IL-6 but not IL-8 (Espagnolle et al., 2014).

This study evaluated changes in the immunomodulation of ADSCs when cultured under stress conditions of hypoxia, serum starvation, hypoxia combined with serum starvation, and TNF-α inflammatory factor treatment (Fig. 1).

2 Materials and Methods

2.1 Cells and Reagents

An ADSC line was obtained from the Stem Cell Institute (SCI) - Cell Bank (Stem Cell Institute, HCMC, VN). ADSC culture medium (ADSCCult I™), serum starvation medium (MSC Cult NF), and a deattachment reagent were bought from Regenmedlab (HCMC, VN). ADSCCult I™ medium was supplemented with TNF-α inflammatory factors (Thermo Fisher Scientific, Waltham, MA), at a concentration of 1000 U/ml.

Peripheral blood samples were collected from a healthy donor with the consent form. Fresh blood samples were drawn in a sterile blood collection tube containing anticoagulants. The blood sample was immediately used after collection (Fig. 2).

RPMI-1640 medium (Thermo Fisher Scientific, Waltham, MA) containing 40 ng/ml IL-4 and 50 ng/ml GM-CSF was prepared to stimulate DC differentiation.

CD133 MicroBead Kit–Tumor Tissue, human and MS Column were purchased from Miltenyi Biotec (Bergisch Gladbach, Germany). CD133/1 (AC133)-PE, human; CD25-APC, human; CD4-FITC, human; CD80-PE, human; and Anti-HLA-DR-FITC, human were also purchased from Miltenyi Biotec (Bergisch Gladbach, Germany).

Ficoll-Paque PREMIUM was purchased from GE Healthcare Bio-Sciences (Uppsala, Sweden). BD Pharm Lyse™ (Lysing buffer) was purchased from BD Biosciences (San Jose, CA).

Human IDO (indoleamine-2,3-dioxygenase) ELISA Kit, Human Prostaglandin E2 (PGE2) ELISA Kit were purchased from MyBioSource, Inc. (San Diego, CA, USA). Human IL-6 ELISA Kit and Human IL-10 ELISA Kit were purchased from Sigma-Aldrich (Saint Louis, MO, USA). VEGF Human ELISA Kit was purchased from Abcam (Cambridge, MA, USA).

The Angiogenesis Starter Kit was purchased from Thermo Fisher Scientific (Waltham, MA) (Fig. 3).

2.2 ADSC Expansion and Characterization

ADSCs were expanded in vitro to get enough cells for further experiments. These cells were cultured in flasks T-75 using an ADSCCult I™ medium. The cells were subcultured using a deattachment reagent. The cells were re-confirmed as MSCs according to the International Society for Cell and Gene Therapy (ISCT).

They were harvested at the 5th passages and stained with the antibodies anti-CD14 (conjugated with FITC), anti-CD34 (conjugated with FITC), anti-CD44 (conjugated with PE), anti-CD45 (conjugated with FITC), anti-CD73 (conjugated with PE), anti-CD90 (conjugated with PE), anti-CD105 (conjugated with PerCP), and anti-HLA-DR (conjugated with APC). All monoclonal antibodies were bought from BD Biosciences. These cells were washed twice with PBS before analyzing the expression of markers using the FACS Calibur flow cytometer

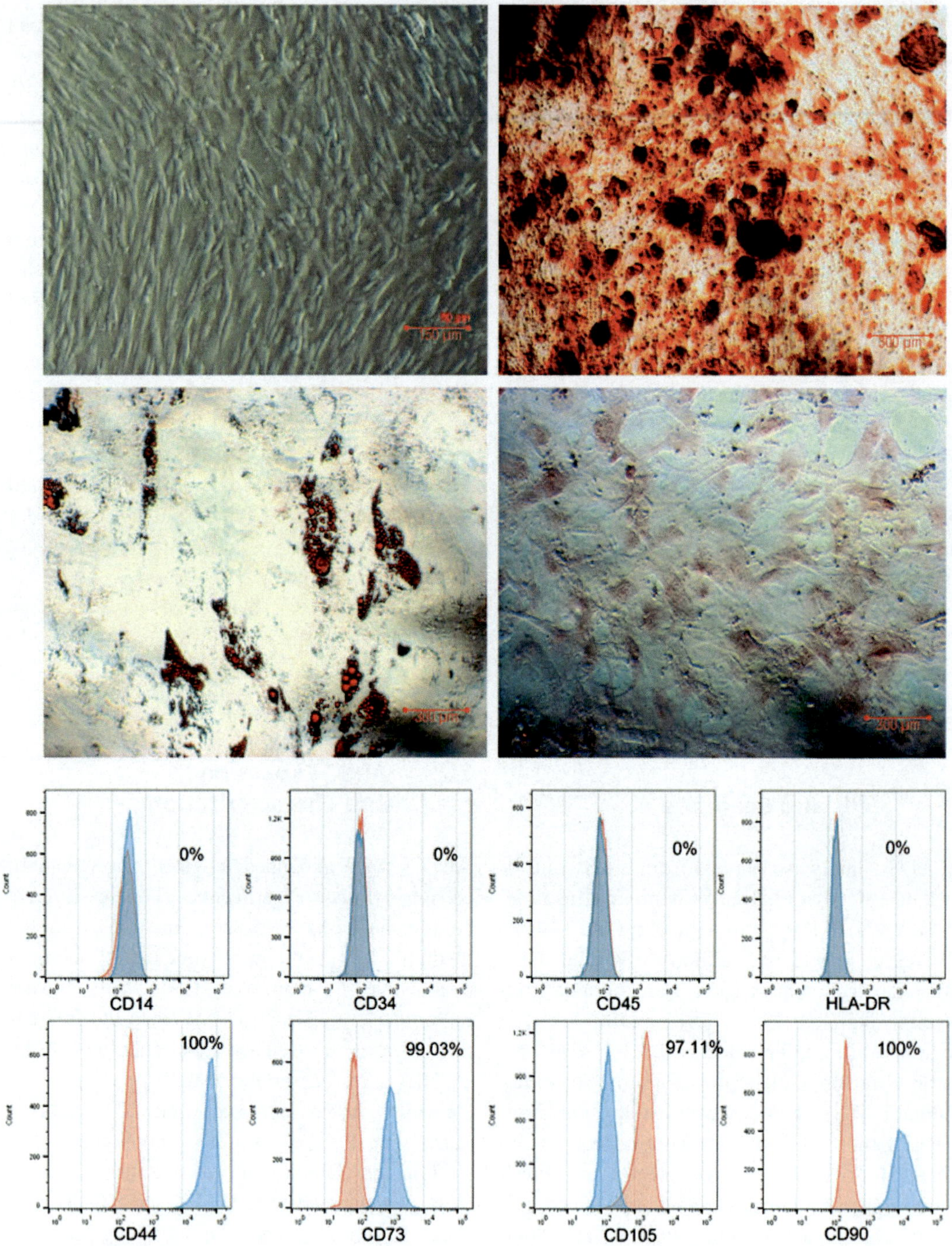

Fig. 1 Expanded ADSCs expressed the MSC phenotypes suggested by the ISCT. These cells expressed a fibroblast-like shape when adhered to the flask surface (**a**); they could be induced into adipocytes (positive with Oil Red), osteoblasts (positive with Alizarin Red), and chondrocytes (positive with Safranin O). They also displayed the MSC marker profiles (positive with CD44, CD73, CD90, and CD105; negative with CD14, CD34, CD45, and HLA-DR). Cell images at 10x magnification

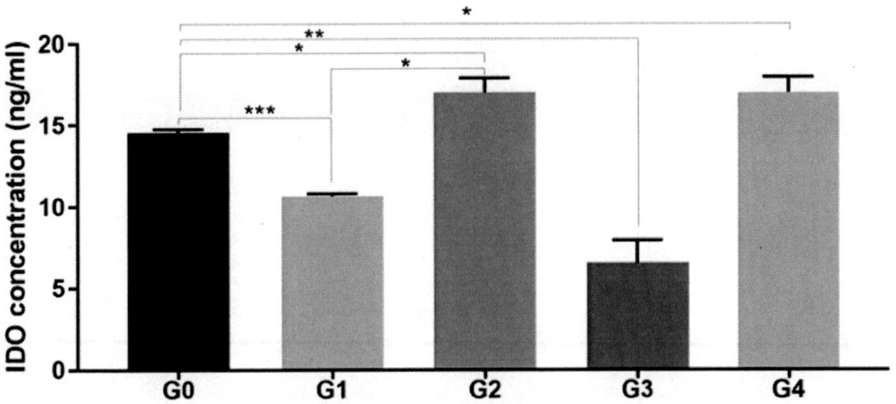

Fig. 2 IDO secretion in ADSC culture supernatants in some stress conditions. ADSCs under the standard in vitro culture condition (G0) secreted IDO at 14.56 ± 0.1229 ng/mL. When faced with hypoxia stress (G1), the inflammatory factor TNF-α (G3) ADSCs significantly reduced IDO secretion. While when cultured in serum starvation alone or in combination with hypoxia, ADSCs significantly increased the secretion of IDO. (*: P < 0.05, **: p < 0.005, ***: p < 0.001)

Fig. 3 PGE2 secretion in ADSC culture supernatants obtained from some conditions of stress. ADSCs strongly reduced PGE2 in the supernatant when faced with stresses of TNF-α (G1), hypoxia (G3) alone, or in combination with serum starvation (G4) compared to normal conditions (G0). However, they produced more PGE2 in serum starvation conditions compared to normal conditions (G0). *: p < 0.05, **: p < 0.005

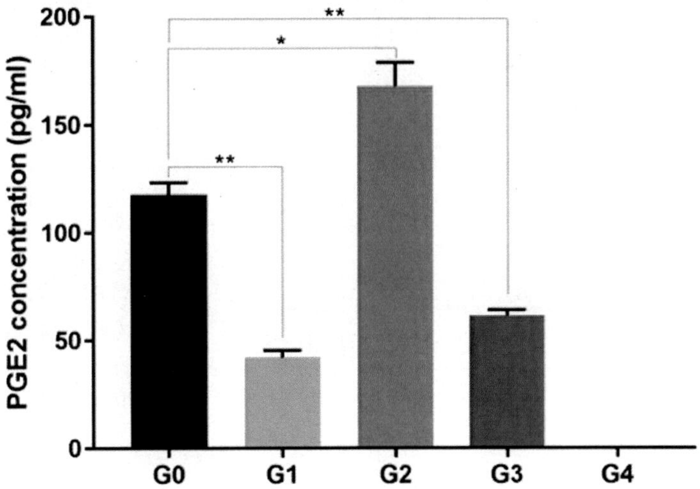

(BD Biosciences). The results were analyzed by FlowJo software (BD Biosciences).

For in vitro differentiation assays, ADSCs were induced into adipocytes, osteoblasts, and chondrocytes using the commercial kits (StemPro Adipogenesis, StemPro Osteogenesis, and StemPro Chondrogenesis, Thermo Fisher). Differentiated cells were stained with Oil Red, Alizarin Red, and Safranin O to confirm the phenotype of adipocytes, osteoblasts, and chondrocytes, respectively (Fig. 4).

2.3 Preconditioning ADSCs with Some Stress Conditions

ADSCs were seeded in the T-75 flasks with 5.000 cells per cm^2 in the ADSCCult I™ medium. Until cell confluency reached 70%, the stress conditions were applied. Eighteen flasks were randomly assigned to five groups with three flasks per group: G1, control (flasks that were put in the standard condition with 37 °C, 5% CO_2); G2, TNF-α treatment; G3, serum starvation, G4,

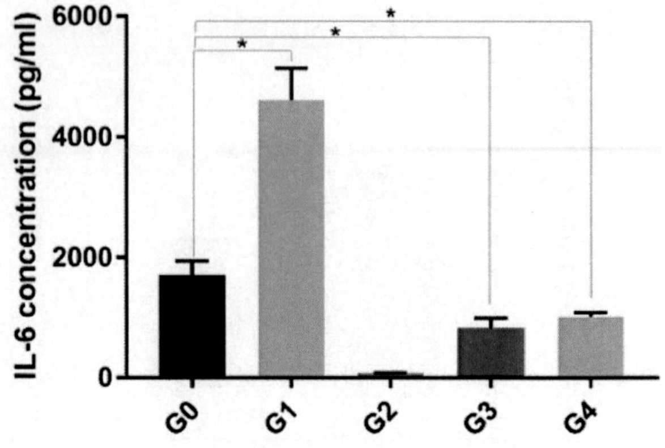

Fig. 4 IL-6 secretion in ADSC in some stress culture conditions. ADSCs significantly increased the production of IL-6 in TNF-α treatment conditions but reduced it in conditions of hypoxia, serum starvation alone, or in combination with hypoxia. *: $p < 0.05$

hypoxia (5% O_2); and G5, a combination of hypoxia (5% O_2) and serum starvation. These flasks were incubated in the stress conditions for 48 h before they were collected for further experiments. After 48 h, both the supernatant (conditioned media) and cells were collected. The conditioned media were collected in 50 centrifuge tubes and then centrifuged at 3500 g for 30 min to remove debris. These centrifuged tubes were put on the −86 °C freezer until they were used. The cells were collected by deattachment with the deattachment reagent from the flask surface.

2.4 Measurement of IDO, PGE2, and IL-6 in the Conditioned Media

IDO, PGE2, and IL-6 concentrations were determined by ELISA methods using the commercial kits (Abcam, Germany). The procedures were conducted following the manufacturer's instructions. Some main steps included he following: (i) preparation of standard protein and samples into wells, (ii) addition of corresponding detected antibodies carrying biotin, (iii) addition of HRP-streptavidin solution, (iv) addition of TMB substrate, and (v) addition of stop solution and immediately observe the result at 450 nm (Beckman Coulter DTX 880 Multimode Detector, Beckman Coulter, USA).

2.5 Cell Assay to Evaluate the Immune Modulations of ADSCs

2.5.1 Collection of Peripheral Blood-Derived MNCs and Induction into Dendritic Cells

Anticoagulated peripheral blood samples were used in this assay. The MNCs were isolated based on the gradient centrifugation with Ficoll 1.077 as per the manufacturer's guidelines, which were as follows: (i) preparation of diluted blood, (ii) collection of mononuclear layer after centrifugation on Ficoll-Paque, and (iii) washing and suspension of MNCs in RPMI-1640 medium with 10% FBS followed by incubation (37 °C, 5% CO_2). After 4 h, the floating cells were transferred to the centrifuge tube (the lymphocyte fraction); attached cells were induced in a DC culture medium (RPMI-1640 supplemented with 10% FBS, 20 ng/mL GMCSF, and 20 ng/mL IL-4) for 7 days to become immature DCs (Fig. 5).

2.5.2 Co-culture ADSCs and MNCs

This essay aimed to evaluate the effects of ADSCs on the presence of Treg cells in MNCs. Firstly, ADSCs were plated in 12-well plates at 10^5 cells per well and cells were allowed to attach to the surface of the plate for 4 h in the ADSCCult I™ medium. Then, the media were removed and replaced with RPMI 1640 10% FBS medium with

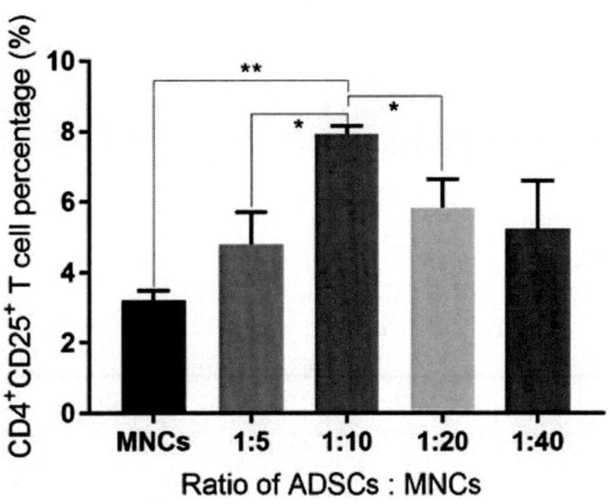

Fig. 5 Treg percentage in MNC populations and after co-culture with ADSCs. ADSCs significantly increased the ratio of Treg when co-cultured with MNC at 1 MNC:10 ADSC. (*: $p < 0.05$, **: $p < 0.005$)

MNCs. The doses of MNCs were added to ADSCs, which included 5.10^5 (5 times of MNCs compared to ADSCs), 10^6 (20 times), and 2.10^6 (40 times). The control wells only contained MNCs without ADSCs. All plates were incubated for 48 h before the supernatants were obtained in a 15-mL centrifuge tube. These tubes were centrifuged at 300 g for 5 min, and the cell pellets were re-suspended in a staining buffer and used in the flow cytometry analysis in the next assay.

2.5.3 Co-culture ADSCs and Lymphocytes

This assay aimed to evaluate the immunomodulation potential of ADSCs on lymphocytes. Firstly, ADSCs were seeded on 6-well plates with 2.10^5 cells per well, and cells were allowed to attach to the plate surface for 4 h. Then, the media were removed and added to the RPMI 1640 medium with lymphocytes (2.10^6 lymphocytes per well). The control wells only contained lymphocytes without ADSCs. The lymphocytes in the control wells were stimulated to divide by phytohaemagglutinin (PHA, 2.5 μg/mL). After 48 h of exposure, supernatants in all wells (included control wells) were collected. The cells and supernatant without cells were separated by centrifugation at 300 g for 5 min. The supernatant (without cells) was used to determine the concentration of IDO, IL-6, and IL-10 by ELISA methods, while the cell pellets were re-suspended in the staining buffers used for further experiments to analyze the presence of Treg cells.

2.5.4 Co-Culture ADSCs and Dendritic Cells

This assay aimed to evaluate the inhibition of maturation of ADSCs on immature DCs in vitro. Firstly, ADSCs were seeded on 6-well plates with 10^5 cells per well, and the cells were allowed to attach to the plate surface for 4 h. Then, the media were removed and replaced with RPMI 1640 with 10^5 cells per well of immature DCs. The control wells only contained immature DCs with RPMI 1640 supplemented with 10 ng/mL of TNF-α. All plates were incubated in 37 °C and 5% CO_2 for 48 h before cells were collected to analyze the maturation of DCs by flow cytometry.

2.6 Flow Cytometry for Immune Cell Markers

The Treg cells were analyzed based on the expression of CD4 and CD25, while the maturation of DCs was based on the expression of CD80 and HLA-DR. The cell suspension was stained with anti-CD4-FITC and anti-CD25-APC antibodies to evaluate the presence of Treg in the cell population. DCs were stained with anti-CD80-PE and anti-HLA-DR-FITC to analyze the maturation of immature DCs. All samples were washed twice with PBS before they were used to analyze the marker expression by FACS Calibur (BD Biosciences) flow cytometer.

2.7 Statistical Analysis

Statistical analysis was performed using GraphPad Prism 7 (GraphPad Software, San Diego, CA, USA). All P values noted were determined using a two-sample method t-test, assuming the same SD, with P values generated for two tails. Specific P values are noted in the figure legends (Figs. 6, 7 and 8).

3 Results

3.1 ADSCs Express the Mesenchymal Stem Cell Phenotypes

After 2 days of culture under selective conditions, ADSCs were observed under a microscope, and their morphology was compared with the ADSCs grown under standard conditions (in ADSC Cult I™ medium, 5% CO_2). ADSCs were relatively homogeneous in most culture conditions and typically had a mesenchymal cell shape and smooth and bright cell surfaces. However, in the TNF-α-supplemented medium, ADSCs were more elongated than those grown in the standard culture.

3.2 IDO Secretion in ADSC Culture Supernatants

After 2 days of culture under stress conditions, the IDO concentration in the supernatants was measured with the ELISA method, and the results showed that the IDO concentration was 14.56 ± 0.1229 ng/mL (G0), 10.64 ± 0.1038 ng/mL (G1), 17.01 ± 0.5037 ng/mL (G2), 6.519 ± 0.8208 ng/mL (G3), and 17 ± 0.5401 ng/mL (G4) in control, TNF-α supplemented medium, serum starvation medium, hypoxia, and serum starvation combined with hypoxia, respectively.

These results showed that under hypoxic stress or TNF-α stress, ADSCs significantly reduced the IDO production and secretion (14.56 ± 0.1229 ng/mL in normal condition vs. 6.519 ± 0.8208 ng/mL and 10.64 ± 0.1038 ng/mL stressed with hypoxia or TND-α, respectively; $p < 0.005$). In contrast to that, ADSCs strongly increased the IDO levels in the supernatant when stressed in the serum starvation medium alone or in combination with hypoxia (14.56 ± 0.1229 ng/mL in normal condition vs. 17.01 ± 0.5037 ng/mL and 17 ± 0.5401 ng/mL for starvation stress and starvation in combination with hypoxia, respectively; $p < 0.05$).

3.3 PGE2 Secretion in ADSC Culture Supernatants

After 2 days of culture under selected conditions, the concentration of PGE2 in the supernatants was measured with the ELISA method; the results showed that the concentration of PGE2 was 117.5 ± 3,993 pg/mL (G0), 42.06 ± 2.342 pg/mL (G1), 167.8 ± 7.696 pg/mL (G2), 61.49 ± 1.747 pg/mL (G3), and 0 pg/mL (G4). ADSCs strongly expressed PGE2 in serum starvation stress compared to normal conditions and other stresses ($p < 0.05$). In hypoxia alone (G3) or hypoxia in combination with serum starvation (G4) and TNF-α treatment, PGE2 significantly reduced ($p < 0.05$).

3.4 IL-6 Secretion in ADSC Culture Supernatants

After 2 days of culture under selected conditions, the IL-6 concentration in the supernatants was measured with the ELISA method, and the results showed that the concentration of IL-6 was 1705 ± 167.6 pg/mL (G0), 4,614 ± 374.7 pg/mL (G1), 75.33 ± 5.728 pg/mL (G2), 828.7 ± 112.2 pg/mL (G3), and 1,014 ± 53.26 pg/mL (G4). ADSCs significantly reduced IL-6 when faced with the stresses of hypoxia, serum starvation alone, or in combination with hypoxia ($p < 0.05$). However, when treated with TNF-α, they clearly increased the production and

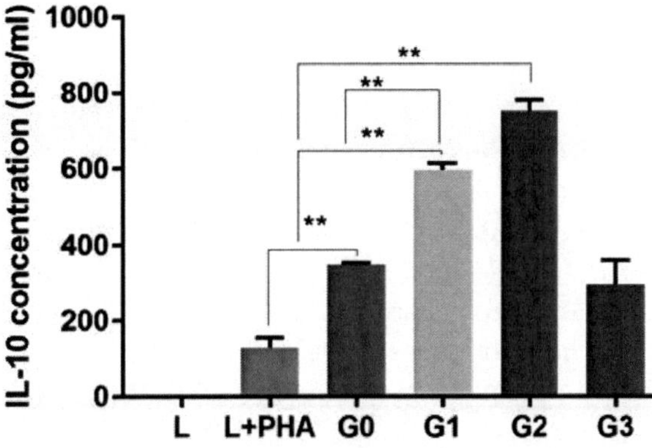

Fig. 6 IL-10 secretion in MNCs in case of co-culture with ADSCs. Lymphocytes did not secrete IL-10 when they were inactivated but secreted IL-10 at 129.1 ± 18.46 pg/mL when stimulated by PHA. After direct contact with ADSCs, MNCs significantly increased IL-10 secretion (**: $p < 0.01$)

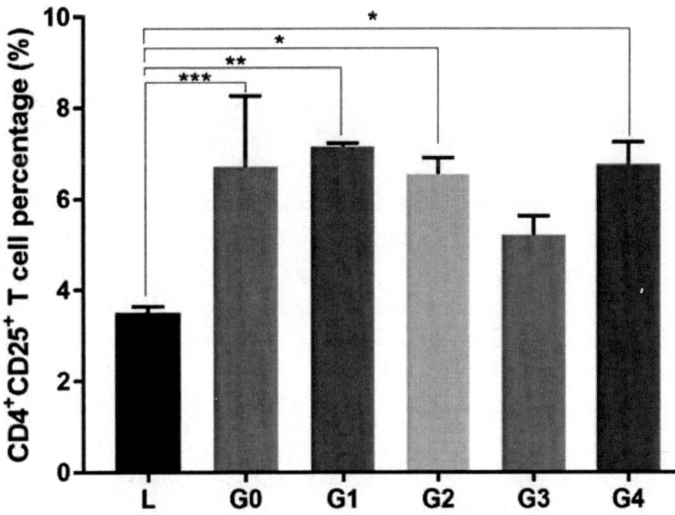

Fig. 7 Treg percentage in lymphocyte populations after co-culture with ADSCs. ADSCs in the presence of TNF-α, serum starvation alone, or in combination with hypoxia significantly stimulated Treg proliferation compared to the without stress group, while with hypoxia, Treg slightly increased. ***: $p < 0.001$, **: $p < 0.01$, *: $p < 0.05$

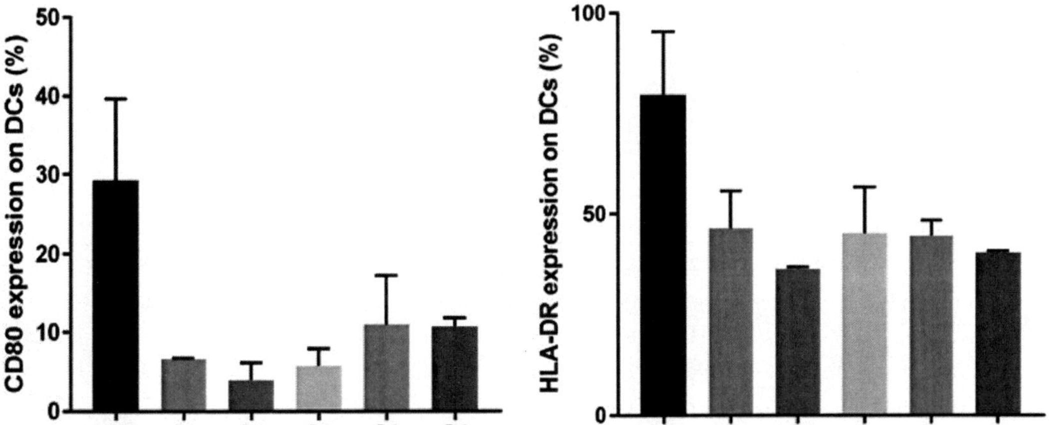

Fig. 8 CD80 and HLA-DR expression on DCs after co-culture with ADSCs. In all stress conditions, both CD80 and HLA-DR expressions on DCs were reduced after co-culture with ADSCs

secretion of IL-6 ($p < 0.05$) (4,614 ± 374.7 pg/mL in G1 vs. 1705 ± 167.6 pg/mL in G0).

3.5 Treg Percentage in MNC Populations After Co-culture with ADSCs

After 2 days of co-culturing ADSCs with MNCs, the percentage of Tregs ($CD4^{+}CD25^{+}$) ADSC-MNCs co-cultures at the ratio of 1:5, 1:10, 1:20, and 1:40 was 4.8 ± 0.5292%, 7.95 ± 0.15%, 5.833 ± 0.4631%, and 5.233 ± 0.7881%, respectively, while the proportion of Tregs in the MNC population was 3.2 ± 0.2%. Thus, ADSCs significantly increased the ratio of Treg when co-culture with MNC was at 1 MNC:10 ADSC ($p < 0.005$) and at 1 MNC:20 ADSC ($p < 0.05$), while 1 MNC:5 ADSC and 1 MNC:40 ADSC did not significantly change the Treg ratio. Moreover, the ratio of 1 MNC:10 ADSC also significantly increased the ratio of Treg compared to the ratio of 1 MNC:5 ADSC and 1 MNC:20 ADSC ($p < 0.05$). Therefore, the following experiments of co-cultured ADSC with immune cells all chose a ratio of 1 immune cell:10 ADSCs.

3.6 IL-10 Secretion in ADSC-Lymphocytes Co-cultured Supernatants

After 2 days of co-cultivating lymphocytes with ADSCs (at the ratio of 1 immune cell:10 ADSCs) induced under stress conditions, the IL-10 concentration in the conditioned media was measured by the ELISA method. The results showed that the IL-10 concentration in the conditioned media was 348.1 ± 3.065 pg/mL (G0), 595.4 ± 12.91 pg/mL (G1), 752 ± 20.67 pg/mL (G2), and 295.6 ± 44.91 pg/mL (G3). The IL-10 concentration in the lymphocytes only and in the lymphocytes supplemented with PHA was 0 pg/mL and 129.1 ± 18.46 pg/mL, respectively.

This meant that the lymphocytes (L) did not secrete IL-10 when they had not been activated; when stimulated by PHA (L + PHA) (the mitogen of T cell division) to the culture medium, the lymphocytes secreted IL-10 at 129.1 ± 18.46 pg/mL. After direct contact with ADSCs, MNCs increased IL-10 secretion to 348.1 ± 3.065 pg/mL ($p < 0.01$) in G0. The IL-10 production was enhanced in stresses of serum starvation and TNF-α treatment ($p < 0.05$) and downregulated in the case of hypoxia ($p < 0.05$).

3.7 Treg Percentage in Lymphocyte Populations After Co-culture with ADSCs

After 2 days of co-cultivating lymphocytes with ADSCs under stress conditions, the ratio of Treg cells ($CD4^{+}CD25^{+}$) were 6.7 ± 1.1% (G0), 7.15 ± 0.05% (G1), 6.55 ± 0.25% (G2), 5.2 ± 0.3% (G3), and 6.75 ± 0.35% (G4), while the proportion of Tregs in the lymphocyte population was 3.5 ± 0.1%.

In almost all stresses, ADSCs produce Treg proliferation. In the stresses of TNF-α, serum starvation, and serum starvation combined with hypoxia, Treg cells significantly increased compared to without stress ($p < 0.05$). In contrast, in the hypoxia, Treg slightly increased compared to control (without stress) ($p > 0.05$).

3.8 The Maturation of Immature DCs after co-Culture with ADSCs

The maturation status of DCs was evaluated based on the expression of CD80 and HLA-DR. Regarding the CD80 expression, after 2 days of co-culture with ADSCs under stress conditions, $CD80^{+}$ expression on DCs was recorded as 6.65 ± 0.05% (G0), 3.967 ± 1.257% (G1), 5.75 ± 1.55% (G2), 510.95 ± 4.45% (G3), and 10.7 ± 0.8% (G4) compared to 29.25 ± 7.35% in the control without stress. The results showed that in the presence of ADSCs in all stress culture conditions, immature DCs reduced the maturation to become mature DCs.

Similarly, after 2 days of co-culturing DCs with ADSCs under stress conditions, $HLA\text{-}DR^{+}$ expression on DCs also reduced by 46.3 ± 6.6%

(G0), 36.45 ± 0.35% (G1), 45.3 ± 8% (G2), 44.65 ± 2.55% (G3), and 40.4 ± 0.3% (G4) compared to 79.6 ± 11.1% in the without-stress condition ($p < 0.05$).

4 Discussion

ADSCs are a potential source of stem cells in regenerative medicine and also have uses beyond the stem cell drug market. To improve the applicability of ADSCs, many strategies are designed to help ADSCs adapt to the conditions of each specific defect. While the potential of ADSCs is clear, their therapeutic effect is not as high as expected because the microenvironments they deal with are harsh. Once tissue damage occurs, the supply from the blood vessel is blocked, leading to a deficiency of oxygen and nutrients. Damaged blood vessels also lead to discontinuous circulation, causing stagnation of inflammatory cytokines. This is why this study investigated the immunomodulation potential of ADSCs stressed by TNF-α-supplemented culture, serum starvation medium, hypoxia alone, and hypoxia in combination with serum starvation compared to the normal culture condition (without stress).

Tryptophan is the rarest and one of the essential amino acids in mammals (Stone & Darlington, 2002). IDO is the primary mechanism of extra-hepatic tryptophan metabolism (Munn, 2011; Gerriets & Rathmell, 2012; Mbongue et al., 2015), which results in the accumulation of degradation products of kynurenine (Trabanelli et al., 2011). IDO is considered to be the main inhibitor of the immune response, leading to the inhibition of T cell proliferation and activated effector T cells and the induction of T, B, and NK cell apoptosis (Munn, 2011; Munn et al., 1998), forming regulatory T cells (Munn & Mellor, 2013; Nguyen et al., 2010).

ADSCs under the standard in vitro culture condition without stress secreted IDO at 14.56 ± 0.1229 ng/mL. After 48 h of culture under the stress conditions, the IDO concentration in the ADSC culture supernatants increased in the starvation medium alone or in combination with hypoxia ($p < 0.05$). In contrast, ADSCs decreased IDO secretion in the TNF-α-supplemented medium ($p < 0.0005$) and hypoxia ($p < 0.005$). Thus, when faced with serum starvation, ADSCs increased IDO secretion, which may increase immunity suppression. These observations are similar to those on cancer cell lines (86HG39-glioblastoma cell line and HeLa cells) and normal cells (foreskin fibroblast), in that they too reduce IDO in hypoxia (Schmidt et al., 2013). Farida Djouad et al. showed that effects of TNF-α and IFN-gamma on ADSCs could depend on their concentration and could increase the IDO activity under low levels; however, in the case of abundant inflammatory cytokines, the immunosuppressive ability of MSCs could be reduced (Djouad et al., 2005). Fallarino et al. pointed out that in the ischemic damage, MSCs can protect the unintentional damage by suppression and induction of neutrophil apoptosis through the increase of IDO production and kynurenine metabolites. Both IDO and kynurenine metabolites are toxic to neutrophils (Fallarino et al., 2006).

PGE2 inhibits T cell proliferation (Goodwin et al., 1977; Goodwin, 1989) and promotes the development of regulatory T cells (Tregs) (Baratelli et al., 2005; Sharma et al., 2005; Bergmann et al., 2007; Soontrapa et al., 2011). PGE2 is required for the development of myeloid-derived suppressor cells (MDSCs) (Obermajer et al., 2011; Ochoa et al., 2007; Sinha et al., 2007; Fujita et al., 2011). It also promotes IL-10 induction, which directly inhibits the production of pro-inflammatory cytokines (Wang et al., 2007; Stolina et al., 2000). PGE2 strongly inhibits the production of Th1 cytokines such as interferon γ (IFN-γ) and IL-2 by Th1 cells (Hilkens et al., 1996), as well as the production of IL-12 and tumor necrosis factor α (TNFα) (Kalinski et al., 2001; van der Pouw Kraan et al., 1995). ADSCs under the standard in vitro culture without stress secreted PGE2 at 117.5 ± 3.993 pg/mL. After 2 days of culture under stress conditions, the concentration of PGE2 in the ADSCs culture supernatant increased in the serum starvation culture ($p < 0.05$). However, ADSCs decreased PGE2 secretion in the other stress conditions. Thus, when faced with starvation stress, ADSCs increased PGE2 secretion, which may increase immunity suppression. Along with the results of IDO expression, ADSCs in the serum starvation culture could give strong immunosuppressive effects. These results showed a correlation between IDO and PGE2 expression and were similar to the research of Deborah Braun et al., who showed that PGE2 induces IDO mRNA expression by transmitting signals via Gs-protein-coupled receptor E prostanoid-2 (EP2), which then activates adenylate cyclase, catalyzes cAMP formation, and activates PKA (Braun et al., 2005). However, when facing starvation and lack of oxygen stress, ADSCs did not secrete PGE2.

IL-6 increases IL-4 production, enhancing the formation of anti-inflammatory Th2 cells (Rincon et al., 1997; Diehl et al., 2002). IL-6 induces anti-inflammatory phenotypes in macrophages, which is recognized by the increase in IL-4 and IL-10 production and reduction in IL-1β secretion (Frisdal et al., 2011). It also converts monocytes into an anti-inflammatory M2 form that produces high levels of IL-10 and TGF-β (Duluc et al., 2007; Roca et al., 2009). ADSCs under in vitro culture without stress yielded IL-6 at 2514 ± 405.1 pg/mL. From the analysis results of IL-6 secretion, we showed that, under the stress of starvation, ADSCs decreased IL-6 secretion. This phenomenon also was observed in the human skin fibroblasts that decreased IL-6 gene expression under hypoxia compared to a concentration of air oxygen (Ewa Przybyt et al., 2016). However, when stressed by the TNF-α inflammatory factor, ADSCs significantly increased IL-6 secretion from 2,514 ± 405.1 pg/mL to 4,427 ± 222.6 pg/mL ($p < 0.05$). The results were similar to the study by Turner et al., which showed that the amount of IL-6 mRNA increased in a concentration-dependent manner after treatment with TNF-α and increased the most by five times when responding to 10 ng/ml TNF-α (Turner et al., 2007). This may be the basis for preconditioning ADSCs cultured in a medium supplemented with TNF-α inflammatory factors to enhance immunosuppression through mechanisms affected by IL-6.

IL-10 is known to be a pleiotropic and potent anti-inflammatory and immunosuppressive cytokine, which is produced by various cells of both innate and adaptive immunity (Moore et al., 2001). IL-10 has been shown to inhibit the production of pro-inflammatory cytokines IL-1β and TNF-α and the expression of major histocompatibility complex (MHC) class II proteins, co-stimulatory and adhesion molecules (CD86 and CD54), and the action mediators of inflammatory immune response IL-12 and IL-23 (McKinstry et al., 2009). Differentiation of Tregs is also increased by IL-10 (Heine et al., 2014). ADSCs under in vitro culture without stress secreted IL-10 at 4.548 ± 2.391 pg/mL. Lymphocytes did not express IL-10; however, when PHA was added, lymphocytes secrete the cytokine at 129.1 ± 18.46 pg/mL. The increase in the secretion of IL-10 is even higher for co-cultured lymphocytes with ADSCs, where co-cultured lymphocytes with ADSCs under in vitro culture without stress measured an IL-10 content of 348.1 ± 3.065 pg/mL. IL-10 concentration in the co-cultured supernatants increased when ADSCs were stressed by either TNF-α supplementation or serum starvation ($p < 0.005$). Moreover, the concentration of IL-10 in a co-culture of lymphocytes and ADSCs in starvation in combination with hypoxia is too high and exceeds the range of the kit. Thus, the direct interaction between lymphocytes and ADSCs strongly stimulates the secretion of IL-10, which may come from the lymphocytes and/or the ADSCs. ADSCs under stress conditions strongly increase IL-10 secretion compared to ADSCs in in vitro culture without stress. The results showed similarities with many studies that have demonstrated that ADSCs can facilitate the production of IL-10 from Tregs, monocytes, and dendritic cells (Nemeth et al., 2009; Zhang et al., 2004).

Co-culturing lymphocytes with ADSCs increased the proportion of Tregs ($CD4^+CD25^+$); specifically, the proportion of Tregs in the lymphocyte population was 3.5 ± 0.1%. This ratio was higher when co-culturing lymphocytes with ADSCs were selected under stress conditions. The results were similar to the research of Bassi et al., who demonstrated that Treg cells were stimulated to proliferate in vitro in the presence of ADSCs (Bassi et al., 2012). In addition, co-cultured DCs with ADSCs reduced the expression of DC maturation markers (including CD80 co-stimulatory molecule and HLA-DR major histocompatibility complex class II). The inhibition of DC maturation by ADSCs can be related to the IL-6 produced by ADSCs (Djouad et al., 2007).

To sum up, stress conditions of hypoxia, starvation, hypoxia combined with starvation, and TNF-α treatment could modify the immunomodulation potency of ADSCs. The immunomodulation potency of ADSCs was clearly enhanced in serum starvation.

5 Conclusion

This study evaluated changes in the immunomodulation potency of ADSCs cultured under stress conditions of hypoxia, serum starvation, hypoxia combined with starvation, and a medium supplemented with TNF-α inflammatory factor. The results showed that ADSCs cultivated under serum starvation strongly secreted IDO and PGE2. In addition, co-cultured lymphocytes with ADSCs showed that ADSCs in all stress conditions increased the proportion of Tregs, especially in the stress of serum starvation. Furthermore, ADSCs under stress conditions strongly increased IL-10 secretion compared to ADSCs in in vitro culture without stress. These results suggested that culture conditions can change the immunomodulation of ADSCs, and further, it can be boosted by serum starvation in vitro. This seems to be an easy way to improve the treatment efficacy for immunity-related disease treatments that use ADSCs.

Acknowledgments The authors thank Van Hanh General Hospital, Ho Chi Minh City, Viet Nam, that approved to use the adipose tissues from donors for the study.

Competing Interests The authors declare that they have no competing interests.

Funding This work is supported by Vietnam National University Ho Chi Minh City, No.TX2019-18-02.

References

Abdolmohammadi, K., Mahmoudi, T., Nojehdehi, S., Tayebi, L., Hashemi, S. M., Noorbakhsh, F., Abdollahi, A., Soleimani, M., Nikbin, B., & Nicknam, M. H. (2020). Effect of hypoxia preconditioned adipose-derived mesenchymal stem cell conditioned medium on cerulein-induced acute pancreatitis in mice. *Advanced Pharmaceutical Bulletin, 10*(2), 297–306.

Al-Ghadban, S., & Bunnell, B. A. (2020). Adipose tissue-derived stem cells: Immunomodulatory effects and therapeutic potential. *Physiology (Bethesda, Md), 35*(2), 125–133.

Anderson, J. D., Johansson, H. J., Graham, C. S., Vesterlund, M., Pham, M. T., Bramlett, C. S., Montgomery, E. N., Mellema, M. S., Bardini, R. L., Contreras, Z., et al. (2016). Comprehensive proteomic analysis of mesenchymal stem cell exosomes reveals modulation of angiogenesis via nuclear factor-KappaB signaling. *Stem Cells, 34*(3), 601–613.

Bacakova, L., Zarubova, J., Travnickova, M., Musilkova, J., Pajorova, J., Slepicka, P., Kasalkova, N. S., Svorcik, V., Kolska, Z., Motarjemi, H., et al. (2018). Stem cells: Their source, potency and use in regenerative therapies with focus on adipose-derived stem cells - a review. *Biotechnology Advances, 36*(4), 1111–1126.

Bader, A. M., Klose, K., Bieback, K., Korinth, D., Schneider, M., Seifert, M., Choi, Y. H., Kurtz, A., Falk, V., & Stamm, C. (2015). Hypoxic preconditioning increases survival and pro-Angiogenic capacity of human cord blood mesenchymal stromal cells in vitro. *PLoS One, 10*(9), e0138477.

Baratelli, F., Lin, Y., Zhu, L., Yang, S. C., Heuze-Vourc'h, N., Zeng, G., Reckamp, K., Dohadwala, M., Sharma, S., & Dubinett, S. M. (2005). Prostaglandin E2 induces FOXP3 gene expression and T regulatory cell function in human CD4+ T cells. *Journal of Immunology, 175*(3), 1483–1490.

Bassi, E. J., Moraes-Vieira, P. M., Moreira-Sa, C. S., Almeida, D. C., Vieira, L. M., Cunha, C. S., Hiyane, M. I., Basso, A. S., Pacheco-Silva, A., & Camara, N. O. (2012). Immune regulatory properties of allogeneic adipose-derived mesenchymal stem cells in the treatment of experimental autoimmune diabetes. *Diabetes, 61*(10), 2534–2545.

Beegle, J., Lakatos, K., Kalomoiris, S., Stewart, H., Isseroff, R. R., Nolta, J. A., & Fierro, F. A. (2015). Hypoxic preconditioning of mesenchymal stromal cells induces metabolic changes, enhances survival, and promotes cell retention in vivo. *Stem Cells, 33*(6), 1818–1828.

Bergmann, C., Strauss, L., Zeidler, R., Lang, S., & Whiteside, T. L. (2007). Expansion of human T regulatory type 1 cells in the microenvironment of cyclooxygenase 2 overexpressing head and neck squamous cell carcinoma. *Cancer Research, 67*(18), 8865–8873.

Braun, D., Longman, R. S., & Albert, M. L. (2005). A two-step induction of indoleamine 2,3 dioxygenase (IDO) activity during dendritic-cell maturation. *Blood, 106*(7), 2375–2381.

Ceccarelli, S., Pontecorvi, P., Anastasiadou, E., Napoli, C., & Marchese, C. (2020). Immunomodulatory effect of adipose-derived stem cells: The cutting edge of clinical application. *Frontiers in Cell and Development Biology, 8*, 236–236.

Chen, L., Xu, Y., Zhao, J., Zhang, Z., Yang, R., Xie, J., Liu, X., & Qi, S. (2014). Conditioned medium from hypoxic bone marrow-derived mesenchymal stem cells enhances wound healing in mice. *PLoS One, 9*(4), e96161.

Dai, R., Wang, Z., Samanipour, R., Koo, K. I., & Kim, K. (2016). Adipose-derived stem cells for tissue engineering and regenerative medicine applications. *Stem Cells International, 2016*, 6737345.

Diehl, S., Chow, C. W., Weiss, L., Palmetshofer, A., Twardzik, T., Rounds, L., Serfling, E., Davis, R. J., Anguita, J., & Rincon, M. (2002). Induction of NFATc2 expression by interleukin 6 promotes T helper type 2 differentiation. *The Journal of Experimental Medicine, 196*(1), 39–49.

Djouad, F., Bony, C., Haupl, T., Uze, G., Lahlou, N., Louis-Plence, P., Apparailly, F., Canovas, F., Reme, T., Sany, J., et al. (2005). Transcriptional profiles discriminate bone marrow-derived and synovium-derived mesenchymal stem cells. *Arthritis Research & Therapy, 7*(6), R1304–R1315.

Djouad, F., Charbonnier, L. M., Bouffi, C., Louis-Plence, P., Bony, C., Apparailly, F., Cantos, C., Jorgensen, C., & Noel, D. (2007). Mesenchymal stem cells inhibit the differentiation of dendritic cells through an interleukin-6-dependent mechanism. *Stem Cells, 25*(8), 2025–2032.

Duluc, D., Delneste, Y., Tan, F., Moles, M. P., Grimaud, L., Lenoir, J., Preisser, L., Anegon, I., Catala, L., Ifrah, N., et al. (2007). Tumor-associated leukemia inhibitory factor and IL-6 skew monocyte differentiation into tumor-associated macrophage-like cells. *Blood, 110*(13), 4319–4330.

Elsässer, A., Suzuki, K., Lorenz-Meyer, S., Bode, C., & Schaper, J. (2001). The role of apoptosis in myocardial ischemia: A critical appraisal. *Basic Research in Cardiology, 96*(3), 219–226.

Espagnolle, N., Balguerie, A., Sensebé, L., & Varin, A. (2014). Important role of the immune environment on BM-MSC and ADSC function: Modulation of immunosuppressive capacities and secretory profiles of MSC by macrophages. *Cytotherapy, 16*(4), S67.

Ewa Przybyt, T. P., Krenning, G., & Harmsen, M. C. (2016). ADSC promote vascular network formation through juxtacrine and paracrine interactions with endothelial cells. In E. Przybyt (Ed.), *Adipose derived stromal cells in cardiovascular regenerative medicine*. Groningen: University of Groningen.

Fallarino, F., Grohmann, U., You, S., McGrath, B. C., Cavener, D. R., Vacca, C., Orabona, C., Bianchi, R., Belladonna, M. L., Volpi, C., et al. (2006). Tryptophan catabolism generates autoimmune-preventive regulatory T cells. *Transplant Immunology, 17*(1), 58–60.

Frisdal, E., Lesnik, P., Olivier, M., Robillard, P., Chapman, M. J., Huby, T., Guerin, M., & Le Goff, W. (2011). Interleukin-6 protects human macrophages from cellular cholesterol accumulation and attenuates the proinflammatory response. *The Journal of Biological Chemistry, 286*(35), 30926–30936.

Fujita, M., Kohanbash, G., Fellows-Mayle, W., Hamilton, R. L., Komohara, Y., Decker, S. A., Ohlfest, J. R., & Okada, H. (2011). COX-2 blockade suppresses gliomagenesis by inhibiting myeloid-derived suppressor cells. *Cancer Research, 71*(7), 2664–2674.

Gerriets, V. A., & Rathmell, J. C. (2012). Metabolic pathways in T cell fate and function. *Trends in Immunology, 33*(4), 168–173.

Goodwin, J. S. (1989). Immunomodulation by eicosanoids and anti-inflammatory drugs. *Current Opinion in Immunology, 2*(2), 264–268.

Goodwin, J. S., Bankhurst, A. D., & Messner, R. P. (1977). Suppression of human T-cell mitogenesis by prostaglandin. Existence of a prostaglandin-producing suppressor cell. *The Journal of Experimental Medicine, 146*(6), 1719–1734.

Gyöngyösi, M., Blanco, J., Marian, T., Trón, L., Petneházy, O., Petrasi, Z., Hemetsberger, R., Rodriguez, J., Font, G., & Pavo, I. J. (2008). Serial noninvasive in vivo positron emission tomographic tracking of percutaneously intramyocardially injected autologous porcine mesenchymal stem cells modified for transgene reporter gene expression. *Circulation: Cardiovascular Imaging, 1*(2), 94–103.

Heine, G., Drozdenko, G., Grun, J. R., Chang, H. D., Radbruch, A., & Worm, M. (2014). Autocrine IL-10 promotes human B-cell differentiation into IgM- or IgG-secreting plasmablasts. *European Journal of Immunology, 44*(6), 1615–1621.

Herrmann, J. L., Wang, Y., Abarbanell, A. M., Weil, B. R., Tan, J., & Meldrum, D. R. (2010). Preconditioning mesenchymal stem cells with transforming growth factor-alpha improves mesenchymal stem cell-mediated cardioprotection. *Shock, 33*(1), 24–30.

Hilkens, C. M., Snijders, A., Snijdewint, F. G., Wierenga, E. A., & Kapsenberg, M. L. (1996). Modulation of T-cell cytokine secretion by accessory cell-derived products. *European Respiratory Journal, 22*, 90s–94s.

Hsiao, S. T., Dilley, R. J., Dusting, G. J., & Lim, S. Y. (2014). Ischemic preconditioning for cell-based therapy and tissue engineering. *Pharmacology & Therapeutics, 142*(2), 141–153.

Hu, X., Yu, S. P., Fraser, J. L., Lu, Z., Ogle, M. E., Wang, J.-A., & Wei, L. (2008). Transplantation of hypoxia-preconditioned mesenchymal stem cells improves infarcted heart function via enhanced survival of implanted cells and angiogenesis. *The Journal of Thoracic and Cardiovascular Surgery, 135*(4), 799–808.

Kalinski, P., Vieira, P. L., Schuitemaker, J. H., de Jong, E. C., & Kapsenberg, M. L. (2001). Prostaglandin E (2) is a selective inducer of interleukin-12 p40 (IL-12p40) production and an inhibitor of bioactive IL-12p70 heterodimer. *Blood, 97*(11), 3466–3469.

Lee, M. J., Kim, J., Kim, M. Y., Bae, Y. S., Ryu, S. H., Lee, T. G., & Kim, J. H. (2010). Proteomic analysis of tumor necrosis factor-alpha-induced secretome of human adipose tissue-derived mesenchymal stem cells. *Journal of Proteome Research, 9*(4), 1754–1762.

Mbongue, J. C., Nicholas, D. A., Torrez, T. W., Kim, N. S., Firek, A. F., & Langridge, W. H. (2015). The role of Indoleamine 2, 3-dioxygenase in immune suppression and autoimmunity. *Vaccines (Basel), 3*(3), 703–729.

McGinley, L. M., McMahon, J., Stocca, A., Duffy, A., Flynn, A., O'Toole, D., & O'Brien, T. (2013). Mesenchymal stem cell survival in the infarcted heart is enhanced by lentivirus vector-mediated heat shock protein 27 expression. *Human Gene Therapy, 24*(10), 840–851.

McKinstry, K. K., Strutt, T. M., Buck, A., Curtis, J. D., Dibble, J. P., Huston, G., Tighe, M., Hamada, H., Sell, S., Dutton, R. W., et al. (2009). IL-10 deficiency unleashes an influenza-specific Th17 response and enhances survival against high-dose challenge. *Journal of Immunology, 182*(12), 7353–7363.

Moore, K. W., de Waal, M. R., Coffman, R. L., & O'Garra, A. (2001). Interleukin-10 and the interleukin-10 receptor. *Annual Review of Immunology, 19*, 683–765.

Munn, D. H. (2011). Indoleamine 2,3-dioxygenase, Tregs and cancer. *Current Medicinal Chemistry, 18*(15), 2240–2246.

Munn, D. H., & Mellor, A. L. (2013). Indoleamine 2,3 dioxygenase and metabolic control of immune responses. *Trends in Immunology, 34*(3), 137–143.

Munn, D. H., Zhou, M., Attwood, J. T., Bondarev, I., Conway, S. J., Marshall, B., Brown, C., & Mellor, A. L. (1998). Prevention of allogeneic fetal rejection by tryptophan catabolism. *Science, 281*(5380), 1191–1193.

Nemeth, K., Leelahavanichkul, A., Yuen, P. S., Mayer, B., Parmelee, A., Doi, K., Robey, P. G., Leelahavanichkul, K., Koller, B. H., Brown, J. M., et al. (2009). Bone marrow stromal cells attenuate sepsis via prostaglandin E(2)-dependent reprogramming of host macrophages to increase their interleukin-10 production. *Nature Medicine, 15*(1), 42–49.

Nguyen, N. T., Kimura, A., Nakahama, T., Chinen, I., Masuda, K., Nohara, K., Fujii-Kuriyama, Y., & Kishimoto, T. (2010). Aryl hydrocarbon receptor negatively regulates dendritic cell immunogenicity via a kynurenine-dependent mechanism. *Proceedings of the National Academy of Sciences of the United States of America, 107*(46), 19961–19966.

Obermajer, N., Muthuswamy, R., Lesnock, J., Edwards, R. P., & Kalinski, P. (2011). Positive feedback between PGE2 and COX2 redirects the differentiation of human dendritic cells toward stable myeloid-derived suppressor cells. *Blood, 118*(20), 5498–5505.

Ochoa, A. C., Zea, A. H., Hernandez, C., & Rodriguez, P. C. (2007). Arginase, prostaglandins, and myeloid-derived suppressor cells in renal cell carcinoma. *Clinical Cancer Research, 13*(2 Pt 2), 721s–726s.

Rincon, M., Anguita, J., Nakamura, T., Fikrig, E., & Flavell, R. A. (1997). Interleukin (IL)-6 directs the differentiation of IL-4-producing CD4+ T cells. *The Journal of Experimental Medicine, 185*(3), 461–469.

Roca, H., Varsos, Z. S., Sud, S., Craig, M. J., Ying, C., & Pienta, K. J. (2009). CCL2 and interleukin-6 promote survival of human CD11b+ peripheral blood mononuclear cells and induce M2-type macrophage polarization. *The Journal of Biological Chemistry, 284*(49), 34342–34354.

Roemeling-Van Rhijn M, Mensah F, Korevaar S, Leijs M, van Osch G, IJzermans J, Betjes M, Baan C, Weimar W, Hoogduijn M: Effects of hypoxia on the immunomodulatory properties of adipose tissue-derived mesenchymal stem cells. Frontiers in Immunology 2013, 4(203).

Rosová, I., Dao, M., Capoccia, B., Link, D., & Nolta, J. A. (2008). Hypoxic preconditioning results in increased motility and improved therapeutic potential of human mesenchymal stem cells. *Stem Cells, 26*(8), 2173–2182.

Saller, M. M., Prall, W. C., Docheva, D., Schönitzer, V., Popov, T., Anz, D., Clausen-Schaumann, H., Mutschler, W., Volkmer, E., Schieker, M., et al. (2012). Increased stemness and migration of human mesenchymal stem cells in hypoxia is associated with altered integrin expression. *Biochemical and Biophysical Research Communications, 423*(2), 379–385.

Schmidt, S. K., Ebel, S., Keil, E., Woite, C., Ernst, J. F., Benzin, A. E., Rupp, J., & Daubener, W. (2013). Regulation of IDO activity by oxygen supply: Inhibitory effects on antimicrobial and immunoregulatory functions. *PLoS One, 8*(5), e63301.

Sharma, S., Yang, S. C., Zhu, L., Reckamp, K., Gardner, B., Baratelli, F., Huang, M., Batra, R. K., & Dubinett, S. M. (2005). Tumor cyclooxygenase-2/prostaglandin E2-dependent promotion of FOXP3 expression and CD4+ CD25+ T regulatory cell activities in lung cancer. *Cancer Research, 65*(12), 5211–5220.

Si, Z., Wang, X., Sun, C., Kang, Y., Xu, J., Wang, X., & Hui, Y. (2019). Adipose-derived stem cells: Sources, potency, and implications for regenerative therapies. *Biomedicine & pharmacotherapy = Biomedecine & pharmacotherapie, 114*, 108765.

Sinha, P., Clements, V. K., Fulton, A. M., & Ostrand-Rosenberg, S. (2007). Prostaglandin E2 promotes tumor progression by inducing myeloid-derived suppressor cells. *Cancer Research, 67*(9), 4507–4513.

Soontrapa, K., Honda, T., Sakata, D., Yao, C., Hirata, T., Hori, S., Matsuoka, T., Kita, Y., Shimizu, T.,

Kabashima, K., et al. (2011). Prostaglandin E2-prostaglandin E receptor subtype 4 (EP4) signaling mediates UV irradiation-induced systemic immunosuppression. *Proceedings of the National Academy of Sciences of the United States of America, 108*(16), 6668–6673.

Stolina, M., Sharma, S., Lin, Y., Dohadwala, M., Gardner, B., Luo, J., Zhu, L., Kronenberg, M., Miller, P. W., Portanova, J., et al. (2000). Specific inhibition of cyclooxygenase 2 restores antitumor reactivity by altering the balance of IL-10 and IL-12 synthesis. *Journal of Immunology, 164*(1), 361–370.

Stone, T. W., & Darlington, L. G. (2002). Endogenous kynurenines as targets for drug discovery and development. *Nature Reviews. Drug Discovery, 1*(8), 609–620.

Toma, C., Pittenger, M. F., Cahill, K. S., Byrne, B. J., & Kessler, P. D. (2002). Human mesenchymal stem cells differentiate to a cardiomyocyte phenotype in the adult murine heart. *Circulation, 105*(1), 93–98.

Trabanelli, S., Ocadlikova, D., Evangelisti, C., Parisi, S., & Curti, A. (2011). Induction of regulatory T cells by dendritic cells through indoleamine 2,3-dioxygenase: A potent mechanism of acquired peripheral tolerance. *Current Medicinal Chemistry, 18*(15), 2234–2239.

Turner, N. A., Mughal, R. S., Warburton, P., O'Regan, D. J., Ball, S. G., & Porter, K. E. (2007). Mechanism of TNFalpha-induced IL-1alpha, IL-1beta and IL-6 expression in human cardiac fibroblasts: Effects of statins and thiazolidinediones. *Cardiovascular Research, 76*(1), 81–90.

van der Pouw Kraan, T. C., Boeije, L. C., Smeenk, R. J., Wijdenes, J., & Aarden, L. A. (1995). Prostaglandin-E2 is a potent inhibitor of human interleukin 12 production. *The Journal of Experimental Medicine, 181*(2), 775–779.

Wang, M. T., Honn, K. V., & Nie, D. (2007). Cyclooxygenases, prostanoids, and tumor progression. *Cancer Metastasis Reviews, 26*(3–4), 525–534.

Wang, L., Hu, X., Zhu, W., Jiang, Z., Zhou, Y., Chen, P., & Wang, J. (2014). Increased leptin by hypoxic-preconditioning promotes autophagy of mesenchymal stem cells and protects them from apoptosis. *Science China Life Sciences, 57*(2), 171–180.

Wells, A., Rodrigues, M., Wells, A., & Nuschke, A. (2016). Starvation as an initiator of mesenchymal stem cell/multipotent stromal cell differentiation. *Journal of Stem Cell Research & Therapy, 1*(3), 00020.

Zhang, W., Ge, W., Li, C., You, S., Liao, L., Han, Q., Deng, W., & Zhao, R. C. (2004). Effects of mesenchymal stem cells on differentiation, maturation, and function of human monocyte-derived dendritic cells. *Stem Cells and Development, 13*(3), 263–271.

Adv Exp Med Biol - Innovations in Cancer Research and Regenerative Medicine (2023) 4: 19–33
https://doi.org/10.1007/5584_2023_776

Published online: 9 June 2023

Interferon-Gamma Increases the Immune Modulation of Umbilical Cord-Derived Mesenchymal Stem Cells but Decreases Their Chondrogenic Potential

Nhat Chau Truong, Thu Ngoc-Minh Phan, Nhi Thao Huynh, Khuong Duy Pham, and Phuc Van Pham

Abstract

Introduction The pro-inflammatory cytokine interferon-gamma (IFN-γ) is reported to be an agent that boosts the immune modulation of mesenchymal stem cells (MSCs). However, the effects of IFN-γ on the chondrogenic potential of treated MSCs have not been evaluated in depth. This study aimed to evaluate the effects of IFN-γ on the immune modulation and chondrogenic potential of human umbilical cord-derived MSCs (hUC-MSCs).

Methods UC-MSCs were isolated and expanded following published protocols. They were characterized as MSCs before their use in further experiments. The UC-MSCs were treated with IFN-γ at 10 ng/mL for 48 h. Changes in phenotype were investigated based on changes in MSC markers, immunomodulatory genes (*TGF-β*, *IL-4*, and *IDO*) for immune modulation, and cartilage-related genes during the induction of differentiation (*Col1a2*, *Col2a1*, *Sox9*, *Runx2*, and *Acan*) for chondrogenic potential.

Results IFN-γ-treated UC-MSCs maintained MSC markers and exhibited decreased expression of transcriptional regulatory factors in chondrogenesis (*Sox9* and *Runx2*) and the extracellular matrix-specific genes *Col1a2* and *Acan* but not *Col2a1* compared to non-treated cells ($p < 0.05$). Furthermore, the immunomodulatory capability of IFN-γ-treated UC-MSCs was clearly revealed through their increased expression of *IDO* and *IL-4* and decreased expression of *TGF-β* compared to non-treated cells ($p < 0.05$).

Nhat Chau Truong and Thu Ngoc-Minh Phan contributed equally to this work.

N. C. Truong and T. N.-M. Phan
Stem Cell Institute, University of Science, Ho Chi Minh City, Viet Nam

Vietnam National University Ho Chi Minh City, Ho Chi Minh City, Viet Nam

N. T. Huynh and K. D. Pham
Vietnam National University Ho Chi Minh City, Ho Chi Minh City, Viet Nam

Laboratory of Stem Cell Research and Application, University of Science, Ho Chi Minh City, Viet Nam

P. Van Pham (✉)
Stem Cell Institute, University of Science, Ho Chi Minh City, Viet Nam

Vietnam National University Ho Chi Minh City, Ho Chi Minh City, Viet Nam

Laboratory of Cancer Research, University of Science, Ho Chi Minh City, Viet Nam
e-mail: phucpham@sci.edu.vn

Conclusion This study demonstrated that UC-MSCs treated with IFN-γ at 10 ng/mL had reduced expression of chondrocyte-specific genes; however, they maintained multi-lineage differentiation and exhibited immunomodulatory properties.

Keywords

Cartilage · Immunomodulation · Inflammation · Interferon-gamma · Umbilical cord mesenchymal stem cells

Abbreviations

Acan	Aggrecan
CD	Cluster of differentiation
Col	Collagen
HLA-DR	Human leukocyte antigen-DR
hUC-MSC	Human umbilical cord-derived mesenchymal stem cell
IDO	Indoleamine 2,3-dioxygenase
IFN-γ	Interferon-gamma
ISCT	International Society for Cellular Therapy
MSC	Mesenchymal stem cell
RT-qPCR	Real-time reverse transcription polymerase chain reaction
Runx2	Runt-related transcription factor 2
Sox9	SRY-box transcription factor 9
TGF-β	Transforming growth factor beta

1 Introduction

Mesenchymal stem cells (MSCs) are the most stem cell type found in the human body. They can be found in and isolated from various tissue sources, including adipose tissue, dental pulp, amniotic fluid, the umbilical cord, cord blood, peripheral blood, menstrual blood, and breast milk (Arutyunyan et al., 2016; Moraes et al., 2016; Ullah et al., 2015). The minimal criteria for defining MSCs suggested by the International Society for Cell and Gene Therapeutics (ISCT) are as follows: (1) must be plastic-adherent when cultured under standard conditions, (2) must be able to differentiate into osteoblasts, chondrocytes, and adipocytes in vitro, (3) must express CD73, CD90, and CD105 and lack CD14, CD34, CD45, and HLA-DR expression (Arutyunyan et al., 2016; Dominici et al., 2006; Lopez-Santalla et al., 2020). The therapeutic effects of these cells are based on mechanisms including immune modulation, multiple differentiation, and angiogenesis.

Theoretically, the enhancement of MSCs' immune modulation can increase treatment efficacy, especially in the case of some inflammatory conditions or diseases. Different approaches have been attempted to increase MSCs' immune modulation, such as the preconditioning of MSCs using cytokines or growth factors (De Witte et al., 2016, 2017; Chinnadurai et al., 2014; Wang et al., 2016), the use of immune receptor agonists (Rashedi et al., 2017; Kim et al., 2018; Qiu et al., 2017), hypoxia treatment (Saller et al., 2012; Kang et al., 2018; Yuan et al., 2019; Liu et al., 2020), and three-dimensional culture (Zimmermann & McDevitt, 2018; Conrad et al., 2021).

Of these approaches, the use of inflammatory factors (IFN-γ and TNF-α) showed potential in boosting immune modulation. Recent publications suggested that MSCs treated with IFN-γ could upregulate the secretion of PGE2, HGF, TGF-β, MCP-1 (De Witte et al., 2016), and PDL-1; cause T cell suppression (Chinnadurai et al., 2014); downregulate the secretion of IFN-γ and TNF-α and Th17 cells (Wang et al., 2016); upregulate the secretion of IL-6 and IL-10; and promote Tregs (Wang et al., 2016).

However, the reverse effects of immune modulation enhancement, especially regarding multiple differentiation potential, have not been carefully evaluated in previous studies. This study aimed to investigate the effects of preconditioning with IFN-γ on both the immunomodulation capacity and chondrogenic potential of human umbilical cord-derived mesenchymal stem cells (hUC-MSCs).

2 Methods

2.1 Expansion of Human Umbilical Cord-Derived Mesenchymal Stem Cells

The hUC-MSCs were provided by SCI Cellbank (Stem Cell Institute, Viet Nam) and expanded in T-75 flasks (Eppendorf, Germany) using MSCCult I medium (Regenmedlab, Viet Nam) with or without 10 ng/mL IFN-γ (Peprotech, US) for 48 h. The cells were cultured as monolayers and under standard conditions (37 °C and 5% CO_2). When the confluence reached 70–80%, the hUC-MSCs were detached using TrypLE™ (Gibco, US) and sub-cultured at a density of 10.10^4 cells/cm^2. All hUC-MSCs used in the following experiments were from passages 5–7.

2.2 Characterization of Human Umbilical Cord-Derived Mesenchymal Stem Cells

After 24 h and 48 h of induction in IFN-γ-supplemented medium, the changes in the morphology of the hUC-MSCs were recorded in comparison with the counterpart. Then, the cells were transferred to a 96-well plate (10^4 cells per well) and induced into osteoblasts, chondroblasts, and adipocytes in vitro using the StemPro™ Osteogenesis Differentiation Kit, StemPro™ Chondrogenesis Differentiation Kit, and StemPro™ Adipogenesis Differentiation Kit (Gibco, US), respectively. After 30 days of osteogenic differentiation and 15 days of chondrogenic and adipogenic differentiation, the differentiated cells were stained with Alizarin Red dye (Sigma-Aldrich, US), Safranin-O dye (Matheson Coleman & Bell, US), and Oil Red dye (Sigma-Aldrich, US) to confirm the osteogenic, chondrogenic, and adipogenic differentiation results, respectively.

The expression of specific surface markers was detected by flow cytometry. The control and IFN-γ-treated hUC-MSCs were harvested, resuspended in Stain Buffer (BD Biosciences, US), and stained overnight (at 4 °C, in the dark) with the following conjugated monoclonal antibodies: CD14-FITC, CD34-FITC, CD44-PE, CD45-APC, CD73-PE, CD90-PE, CD105-FITC, and HLA-DR-FITC (BD Biosciences, US). Then, excess antibodies were removed by washing twice with Stain Buffer and the samples were analyzed using a FACS Melody system (BD Biosciences, US). The results were processed using Flowjo software (BD Biosciences, US).

2.3 Chondrogenic Differentiation for Gene Expression Analysis

After 48 h of treatment with or without IFN-γ, hUC-MSCs were transferred into a 6-well plate at a density of 10^5 cells per well and incubated overnight. Then, the cells were induced using the StemPro™ Chondrogenesis Differentiation Kit for 15 days at 37 °C, with 5% CO_2. The medium was refreshed every 5 days and the formation of chondrogenic spheroids was observed under an inverted microscope on days 0, 5, 10, and 15.

2.4 Quantitation of the Relative Expression of Chondrocyte-Related Genes and Immunomodulatory Genes

Chondrocyte-related genes, including *Col1*, *Col2*, *Sox9*, *Runx2*, and *Acan*, were quantified before (day 0) and after (day 15) chondrogenic differentiation. In parallel, immunomodulatory genes, including *TGF-β*, *IDO*, and *IL-4*, were quantified before and after IFN-γ treatment. Quantitative RT-PCR was performed using the Luna® Universal One-Step RT-qPCR Kit (New England Biolabs, US). Total RNA was extracted using the Easy Blue Total RNA Extraction Kit (iNtRON Biotechnology, Korea). The RT-PCR reaction consisted of a cycle at 45 °C for 10 min and 95 °C for 1 min, followed by 40 cycles at 95 °C for 10 s and 60 °C for 30 s. Gene expression was normalized to GAPDH and the relative

Table 1 Primers used for the quantification of chondrogenesis- and immunomodulation-related gene expression

Gene	Primer sequence (5′-3′)		Accession no.	References
hCol1a2	F	AGCAGGAGGTTTCGGCTAAG	NM_000089.4	Le et al. (2021)
	R	GCAACAAAGTCCGCGTATCC		
hCol2a1	F	ATCCGGTATTAGGGTCGCTTG	XM_017018831.1	Le et al. (2021)
	R	GAGGAGCGACTGGAAGGTTT		
hSox9	F	AGGACCACCCGGATTACAAG	NM_000346.4	Le et al. (2021)
	R	CCTTGAAGATGGCGTTGGG		
hRunx2	F	GGAGTGGACGAGGCAAGAGTTT	NM_001015051.4	Le et al. (2021)
	R	AGCTTCTGTCTGTGCCTTCTGG		
hAcan	F	TCGAGGACAGCGAGGCC	NM_001135.4	Le et al. (2021)
	R	TCGAGGGTGTAGCGTGTAGAGA		
hTGF-β	F	CAGCAACAATTCCTGGCGATA	NM_000660.7	Strong et al. (2015)
	R	AAGGCGAAAGCCCTCAATTT		
hIDO	F	GGCTTTGCTCTGCCAAATCC	NM_002164.6	Croitoru-Lamoury et al. (2011)
	R	TTCTCAACTCTTTCTCGAAGCTG		
hIL-4	F	AACAGCCTCACAGAGCAGAAGAC	NM_001354990.2	Boeuf et al. (2005)
	R	GCCCTGCAGAAGGTTTCCTT		
hGAPDH	F	TGAAGGTCGGAGTCAACGGATTTGGT	NM_001357943.2	Glare et al. (2002)
	R	CATGTGGGCCATGAGGTCCACCAC		

gene expression was analyzed by the Livak method (2− ΔΔCT) (Livak & Schmittgen, 2001). The primers used for this assay are listed in Table 1.

2.5 Statistical Analysis

The data were statistically analyzed by the t-test using Graphpad Prism 9.0 software (GraphPad Software, Inc., US). The difference between groups was considered statistically significant when the p-value was less than 0.05.

3 Results

3.1 Human Umbilical Cord-Derived Mesenchymal Stem Cells Gradually Lost Their Morphological Features Under the Influence of IFN-γ

After 24 h of culture, IFN-γ-treated hUC-MSCs retained the typical MSC morphology, fibroblast-like elongated shape, and smooth cell surface. After 48 h of induction, the cell density and proliferative capacity of the hUC-MSCs reduced; cells were clustered (Fig. 1, red arrow) and tended to spread flat on the culture surface, while the control sample retained MSC-specific morphology and maintained a stable proliferation rate. Furthermore, the IFN-γ-treated hUC-MSCs gradually lost their characteristic MSC morphology, with the appearance of many inclusions inside the cytoplasm, leading to the cell surface no longer being smooth compared to the control sample.

3.2 Human Umbilical Cord-Derived Mesenchymal Stem Cells Maintained the Capacity for Multi-Lineage Differentiation Under the Influence of IFN-γ

After 30 days of osteogenic induction, the hUC-MSCs accumulated calcium in their extracellular matrix, so the cells turned orange-red to deep red when stained with Alizarin Red dye. The greater the amount of calcium deposited in the matrix, the more intense the stain. The results showed that both the control and IFN-γ-treated

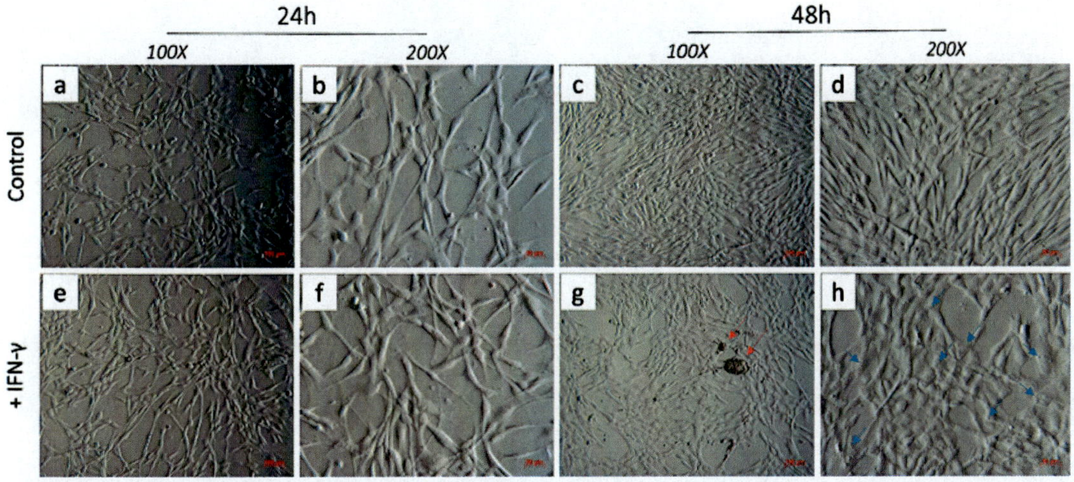

Fig. 1 Human umbilical cord-derived mesenchymal stem cells (hUC-MSCs) change their morphology of under the influence of IFN-γ. (**a**–**d**): Control hUC-MSCs after 24 h (**a**, **b**) and 48 h (**c**, **d**); (**e**–**h**): IFN-γ-treated hUC-MSCs after 24 h (**e**, **f**) and 48 h (**g**, **h**) of subculture; IFN-γ-treated hUC-MSCs reduced proliferative capacity, gradually lost the characteristic morphology of MSCs, clustered (red arrows) and appeared many inclusions when compared with control

hUC-MSCs could maintain a stable potential for differentiation into osteoblasts; however, the control sample showed better osteogenic differentiation capacity (Fig. 2a, b).

After 15 days of chondrogenic induction, the hUC-MSCs accumulated proteoglycans in their chondrogenic matrix, forming orange-red to dark-red spheroids when stained with Safranin-O dye. The central region of the spheroid appeared dark red because of the thick layer of chondrocytes that prevented light from passing through. At the edge of the spheroid, the cell layer was thinner and thus stained orange-red. The results showed that the control samples and IFN-γ-treated hUC-MSCs both differentiated into spheroids, secreted proteoglycans in the chondrogenic matrix, and maintained their potential for differentiation into chondrocytes (Fig. 2c, d).

After 15 days of adipogenic induction, lipid droplets were formed inside the cells. These lipid droplets stained positive with Oil Red O. The results showed that the control samples and IFN-γ-treated hUC-MSCs both contained accumulated lipid droplets. The control cells formed more lipid droplets than the IFN-γ-treated hUC-MSCs; the lipid droplets were also larger in size. Visually, IFN-γ aggravated the adipocyte differentiation potential of hUC-MSCs (Fig. 2e, f).

3.3 The Immunophenotype of Human Umbilical Cord-Derived Mesenchymal Stem Cells Was Almost Unchanged After IFN-γ Treatment

The results showed that the control samples strongly expressed the positive markers CD44 (99.994% ± 0.005%), CD73 (99.985% ± 0.005%), CD90 (99.992% ± 0%), and CD105 (94.070% ± 0.488%; Fig. 3e–h) and showed almost no expression of the negative markers CD14 (0.003% ± 0.005%), CD34 (0.003% ± 0.005%), CD45 (0.015% ± 0.01%), and HLA-DR (0.014% ± 0.024%; Fig. 3a–d). IFN-γ-treated hUC-MSCs also expressed high levels of CD44 (99.905% ± 0.039%), CD73 (99.840% ± 0.026%), CD90 (99.893% ± 0.067%), and CD105 (91.850% ± 0.921%; Fig. 3m–p) and did not express negative markers, specifically CD14 (0.027% ± 0.025%), CD34

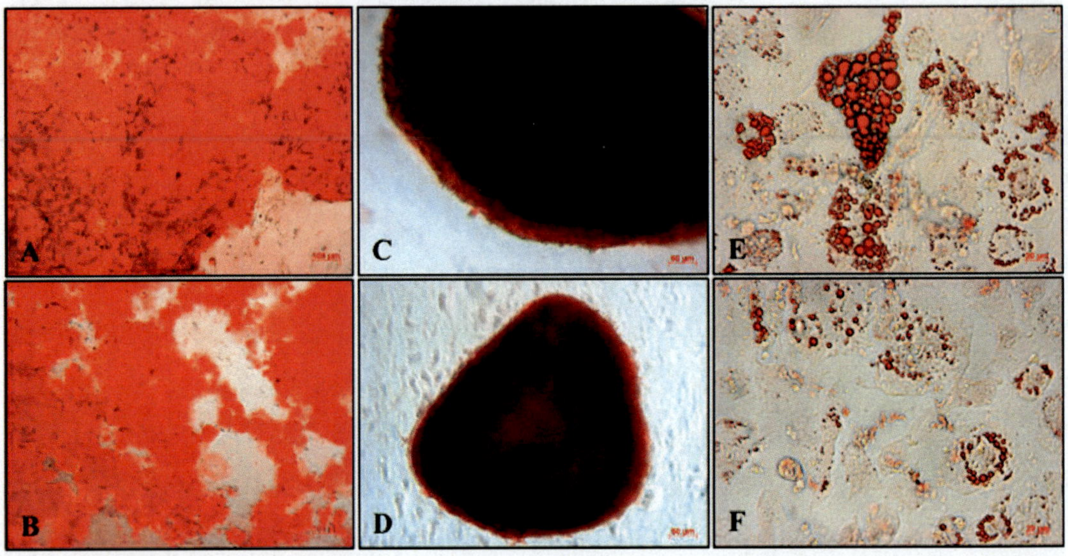

Fig. 2 The differentiation potential of human umbilical cord-derived mesenchymal stem cells (hUC-MSCs) under the influence of IFN-γ. (**a**, **c**, **e**): Control hUC-MSCs after osteogenic, chondrogenic, and adipogenic induction. (**b**, **d**, **e**): Those of IFN-γ-treated hUC-MSCs

(0.006% ± 0.01%), CD45 (0.01% ± 0.017%), and HLA-DR (0.021% ± 0.02%; Fig. 3i–l).

3.4 Human Umbilical Cord-Derived Mesenchymal Stem Cells Showed Increased Expression of *IDO* and *IL-4* and Decreased Expression of *TGF-β* Under the Influence of IFN-γ

Both untreated hUC-MSCs and those treated with IFN-γ expressed genes related to immunomodulation, such as *TGF-β*, *IDO*, and *IL-4*. IFN-γ-treated hUC-MSCs, however, increased the transcriptional activity of *IDO* and *IL-4* and decreased that of *TGF-β*. The expression of *TGF-β* was downregulated 0.31 ± 0.08-fold ($p < 0.01$), while that of *IDO* and *IL-4* was upregulated 203,781.94 ± 33,185.63-fold ($p < 0.01$) and 11.63 ± 2.56-fold, respectively ($p < 0.05$; Fig. 4).

3.5 Human Umbilical Cord-Derived Mesenchymal Stem Cells Showed Decreased Expression of Genes Related to Procollagen and Proteoglycan Under the Influence of IFN-γ

hUC-MSCs treated with or without IFN-γ both tended to clump together to form clusters. Initially, IFN-γ-treated hUC-MSCs clumped together later than their counterparts (Fig. 5b, f). On day 10, the cells agglomerated into a spheroid-like structure, but some spheroids with irregular fringes that incompletely formed into a unified spheroid due to a few discrete clusters of cells around the mass remained (Fig. 5c, g). On day 15, the number of spheroids formed was higher, and the fringes of the spheroids were more uniformly rounded in shape, in the untreated hUC-MSC sample than in IFN-γ-treated hUC-MSCs (Fig. 5d, h).

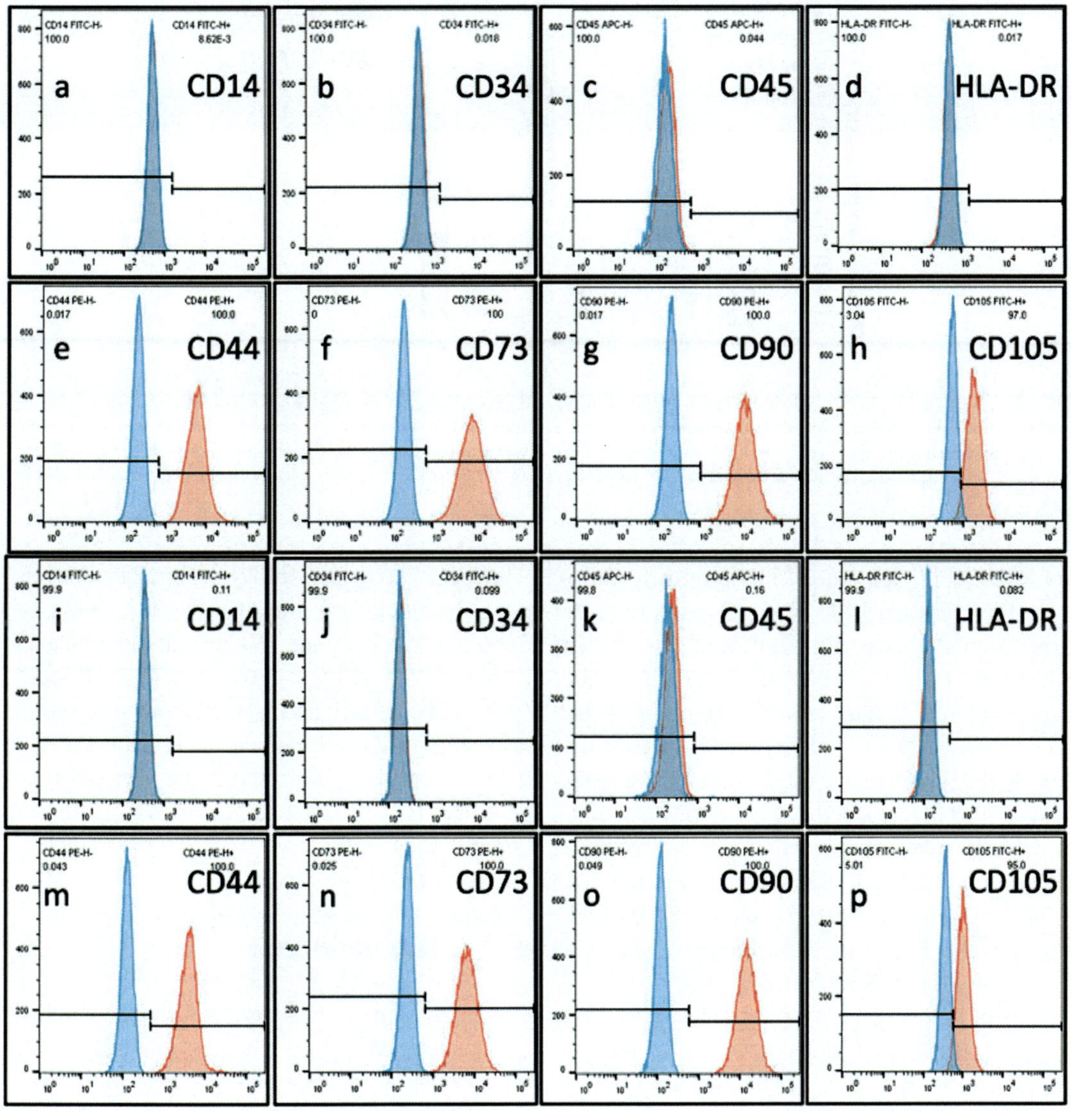

Fig. 3 Immunophenotype of human umbilical cord-derived mesenchymal stem cells (hUC-MSCs) under the influence of IFN-γ. (**a**–**h**): Immunophenotype of control hUC-MSCs. (**i**–**p**): Immunophenotype of IFN-γ-treated hUC-MSCs. hUC-MSCs did not expressed negative markers and strongly expressed positive markers even if induced 48 h by 10 ng/mL IFN-γ

3.6 Changes in Chondrogenic Gene Expression Under the Influence of IFN-γ

The control and IFN-γ-treated hUC-MSCs both expressed chondrogenic genes such as *Col1a2*, *Col2a1*, *Sox9*, *Runx2*, and *Acan* on day 0 of differentiation. IFN-γ-treated hUC-MSCs showed increased expression of *Acan*, while the expression of *Col1a2*, *Sox9*, and *Runx2* decreased 0.09 ± 0.02-fold ($p < 0.0001$), 0.47 ± 0.07-fold ($p < 0.001$), and 0.07 ± 0.01-fold ($p < 0.0001$), respectively, of those without treatment. The expression of *Col2a1* increased slightly but was not significant (1.31 ± 0.27-fold, $p > 0.05$). *Acan* was dramatically upregulated 4.02 ± 0.93-fold

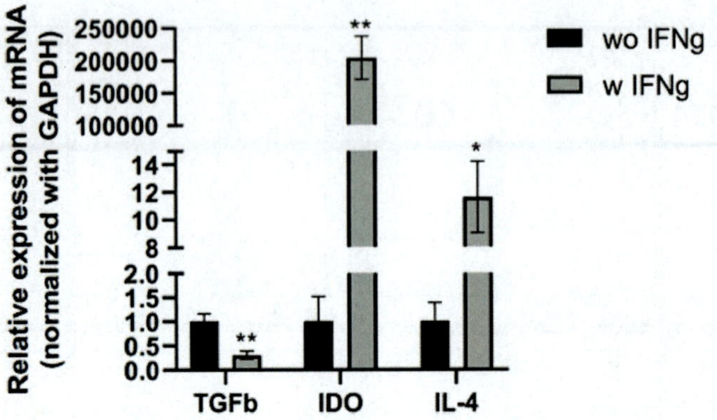

Fig. 4 Changes in expression of genes related to the immunomodulatory capability of human umbilical cord-derived mesenchymal stem cells (hUC-MSCs) under the influence of IFN-γ. TGF-β expression levels of IFN-γ-treated hUC-MSCs were decreased compared to the control while the expression of IDO and IL-4 were increased strongly compared to the control. *: $p < 0.05$, **: $p < 0.01$

($p < 0.05$; Fig. 6a). After 15 days of differentiation, IFN-γ-treated hUC-MSCs showed reduced expression of almost all investigated genes when compared with the control. Specifically, the expression of *Col1a2*, *Runx2*, and *Acan* was downregulated 0.17 ± 0.02-fold ($p < 0.0001$), 0.32 ± 0.02-fold ($p < 0.0001$), and 0.08 ± 0.02-fold ($p < 0.01$), respectively, in these cells compared to the cells without IFN-γ treatment. The expression of *Col2a* and *Sox9* was upregulated 1.04 ± 0.07-fold ($p > 0.05$) and 0.92 ± 0.05-fold, respectively; however, these changes were not statistically significant ($p > 0.05$; Fig. 6b).

The control both before (day 0) and after (day 15) the induction of chondrocyte differentiation expressed chondrogenic genes such as *Col1a2*, *Col2a1*, *Sox9*, *Runx2*, and *Acan*. After 15 days of normal differentiation, the expression levels of *Col2a1* and *Acan* increased dramatically, while the expression levels of *Sox9* and *Runx2* decreased compared to day 0. The expression of *Runx2* and *Sox9* on day 15 was downregulated 0.78 ± 0.05-fold ($p < 0.01$) and 0.17 ± 0.01-fold ($p < 0.0001$), respectively, on day 0. The expression of *Col2a1* and *Acan* was upregulated 106.89 ± 6.28-fold and 36.38 ± 7.65-fold, respectively ($p > 0.05$). *Col1a2*'s increase was not statistically significant (Fig. 6c). After 15 days of differentiation, IFN-γ-treated hUC-MSCs showed increased expression of the *Col1a2*, *Col2a1*, and *Sox9* genes compared to day 0; other gene levels were decreased. Specifically, the expression of *Col1a2*, *Col2a1*, and *Sox9* was upregulated 2.16 ± 0.20-fold ($p < 0.01$), 85.04 ± 6.05-fold ($p < 0.0001$), and 1.53 ± 0.09-fold ($p < 0.01$), respectively. The expression of *Runx2* decreased 0.72 ± 0.04-fold ($p < 0.05$) and that of *Acan* decreased 0.77 ± 0.21-fold but was not significant ($p > 0.05$) (Fig. 6).

4 Discussion

Currently, stem cell therapy is being researched and developed to treat many diseases. MSC transplantation has been proven to exert a therapeutic effect by reducing inflammation through the immunomodulatory and anti-inflammatory properties and differentiation potential of MSCs (Shabgah et al., 2020; Yu et al., 2019). However, it is still unclear whether pro-inflammatory cytokines reduce or increase the effectiveness of therapy (Abarbanell et al., 2009; Luque-Campos et al., 2019; Zhu et al., 2020). In this study, we evaluated the chondrocyte differentiation and immunomodulatory capability of hUC-MSCs in an inflammatory microenvironment in vitro to initially investigate the impact of pro-inflammatory cytokines on transplanted or endogenous hUC-MSCs.

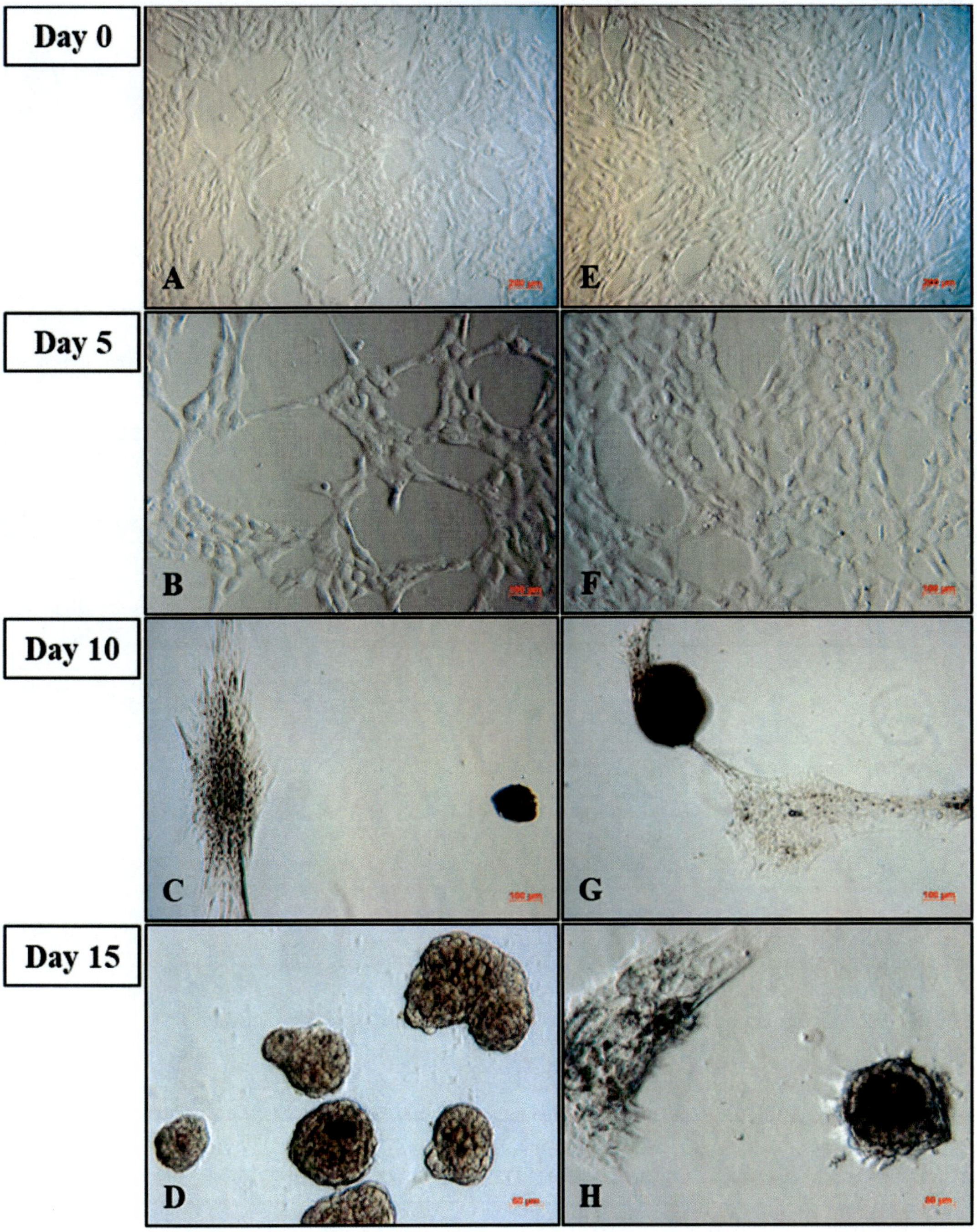

Fig. 5 Changes in human umbilical cord-derived mesenchymal stem cells (hUC-MSCs) morphology during the differentiation of chondrocytes under the influence of IFN-γ. (**a**–**d**) Control sample, (**e**–**h**) IFN-γ-treated hUC-MSCs. (**a**, **e**) Day 0; (**b**, **f**) day 5; (**c**, **g**) day 10; (**d**, **h**) day 15

First, we investigated the morphological changes in cells under an inflammatory condition induced by 10 ng/mL IFN-γ. After 24 h of treatment, hUC-MSCs continued to proliferate well and maintained their spindle shape. After another 24 h, IFN-γ-treated hUC-MSCs showed

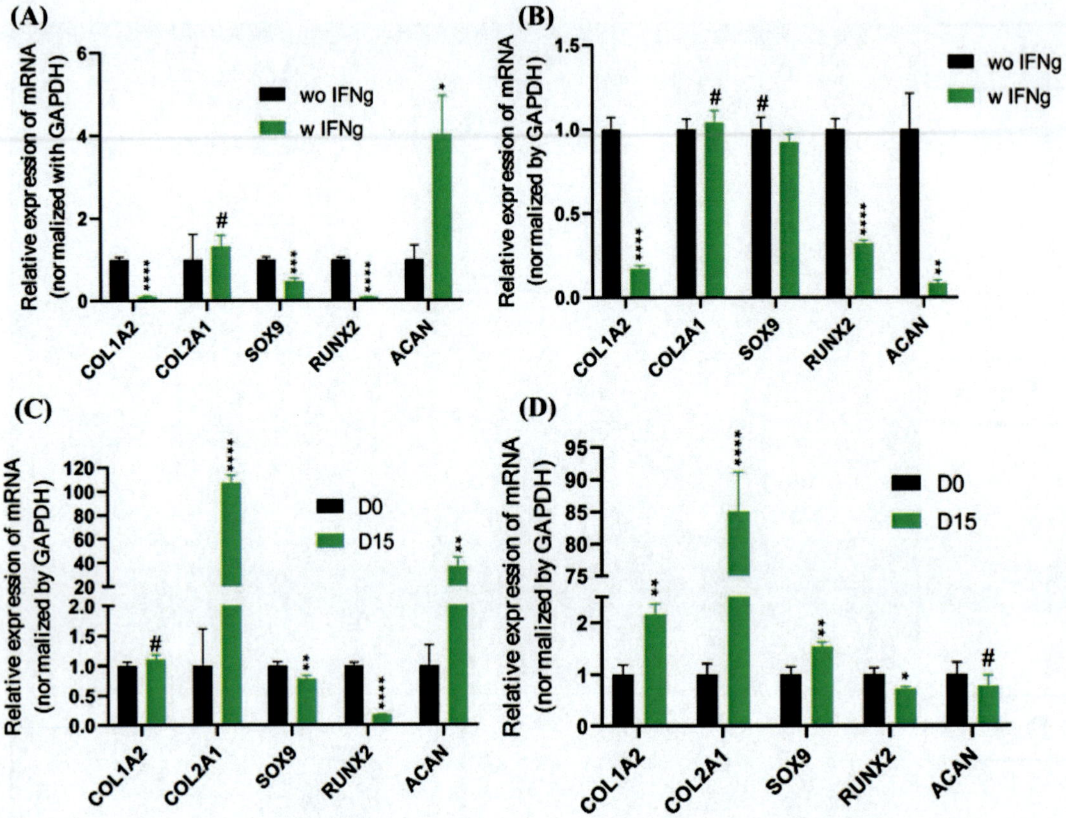

Fig. 6 Changes in chondrogenic-related gene expression under the influence of IFN-γ. (**a**) The difference in gene expression at day 0 between the control and IFN-γ-treated hUC-MSCs. (**b**) The difference in gene expression at day 15 between the control and IFN-γ-treated hUC-MSCs. (**c**) The difference in gene expression of the control between day 0 and day 15. (**d**) The difference in gene expression of IFN-γ-treated hUC-MSCs between day 0 and day 15

decreased proliferation, tended to cluster, became flattened, and spread over the culture surface. The appearance of cytoplasmic inclusions made these cells larger than the control, and they were no longer smooth. The above changes might be triggered by the accumulation of a variety of cellular lesions, including DNA damage (Kawanishi et al., 2017; Lopez-Otin et al., 2013). This leads to senescence and a decrease in the regenerative potential of MSCs (Lopez-Otin et al., 2013; Rossi et al., 2007). Senescent cells can also induce senescence in neighbors through cell-cell interactions (Nelson et al., 2012). Inflammation not only accumulates DNA damage but also causes oxidative stress. The overproduction of free radicals impairs the function of cell membranes (Khansari et al., 2009); thus, MSCs gradually lose their characteristic shape. Oxidative imbalances accompanied by DNA damage can also lead to autophagy dysfunction (Wang et al., 2019). When the autophagy of cells is impaired, the concentration of free radicals continues to increase due to the accumulation of many damaged cells (Wang & Levine, 2010). A high intracellular free radical concentration is also one of the causes of cell flattening and spreading (Gu et al., 2016). In brief, the control maintained better cellular morphology than the IFN-γ-treated cells.

The control and treated hUC-MSCs both lacked expression of CD14, CD34, CD45, and HLA-DR (less than 2%) and strongly expressed CD44, CD73, CD90, and CD105. The CD105 expression of the treated cells was a little lower than that of the control due to their tendency to cluster, which affected the expression of this marker (Hagmann et al., 2013). Furthermore, stem cells are widely used in clinical treatments because of their multilineage differentiation, comprising osteoblasts, chondroblasts, and adipocytes. The osteoblast, chondroblast, and adipocyte differentiation of the control was higher than that of IFN-γ-treated hUC-MSCs; however, the difference between these two samples was not significant. Although the morphology of IFN-γ-treated hUC-MSCs changed after 48 h of culture, their capacity to maintain their differentiation potential was virtually unaffected by IFN-γ in our experiment setting; in other words, 10 ng/mL IFN-γ did not completely inhibit the osteogenesis, chondrogenesis, and adipogenesis of hUC-MSCs.

Continuously, IFN-γ-treated hUC-MSCs showed increased expression of *IDO* and *IL-4* and decreased expression of *TGF-β*. Several reports showed that hUC-MSCs induced with IFN-γ or other pro-inflammatory cytokines express large amounts of immunomodulatory mediators such as *IDO* and *IL-4* (de Witte et al., 2017). These anti-inflammatory cytokines promote the growth of fibroblasts, endothelial cells, and tissue progenitor cells to repair or regenerate damaged cells (de Cássia Noronha et al., 2019; Ma et al., 2014). Meanwhile, *TGF-β* has complex pro- and anti-inflammatory effects (Li & Flavell, 2008a, b), with its dual effect being dependent on the cell and ambient environment (Li et al., 2006). Under inflammatory conditions, *TGF-β* plays a role in promoting inflammation and increasing autoimmunity (Sanjabi et al., 2009). Thus, IFN-γ-treated hUC-MSCs might be immunomodulatory through the decreased expression of *TGF-β*. Furthermore, many studies have shown that IFN-γ can reduce the expression of *TGF-β* by increasing the expression of the protein Smad7 (Kuga et al., 2003; Massagué & Chen, 2000; Nakao et al., 2000). When IFN-γ activates the Smad7 promoter, Smad7 protein expression is increased (Weng et al., 2007) and inhibits TGF-β signaling by blocking the phosphorylation of the receptor-regulated SMADs (R-Smads) (Macias et al., 2015; Wen et al., 2004; Yan et al., 2009). *TGF-β* is also known to regulate cell proliferation and differentiation (Moustakas et al., 2002; Wang et al., 2012; Zhang et al., 2017); hence, a decrease in *TGF-β* expression might lead to a decrease in proliferation and differentiation capability. This is consistent with the above results showing that the proliferation and density of IFN-γ-treated hUC-MSCs reduced after 48 h of culture and that their potential for differentiation into osteoblasts, chondroblasts, and adipocytes was inferior to that of the control.

On day 15 of chondrocyte differentiation, hUC-MSCs showed increased expression of *Col2a1* and *Acan*, decreased expression of the transcription factors *Sox9* and *Runx2*, and unchanged expression of *Col1a2*. After treatment with IFN-γ, the hUC-MSCs showed the same expression trends for *Col2a1* and *Runx2* while increasing *Col1a2* and *Sox9* and decreasing *Acan* during differentiation (Fig. 6c, d). *Sox9* is an essential transcription factor for chondrogenesis, strongly expressed to promote chondrogenesis in the early stage but having reduced expression in the late stage. This transcription factor, together with *Sox5* and *Sox6*, regulates the synthesis of extracellular proteins during chondrocyte development, including *Col1a2* and *Col2a1* (Akiyama et al., 2002; Ikeda et al., 2004; Sekiya et al., 2000; Tew et al., 2005). Type II collagen is recognized to express early and increase continuously throughout the process of articular cartilage differentiation (Goldring et al., 2006; Luo et al., 2017), while type I forms the structure of bone and some other connective tissues (Henriksen & Karsdal, 2019). In our study, the increased expression of both collagen types with IFN-γ treatment may be related to *Sox9* expression when the process of cartilage development has not ended. Therefore, the expression of *Col1a2* in the treated cells increased but was still lower than that in the control at all time points. *Col2a1* expression increased very strongly after 15 days of differentiation regardless of IFN-γ treatment, consistent with the fact that this type of collagen has the highest proportion in cartilage tissue (Adolphe et al., 1997).

Along with the regulation of chondrocyte proliferation through *Sox9*, the expression of *Runx2* is also required (Zhou et al., 2006). This transcription factor is essential for chondrocyte maturation or in the early stage of differentiation and is known to regulate RANKL expression in chondrocytes (Usui et al., 2008) and promote the expression of catabolic factors in the chondrocyte extracellular matrix (Catheline et al., 2018). On day 15, both the control and IFN-γ-treated hUC-MSCs showed decreased expression of *Runx2* (Fig. 6c, d); this decrease was greater in treated cells (Fig. 6a, b). This result corresponds to the morphology of chondrogenic spheroids described above (Fig. 5). More chondrogenic spheroids were produced in the control; the spheroids were also larger and thicker. Although the mechanism by which IFN-γ reduces collagen and proteoglycan synthesis (*Col1a2* and *Acan*) has not been fully elucidated (Mallat et al., 1995), data have shown that IFN-γ reduces procollagen mRNA expression (Goldring et al., 1986) and inhibits proteoglycan synthesis (Dodge et al., 1998) at the transcriptional level. Moreover, the inhibition of collagen and proteoglycan synthesis was dependent on IFN-γ concentration (Jimenez et al., 1984). Additionally, hUC-MSCs, after differentiation into chondrocytes, are responsible for the maintenance and repair of the extracellular matrix. However, differentiated hUC-MSCs are unstable (Pelttari et al., 2006) and may differentiate into hypertrophic chondrocytes (Fong et al., 2012; Goldring & Marcu, 2009). This will increase *Runx2* expression (Lu et al., 2016), as *Runx2* is involved in activating hypertrophic differentiation (Hunter & Felson, 2006). The results in this study showed a decrease in the *Runx2* mRNA level on day 15, indicating that the differentiated hUC-MSCs did not reach hyper-differentiation.

5 Conclusion

hUC-MSCs induced for 48 h with 10 ng/mL IFN-γ could proliferate and maintain MSC morphology, immunophenotype, and differentiation potential. The cells still exhibited immunomodulatory and chondrocyte differentiation capabilities; however, there was a decrease in the expression of genes related to procollagen and proteoglycan.

Acknowledgments This research is funded by the National University Ho Chi Minh City (VNU-HCM) under grant number NV2018-18-2.

Author's Contributions All authors read and approved the final manuscript.

Competing Interests The authors declare that they have no competing interests.

References

Abarbanell, A. M., et al. (2009). Proinflammatory cytokine effects on mesenchymal stem cell therapy for the ischemic heart. *The Annals of Thoracic Surgery, 88*(3), 1036–1043.

Adolphe, M., Thenet-Gauci, S., & Demignot, S. (1997). Chondrocyte culture: A target system to evaluate: Pharmacotoxicological effects of drugs. In *In vitro methods in pharmaceutical research* (pp. 181–207). Elsevier.

Akiyama, H., et al. (2002). The transcription factor Sox9 has essential roles in successive steps of the chondrocyte differentiation pathway and is required for expression of Sox5 and Sox6. *Genes & Development, 16*(21), 2813–2828.

Arutyunyan, I., et al. (2016). Umbilical cord as prospective source for mesenchymal stem cell-based therapy. *Stem Cells International, 2016*, 6901286.

Boeuf, P., et al. (2005). CyProQuant-PCR: A real time RT-PCR technique for profiling human cytokines, based on external RNA standards, readily automatable for clinical use. *BMC Immunology, 6*, 5.

Catheline, S. E., et al. (2018). Chondrocyte-specific RUNX2 overexpression causes chondrodysplasia during development, but is not sufficient to induce OA-like articular cartilage degeneration in adult mice without injury. bioRxiv, 470005.

Chinnadurai, R., et al. (2014). IDO-independent suppression of T cell effector function by IFN-γ–licensed human mesenchymal stromal cells. *The Journal of Immunology, 192*(4), 1491–1501.

Conrad, B., Hayashi, C., & Yang, F. (2021). Gelatin-based microribbon hydrogels guided mesenchymal stem cells to undergo endochondral ossification in vivo with bone-mimicking mechanical strength. *Regenerative Engineering and Translational Medicine, 7*, 301–311.

Croitoru-Lamoury, J., et al. (2011). Interferon-γ regulates the proliferation and differentiation of mesenchymal stem cells via activation of indoleamine 2, 3 dioxygenase (IDO). *PLoS One, 6*(2), e14698.

de Cássia Noronha, N., et al. (2019). Priming approaches to improve the efficacy of mesenchymal stromal cell-based therapies. *Stem Cell Research & Therapy, 10*(1), 1–21.

De Witte, S. F., et al. (2016). Toward development of iMesenchymal stem cells for immunomodulatory therapy. *Frontiers in Immunology, 6*, 648.

de Witte, S. F., et al. (2017). Cytokine treatment optimises the immunotherapeutic effects of umbilical cord-derived MSC for treatment of inflammatory liver disease. *Stem Cell Research & Therapy, 8*(1), 1–12.

Dodge, G. R., et al. (1998). Effects of interferon-γ and tumor necrosis factor α on the expression of the genes encoding aggrecan, biglycan, and decorin core proteins in cultured human chondrocytes. *Arthritis & Rheumatism: Official Journal of the American College of Rheumatology, 41*(2), 274–283.

Dominici, M., et al. (2006). Minimal criteria for defining multipotent mesenchymal stromal cells. The International Society for Cellular Therapy position statement. *Cytotherapy, 8*(4), 315–317.

Fong, C., et al. (2012). Human umbilical cord Wharton's jelly stem cells and its conditioned medium support hematopoietic stem cell expansion ex vivo. *Journal of Cellular Biochemistry, 113*(2), 658–668.

Glare, E. M., et al. (2002). Beta-actin and GAPDH housekeeping gene expression in asthmatic airways is variable and not suitable for normalising mRNA levels. *Thorax, 57*(9), 765–770.

Goldring, M. B., & Marcu, K. B. (2009). Cartilage homeostasis in health and rheumatic diseases. *Arthritis Research & Therapy, 11*(3), 1–16.

Goldring, M. B., et al. (1986). Immune interferon suppresses levels of procollagen mRNA and type II collagen synthesis in cultured human articular and costal chondrocytes. *Journal of Biological Chemistry, 261*(19), 9049–9055.

Goldring, M. B., Tsuchimochi, K., & Ijiri, K. (2006). The control of chondrogenesis. *Journal of Cellular Biochemistry, 97*(1), 33–44.

Gu, Y., et al. (2016). Changes in mesenchymal stem cells following long-term culture in vitro. *Molecular Medicine Reports, 13*(6), 5207–5215.

Hagmann, S., et al. (2013). Different culture media affect growth characteristics, surface marker distribution and chondrogenic differentiation of human bone marrow-derived mesenchymal stromal cells. *BMC Musculoskeletal Disorders, 14*(1), 1–11.

Henriksen, K., & Karsdal, M. A. (2019). Chapter 1: Type I collagen. In M. A. Karsdal (Ed.), *Biochemistry of collagens, laminins and elastin* (2nd ed., pp. 1–12). Academic Press.

Hunter, D. J., & Felson, D. T. (2006). Clinical review-osteoarthritis. *BMJ-British Medical Journal-International Edition, 332*(7542), 639–642.

Ikeda, T., et al. (2004). The combination of SOX5, SOX6, and SOX9 (the SOX trio) provides signals sufficient for induction of permanent cartilage. *Arthritis and Rheumatism, 50*(11), 3561–3573.

Jimenez, S., Freundlich, B., & Rosenbloom, J. (1984). Selective inhibition of human diploid fibroblast collagen synthesis by interferons. *The Journal of Clinical Investigation, 74*(3), 1112–1116.

Kang, I., et al. (2018). Donor-dependent variation of human umbilical cord blood mesenchymal stem cells in response to hypoxic preconditioning and amelioration of limb ischemia. *Experimental & Molecular Medicine, 50*(4), 1–15.

Kawanishi, S., et al. (2017). Crosstalk between DNA damage and inflammation in the multiple steps of carcinogenesis. *International Journal of Molecular Sciences, 18*(8), 1808.

Khansari, N., Shakiba, Y., & Mahmoudi, M. (2009). Chronic inflammation and oxidative stress as a major cause of age-related diseases and cancer. *Recent Patents on Inflammation & Allergy Drug Discovery, 3*(1), 73–80.

Kim, D. S., et al. (2018). Involvement of TLR3-dependent PGES expression in immunosuppression by human bone marrow mesenchymal stem cells. *Stem Cell Reviews and Reports, 14*, 286–293.

Kuga, H., et al. (2003). Interferon-γ suppresses transforming growth factor-β-induced invasion of gastric carcinoma cells through cross-talk of Smad pathway in a three-dimensional culture model. *Oncogene, 22*(49), 7838–7847.

Le, H. T.-N., et al. (2021). Production of engineered cartilage from mesenchymal stem cell spheroids. *Frontiers in Bioscience-Landmark, 26*(2), 266–285.

Li, M. O., & Flavell, R. A. (2008a). TGF-β: A master of all T cell trades. *Cell, 134*(3), 392–404.

Li, M. O., & Flavell, R. A. (2008b). Contextual regulation of inflammation: A duet by transforming growth factor-β and interleukin-10. *Immunity, 28*(4), 468–476.

Li, M. O., et al. (2006). Transforming growth factor-β regulation of immune responses. *Annual Review of Immunology, 24*, 99–146.

Liu, W., et al. (2020). Hypoxic mesenchymal stem cell-derived exosomes promote bone fracture healing by the transfer of miR-126. *Acta Biomaterialia, 103*, 196–212.

Livak, K. J., & Schmittgen, T. D. (2001). Analysis of relative gene expression data using real-time quantitative PCR and the 2− ΔΔCT method. *Methods, 25*(4), 402–408.

Lopez-Otin, C., et al. (2013). The hallmarks of aging. *Cell, 153*(6), 1194–1217.

Lopez-Santalla, M., Fernandez-Perez, R., & Garin, M. I. (2020). Mesenchymal stem/stromal cells for rheumatoid arthritis treatment: An update on clinical applications. *Cell, 9*(8), 1852.

Lu, H., et al. (2016). Inhibition of RUNX2 expression promotes differentiation of MSCs correlated with SDF-1 up-regulation in rats. *International Journal of Clinical and Experimental Pathology, 9*(11), 11388–11395.

Luo, Y., et al. (2017). The minor collagens in articular cartilage. *Protein & Cell, 8*(8), 560–572.

Luque-Campos, N., et al. (2019). Mesenchymal stem cells improve rheumatoid arthritis progression by controlling memory T cell response. *Frontiers in Immunology, 10*, 798.

Ma, S., et al. (2014). Immunobiology of mesenchymal stem cells. *Cell Death & Differentiation, 21*(2), 216–225.

Macias, M. J., Martin-Malpartida, P., & Massagué, J. (2015). Structural determinants of Smad function in TGF-β signaling. *Trends in Biochemical Sciences, 40*(6), 296–308.

Mallat, A., et al. (1995). Interferon alfa and gamma inhibit proliferation and collagen synthesis of human Ito cells in culture. *Hepatology, 21*(4), 1003–1010.

Massagué, J., & Chen, Y.-G. (2000). Controlling TGF-β signaling. *Genes & Development, 14*(6), 627–644.

Moraes, D. A., et al. (2016). A reduction in CD90 (THY-1) expression results in increased differentiation of mesenchymal stromal cells. *Stem Cell Research & Therapy, 7*(1), 1–14.

Moustakas, A., et al. (2002). Mechanisms of TGF-β signaling in regulation of cell growth and differentiation. *Immunology Letters, 82*(1–2), 85–91.

Nakao, A., et al. (2000). Blockade of transforming growth factor β/Smad signaling in T cells by overexpression of Smad7 enhances antigen-induced airway inflammation and airway reactivity. *The Journal of Experimental Medicine, 192*(2), 151–158.

Nelson, G., et al. (2012). A senescent cell bystander effect: Senescence-induced senescence. *Aging Cell, 11*(2), 345–349.

Pelttari, K., et al. (2006). Premature induction of hypertrophy during in vitro chondrogenesis of human mesenchymal stem cells correlates with calcification and vascular invasion after ectopic transplantation in SCID mice. *Arthritis & Rheumatism: Official Journal of the American College of Rheumatology, 54*(10), 3254–3266.

Qiu, Y., et al. (2017). TLR3 preconditioning enhances the therapeutic efficacy of umbilical cord mesenchymal stem cells in TNBS-induced colitis via the TLR3-Jagged-1-Notch-1 pathway. *Mucosal Immunology, 10*(3), 727–742.

Rashedi, I., et al. (2017). TLR3 or TLR4 activation enhances mesenchymal stromal cell-mediated Treg induction via notch signaling. *Stem Cells, 35*(1), 265–275.

Rossi, D. J., et al. (2007). Deficiencies in DNA damage repair limit the function of haematopoietic stem cells with age. *Nature, 447*(7145), 725–729.

Saller, M. M., et al. (2012). Increased stemness and migration of human mesenchymal stem cells in hypoxia is associated with altered integrin expression. *Biochemical and Biophysical Research Communications, 423*(2), 379–385.

Sanjabi, S., et al. (2009). Anti-inflammatory and pro-inflammatory roles of TGF-β, IL-10, and IL-22 in immunity and autoimmunity. *Current Opinion in Pharmacology, 9*(4), 447–453.

Sekiya, I., et al. (2000). SOX9 enhances aggrecan gene promoter/enhancer activity and is up-regulated by retinoic acid in a cartilage-derived cell line, TC6. *Journal of Biological Chemistry, 275*(15), 10738–10744.

Shabgah, A. G., et al. (2020). A significant decrease of BAFF, APRIL, and BAFF receptors following mesenchymal stem cell transplantation in patients with refractory rheumatoid arthritis. *Gene, 732*, 144336.

Strong, A. L., Gimble, J. M., & Bunnell, B. A. (2015). Analysis of the pro- and anti-inflammatory cytokines secreted by adult stem cells during differentiation. *Stem Cells International, 2015*, 412467–412467.

Tew, S. R., et al. (2005). Retroviral transduction with SOX9 enhances re-expression of the chondrocyte phenotype in passaged osteoarthritic human articular chondrocytes. *Osteoarthritis and Cartilage, 13*(1), 80–89.

Ullah, I., Subbarao, R. B., & Rho, G. J. (2015). Human mesenchymal stem cells-current trends and future prospective. *Bioscience Reports, 35*(2).

Usui, M., et al. (2008). Murine and chicken chondrocytes regulate osteoclastogenesis by producing RANKL in response to BMP2. *Journal of Bone and Mineral Research, 23*(3), 314–325.

Wang, R. C., & Levine, B. (2010). Autophagy in cellular growth control. *FEBS Letters, 584*(7), 1417–1426.

Wang, M.-K., et al. (2012). Different roles of TGF-β in the multi-lineage differentiation of stem cells. *World Journal of Stem Cells, 4*(5), 28.

Wang, Q., et al. (2016). Comparative analysis of human mesenchymal stem cells from fetal-bone marrow, adipose tissue, and Warton's jelly as sources of cell immunomodulatory therapy. *Human Vaccines & Immunotherapeutics, 12*(1), 85–96.

Wang, S., et al. (2019). Autophagy dysfunction, cellular senescence, and abnormal immune-inflammatory responses in AMD: From mechanisms to therapeutic potential. *Oxidative Medicine and Cellular Longevity, 2019*.

Wen, F.-Q., et al. (2004). Interferon-γ inhibits transforming growth factor-β production in human airway epithelial cells by targeting Smads. *American Journal of Respiratory Cell and Molecular Biology, 30*(6), 816–822.

Weng, H., et al. (2007). IFN-γ abrogates profibrogenic TGF-β signaling in liver by targeting expression of inhibitory and receptor Smads. *Journal of Hepatology, 46*(2), 295–303.

Yan, X., Liu, Z., & Chen, Y. (2009). Regulation of TGF-β signaling by Smad7. *Acta Biochimica et Biophysica Sinica, 41*(4), 263–272.

Yu, Y., et al. (2019). Therapeutic effect of long-interval repeated intravenous administration of human umbilical cord blood-derived mesenchymal stem cells in DBA/1 mice with collagen-induced arthritis. *Journal of Tissue Engineering and Regenerative Medicine, 13*(7), 1134–1142.

Yuan, O., et al. (2019). Exosomes derived from human primed mesenchymal stem cells induce mitosis and potentiate growth factor secretion. *Stem Cells and Development, 28*(6), 398–409.

Zhang, Y., Alexander, P. B., & Wang, X.-F. (2017). TGF-β family signaling in the control of cell

proliferation and survival. *Cold Spring Harbor Perspectives in Biology, 9*(4), a022145.
Zhou, G., et al. (2006). Dominance of SOX9 function over RUNX2 during skeletogenesis. *Proceedings of the National Academy of Sciences, 103*(50), 19004–19009.
Zhu, D., et al. (2020). Inflammatory cytokines alter mesenchymal stem cell mechano-sensing and adhesion on stiffened infarct heart tissue after myocardial infarction. *Frontiers in Cell and Developmental Biology, 8*, 1183.
Zimmermann, J., & McDevitt, T. (2018). Engineering the 3D MSC spheroid microenvironment to enhance immunomodulation. *Cytotherapy, 20*(5), S106.

Adv Exp Med Biol - Innovations in Cancer Research and Regenerative Medicine (2023) 4: 35–45
https://doi.org/10.1007/5584_2023_768

Published online: 5 April 2023

Exosomes Derived from Human Umbilical Cord Mesenchymal Stem Cells Enhance Angiogenesis Through Upregulation of the *VWF* and *Flk1* Genes in Endothelial Cells

Phat Duc Huynh, Phuc Van Pham, and Ngoc Bich Vu

Abstract

Introduction Exosomes derived from mesenchymal stem cells (MSCs) are crucial mediators of the paracrine effects as well as tissue repair and have promising clinical applications. They enhance tissue regeneration by reducing inflammatory responses, enhancing proliferation, inhibiting apoptosis, and stimulating angiogenesis. This study aimed to evaluate the mechanism of angiogenesis supported by exosomes derived from MSCs.

Methods Exosomes were isolated via ultracentrifugation of a conditioned medium collected from human umbilical cord MSC (hUCMSC) cultures. These exosomes were characterized using transmission electron microscopy, and the expression of specific markers (CD9, CD81, and CD63) was evaluated. To understand the mechanism of angiogenesis, we evaluated the effects of exosomes in endothelial cells (HUVECs). The obtained exosomes were supplemented at a dose of 20 μg/mL into two kinds of culture media for HUVECs (M200 medium and endothelial cell growth medium), while phosphate-buffered saline was added to these media as a control. The effects of the exosomes were evaluated based on the formation of a tubular structure in the culture and the expression of angiogenic genes (*MMP-2*, *Ephrin B2*, *Ephrin B4*, *Flk1*, *Flt1*, *VWF*, *VE-cadherin*, *CD31*, *ANG1*, *ANG2*, and *HGF*) via RT-PCR.

Results The exosomes were obtained from the hUCMSCs at a concentration of 0.7 ± 0.029 μg/mL. They accelerated the formation of new blood vessels by upregulating *HGF*, *VWF*, *CD31*, *Flt1*, and *Flk1* (especially *VWF* and *Flt1*).

P. D. Huynh
Laboratory of Stem Cell Research and Application, University of Science, Ho Chi Minh City, Vietnam

Vietnam National University, Ho Chi Minh City, Vietnam
e-mail: hdphat@hcmus.edu.vn

P. Van Pham
Laboratory of Stem Cell Research and Application, University of Science, Ho Chi Minh City, Vietnam

Vietnam National University, Ho Chi Minh City, Vietnam

Stem Cell Institute, University of Science, Ho Chi Minh City, Vietnam
e-mail: phucpham@sci.edu.vn

N. B. Vu (✉)
University of Science, Ho Chi Minh City, Vietnam

Stem Cell Institute, University of Science, Ho Chi Minh City, Vietnam
e-mail: ngocvu@sci.edu.vn; vbngoc@hcmus.edu.vn

Conclusion Exosomes derived from hUCMSCs can promote angiogenesis through upregulation of *VWF* and *Flt1* in endothelial cells.

Keywords

Angiogenesis · Endothelial cells · Exosomes · Mesenchymal stem cells · Umbilical cord mesenchymal stem cells

Abbreviations

ANG1	Angiopoietin 1
ANG2	Angiopoietin 2
CD	Cluster of differentiation
GFP	Green fluorescent protein
HGF	Hepatic growth factor
hUCMSC	Human umbilical cord mesenchymal stem cells
HUVEC	Human umbilical vein endothelial cells
PBS	Phosphate buffer saline
PIGF	Placental growth factor
TGF-β	Transforming growth factor beta
VEGF	Vascular endothelial growth factor

1 Introduction

Mesenchymal stem cells (MSCs) are potent candidates in regenerative medicine and can be derived from multiple human tissues or organs, such as bone marrow, umbilical cord, and adipose tissue (Panda et al., 2021). The therapeutic effects of MSCs are based on their immunomodulatory, anti-inflammatory, and angiogenic properties. Although with their large size, a large proportion of transplanted MSCs could migrate and be entrapped within the lungs after intravenous administration (Assis et al., 2010; Masterson et al., 2021), the intravenous administration of MSCs can yield therapeutic effects on some diseases of the liver and kidney (Eggenhofer et al., 2012; Cao et al., 2020; Wu & Meng, 2021; Li et al., 2021; Akan et al., 2021). These observations suggest that MSCs can deliver some factors that target injured tissues. Recent reports have described exosomes derived from MSCs as crucial mediators in tissue repair and regeneration (Ahmed & Al-Massri, 2022; Heo & Kim, 2022; Liang et al., 2021; Yin et al., 2019).

Exosomes are classified as a subset of extracellular vesicles (EVs) with a size ranging from 30 to 150 nm (Panda et al., 2021). They were previously considered the "trash bags" of cells. However, exosomes have been discovered to play roles in cell–cell communication, intracellular signaling, and different biological processes. They are also master transporters of microRNAs, mRNAs, and proteins, which are shuttled from their parental cells to recipient cells (Valadi et al., 2007). Exosomes derived from MSCs could enhance the proliferation and migratory capacity of fibroblasts derived from patients with chronic wounds and healthy donors and stimulate angiogenesis in vitro (Shabbir et al., 2015). Moreover, exosomes were identified as a cardioprotective component in the paracrine effects of MSCs. Particularly, they were previously shown to reduce the infarct size (Lai et al., 2010), increase the ATP level, decrease oxidative stress, and activate pro-surviving signaling to enhance myocardial viability and prevent adverse remodeling after myocardial ischemia or reperfusion injury (Arslan et al., 2013). In recent studies, exosomes were proven to stimulate angiogenesis (Komaki et al., 2017; Qiu et al., 2020; Sun et al., 2022; Takeuchi et al., 2019; Yan et al., 2022). Some mechanisms of exosomes that can trigger angiogenesis have been discovered, including those via VEGF/VEGFR (Han et al., 2019), HIF-1 alpha (Zhang et al., 2019), miR-612 (Ge et al., 2021), and miR-21-5p (Huang et al., 2021). However, almost all current studies focus on some molecules inside exosomes that can promote angiogenesis, and the effects of these molecules on the target cells (endothelial cells) for angiogenesis have not been studied. Therefore, this study aimed to determine the effects of exosomes in endothelial cells for angiogenesis.

2 Materials and Methods

2.1 Isolation and Characterization of Human Umbilical Cord MSCs

Frozen human umbilical cord MSCs (hUCMSCs) were supplied by a stem cell bank (SCI Biobank, Stem Cell Institute, HCMC, VN). The cells were rapidly thawed at 37 °C in a water bath. Thereafter, a similar amount of ThawBest solution (Regenmedlab, VN) was added and centrifuged at 500 × g for 5 min. The supernatant was removed, and the pellet was reconstituted using a culture medium (MSCCult I, Regenmedlab). The cells were then cultured in a T25 flask at 37 °C and 5% CO_2. On the next day of culture, the medium was refreshed.

To confirm the MSC phenotype after expansion, we analyzed the surface markers of the hUCMSCs using flow cytometry with anti-human antibodies conjugated with phycoerythrin (PE), fluorescein isothiocyanate (FITC), or PerCP-Cy5.5-conjugated or allophycocyanin (APC): CD14-FITC, CD19-PerCP, CD34-FITC, CD45-APC, CD73-PE, CD90-PerCP, CD105-PerCP, and HLA-DR-FITC (Santa Cruz Biotechnology, Dallas, TX, USA). After incubation at room temperature for 20 min, the cells were examined using the FACSCalibur flow cytometer (BD Bioscience, Franklin, Lakes, NJ, USA). The results were analyzed using the BD CellQuest Pro software (10,000 events).

The osteogenic, chondrogenic, and adipogenic differentiation capacities of the hUCMSCs were assessed using the StemPro® Osteogenesis Differentiation Kit, StemPro® Chondrogenesis Differentiation Kit, and StemPro® Adipogenesis Differentiation Kit (Thermo Fisher Scientific, USA), respectively. After 14–21 days of the incubation period, differentiation was detected by staining with Alizarin Red S (Sigma-Aldrich, USA) for osteoblasts, Oil Red O (Sigma-Aldrich) for adipocytes, and Alcian Blue (Sigma-Aldrich) for chondroblasts.

2.2 Isolation and Characterization of Exosomes Derived from hUCMSCs

The hUCMSCs were cultured in a T175 flask in MSCCult I until 70% of the confluent flasks were replaced by an exosome-free medium (MSCCult MV, Regenmedlab). The cells were incubated in MSCCult MV for 48 h before the conditioned media were collected for exosome isolation.

The conditioned medium was collected and subjected to centrifugation at 300 × g for 10 min, 2000 × g for 10 min, and 10,000 × g for 30 min to remove dead cells and cell debris. Next, the medium was ultracentrifuged at 100,000 × g for 70 min. The supernatant was discarded; the residue was mixed with phosphate-buffered saline (PBS); and centrifugation at 100,000 × g for 70 min was repeated once. The obtained pellets were then incubated again with PBS and stored at −80 °C for further experiments.

The morphology of the exosomes was observed using transmission electron microscopy (TEM). The existence of tetraspanins, including CD9, CD63, and CD81, on the surface of the exosomes was confirmed via flow cytometry. The exosomes were adsorbed to solid 3.9 μm latex beads (Invitrogen, USA). The protocol was conducted according to the manufacturer's instructions. In particular, 5 μg of exosomes was incubated with 10 μL of latex beads for 15 min in a 1.5 mL centrifuge tube at room temperature. PBS^- was added to make a volume of 1 mL and incubated overnight at 4 °C. Next, 110 μL of 1 M glycine (Sigma-Aldrich) was added, gently mixed, and incubated at room temperature for 30 min. The sample was centrifuged at 4000 rpm for 3 min to remove the supernatant, and the bead cluster was washed twice in 1 mL PBS/0.5% BSA. Thereafter, the bead cluster was mixed with 500 mL PBS/0.5% BSA and incubated with the antibody diluted in PBS/0.5% BSA for 30 min at 4 °C. The samples were again washed twice with 150 μL PBS/0.5% BSA and analyzed using the FACSMelody flow cytometer (BD Bioscience), and the data were analyzed using the FlowJo software (BD Bioscience).

2.3 Protein Concentration

Six different concentrations of BSA (Sigma-Aldrich) standards—0, 0.25, 0.5, 0.75, 1, and 1.5 mg/mL—were prepared in PBS. Subsequently, 10 μL of each standard was added to a 96-well plate (SPL, Korea). Thereafter, 200 μL of Coomassie Blue G-250 dye was added to the loaded wells. Each reaction was set up in triplicate. The exosome samples were also added to the plate with Coomassie Blue G-250 dye, similar to the standard samples. The samples were then analyzed using DTX 880 (Beckman Coulter 880, USA) to determine their absorbance at 595 nm.

2.4 Tube Formation Assay

Briefly, 96-well microplates were coated with Matrigel (50 μL/well) and allowed to polymerize for 30 min at 37 °C. The HUVECs were then seeded at a density of 1×10^4 cells/well in an endothelial cell growth medium (ECGM) (Sigma-Aldrich) (positive control), an M200 medium (negative control), or 20 μg/mL exosomes. Images were captured using an inverted microscope after 6 h of seeding.

2.5 RNA Extraction and qRT-PCR Analysis

Approximately 3.10^5 hUCMSCs were seeded in a T25 flask and incubated in an ECGM (Sigma-Aldrich) and another ECGM supplemented with 20 μg/mL exosomes. After 12 h of incubation, the total RNA from the cells was isolated using the easy-BLUETM Total RNA Extraction Kit (Intron Biotechnology, Korea). qRT-PCR was performed using the RealPlex4 (Eppendorf) and Luna® Universal One-Step RT-qPCR Kit (Biolabs Inc., New England). Relative expression was calculated using the Linvar formula: $2^{-(\Delta\Delta CT)}$. Primer sequences are listed in Table 1.

Table 1 Oligonucleotide primers

No.	Genes	Size (bp)	Sequence (3′ – 5′)	NCBI gene ID
1.	GAPDH	131	F: GTCTCCTCTGACTTCAACAGCG R: ACCACCCTGTTGCTGTAGCCAA	NM_001357943.2
2.	ANG1	150	F:CAACAGTGTCCTTCAGAAGCAGC R: CCAGCTTGATATACATCTGCACAG	NM_001314051.2
3.	ANG 2	139	F: ATTCAGCGACGTGAGGATGGCA R: GCACATAGCGTTGCTGATTAGTC	NM_001386337.1
4.	CD31	133	F: AAGTGGAGTCCAGCCGCATATC R: ATGGAGCAGGACAGGTTCAGTC	NM_000442.5
5.	Ephrin B2	135	F:GCAAGTTCTGCTGGATCAACCAG R: GCTGTTGCCGTCTGTGCTAGAA	NM_001372056.1
6.	Ephrin B4	126	F: ACATCACAGCCAGACCCAACTG R: AGGCAGAGAACTGCGACCACAA	NM_004444.5
7.	Flk-1	132	F: GGAACCTCACTATCCGCAGAGT R: CCAAGTTCGTCTTTTCCTGGGC	NM_002253.4
8.	Flt-1	118	F: CCTGCAAGATTCAGGCACCTATG R: GTTTCGCAGGAGGTATGGTGCT	NM_001160030.2
9.	HGF	102	F: GAGAGTTGGGTTCTTACTGCACG R: CTCATCTCCTCTTCCGTGGACA	NM_001010932.3
10.	MMP-2	138	F: AGCGAGTGGATGCCGCCTTTAA R: CATTCCAGGCATCTGCGATGAG	NM_001302510.2
11.	TGF-beta	70	F:CATGGAGCTGGTGAAACGGA R: GGCGAGCCTTAGTTTGGACA	NM_011577.2
12.	VE cadherin	112	F: GAAGCCTCTGATTGGCACAGTG R: TTTTGTGACTCGGAAGAACTGGC	NM_001795.5
13.	Vwf	157	F: CCTTGAATCCCAGTGACCCTGA R: GGTTCCGAGATGTCCTCCACAT	NM_000552.5

2.6 Statistical Analysis

Data were analyzed using Student's t test. Statistical analysis was performed using the GraphPad Prism 8.0 software, and differences were considered significant when the p values were < 0.05.

3 Results

3.1 Characterization of hUCMSCs

After 1 day of thawing, the cells adhered and proliferated on the culture surface with an increasing number of days. The hUCMSCs showed a morphology similar to that of fibroblasts, which have elongated shapes. At the higher objective, the prominent large, round nucleus in the center of the cell could be observed (Fig. 1a).

Osteodifferentiation was characterized by staining calcium deposits using Alizarin Red. The osteoblasts differentiated from the hUCMSCs showed intense red–orange staining of extracellular calcium deposits (Fig. 1b). After 14–21 days of culture, chondrogenic differentiation was observed in the sections stained with Alcian Blue (Fig. 1c). This dye is commonly used to detect chondrocyte-associated extracellular proteoglycans. In this study, strong Alcian Blue staining was observed in the hUCMSCs that demonstrated differentiation toward the chondrogenic lineage. After 3–5 days of culture in the adipogenic differentiation medium, lipid droplets were observed clearly under the microscope. Following days of differentiation, a

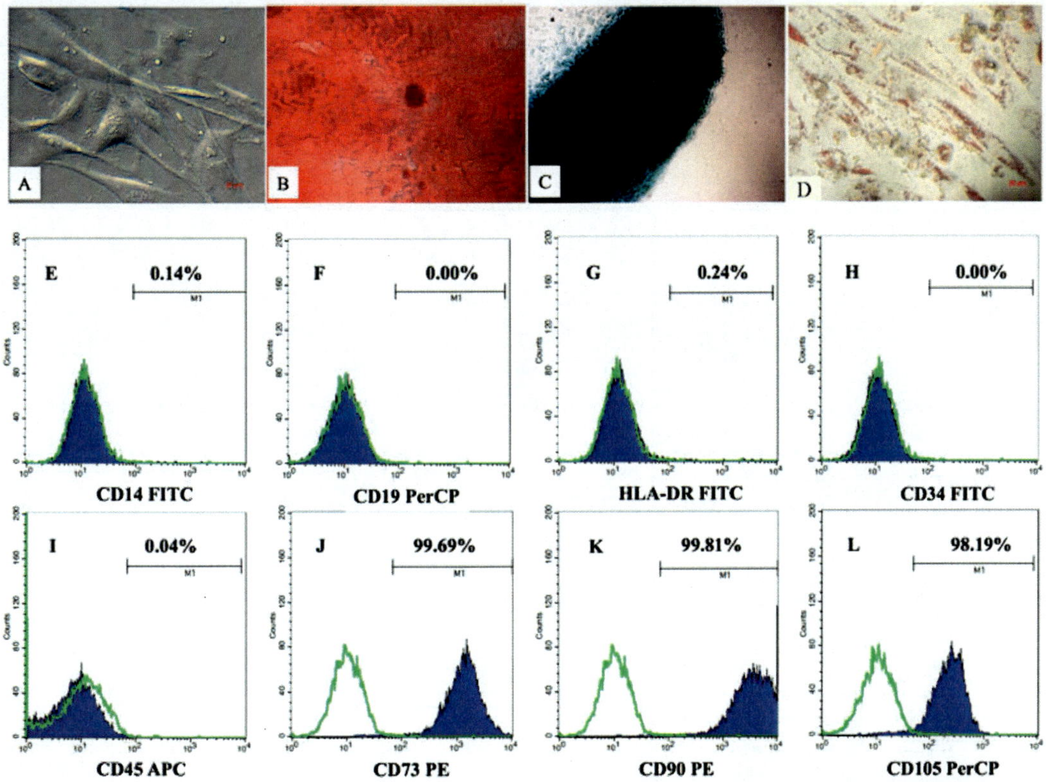

Fig. 1 Phenotype of UCMSCs after thawing and proliferation. (**a**) Morphology of hUCMSC after 1 day of thawing; (**b**) induced cells differentiated into osteoblasts were stained red with Alizarin Red; (**c**) induced cells differentiated into chondrocyte were stained blue with Alcian blue; (**d**) induced cells differentiated into adipocyte were stained red with Oil Red; hUCMSC was negative for CD14 (**e**), CD19 (**f**), HLA-DR (**g**), CD34 (**h**), and CD45 (**i**) and positive for CD73 (**j**), CD90 (**k**), and CD105 (**l**)

large number of lipid droplets accumulated in these cells, which could be seen through positive Oil Red O staining (Fig. 1d).

The surface markers of the target cells were analyzed via flow cytometry. The cells were all positive for CD73 (99.69%), CD90 (99.81%), and CD105 (98.19%) and negative for CD14 (0.14%), CD19 (0%), CD34 (0%), CD45 (0.04%), and HLA-DR (0.24%) (Fig. 1e–l).

3.2 Efficacy of Exosome Isolation from Cell Culture Media

After ultracentrifugation, the obtained exosome amount was 1.128 ± 0.091 μg/mL protein (per milliliter) of the conditioned medium. TEM and flow cytometry were performed to identify the harvested exosomes derived from the hUCMSCs. As shown in Fig. 2, the typical structure of a homogeneous, spherical, and enveloped vesicle was observed on TEM. Flow cytometry also showed that the exosome-bound bead cluster exhibited 84.27% positivity for CD63 and CD81 (Fig. 2a–c, e–f). Among the CD63- and CD81-positive beads, 36.18% were also positive for CD9 (Fig. 2a–c, e–f).

3.3 Facilitation of Endothelial Cell Angiogenesis In Vitro by Exosomes Derived from hUCMSCs

During seeding of the HUVECs on the plates without Matrigel coating, all four media were observed to have spreading cells (Fig. 3). However, on the plates with Matrigel coating, the HUVECs displayed a different tube formation capacity. In the M200 medium without growth factors, the HUVECs did not form tubular structures, while in the medium supplemented with exosomes, the HUVECs formed tubular structures. In the ECGM (as a positive control), the HUVECs easily formed tubular structures. The tubular structure was not clear between the ECGM and ECGM with exosomes (Fig. 3).

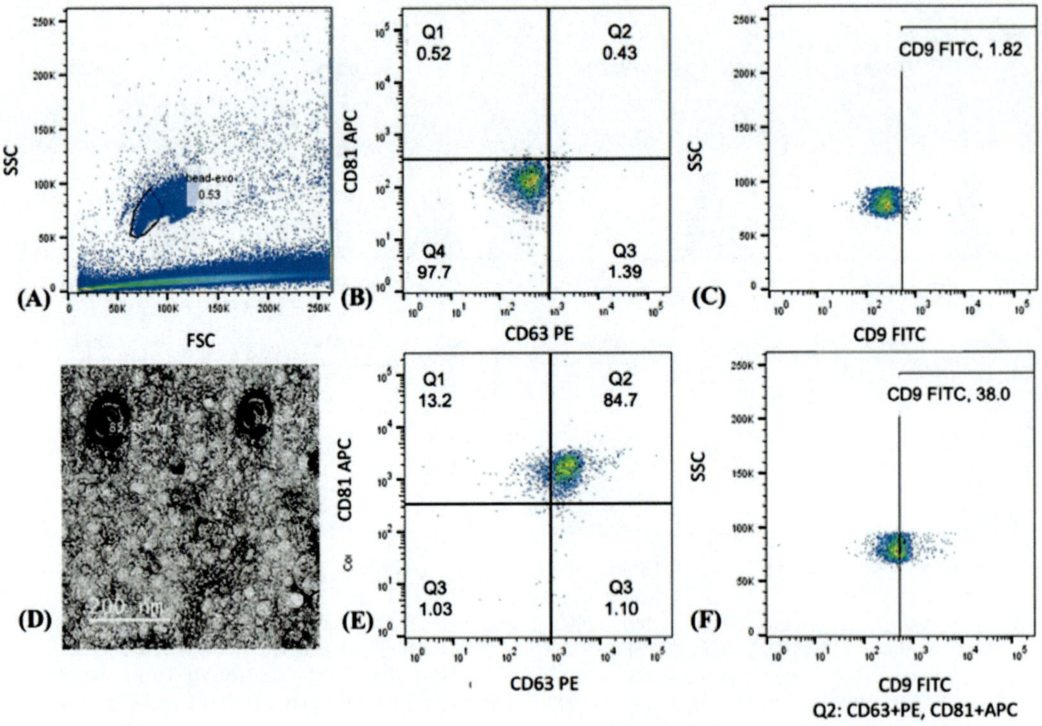

Fig. 2 Characterization of hUCMSC-derived exosomes. The population of exosome-bead (**a**); the expression of CD63, CD81 (**b**), and CD9 (**c**) of the control sample; the morphology of Exos was examined by TEM (**d**); CD63, CD81 (**e**), and CD9 (**f**) of the stained sample

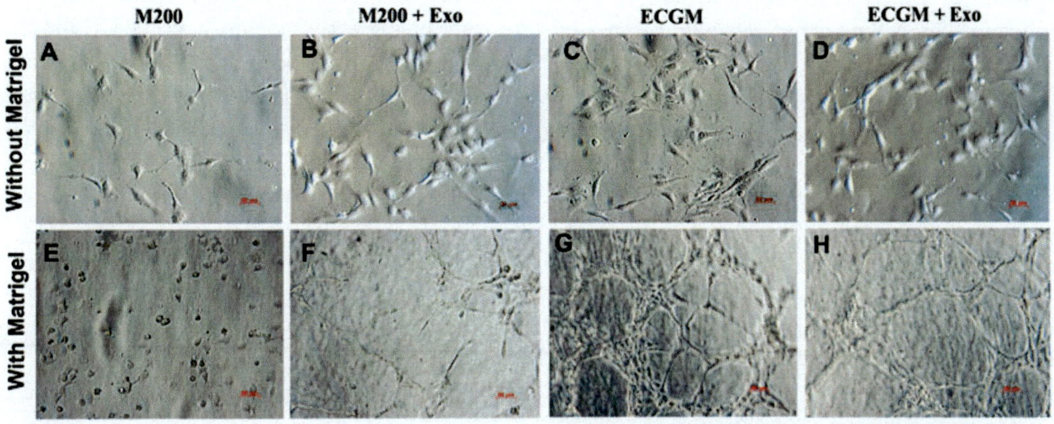

Fig. 3 Tubular structure formation of endothelial cells. In wells without Matrigel coating, HUVECs could not form the tubular structure when cultured in M200 with (**b**) or without exosomes (**a**) and in ECGM medium with (**d**) or without exosomes (**c**). In Matrigel-coated wells, HUVECs could not form tubular structure in M200 medium (**e**) but could form tubular structure in M200 medium supplemented with exosomes (**f**). HUVECs could form tubular structures when cultured in Matrigel-coated wells in ECGM without (**g**) or with exosome supplement (**h**)

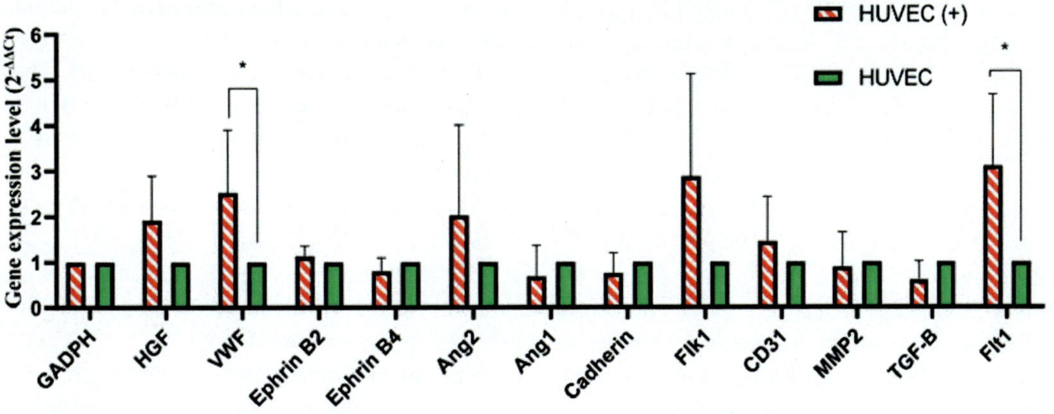

Fig. 4 Gene expression of some angiogenic factors in HUVECs when cultured in the medium with exosome supplement. *HGF*, *VWF*, *CD31*, *Flt1*, *Flk1*, *VWF*, and *Flt1* genes were upregulated after treatment. *Flt1* increased 3.1 ± 1.56 times and *VWF* increased 2.53 ± 1.38 times compared to control (*p < 0.05)

3.4 Expression of Angiogenic Genes in HUVECs

As shown in Fig. 4, the effects of the exosomes and HUVECs increased the expression of some genes related to angiogenesis, including *HGF*, *VWF*, *CD31*, *Flt1*, *Flk1*, *VWF*, and *Flt1*. Specifically, the expression of Flt1 increased 3.1 ± 1.56 times, while that of VWF increased 2.53 ± 1.38 times compared with that of the control (Fig. 4).

4 Discussion

Exosomes have been used in both preclinical and clinical settings to treat some ischemic diseases. They can promote angiogenesis via the main mechanisms related to some microRNAs. This study aimed to explore the effects of exosomes in endothelial cells that trigger angiogenesis and found that exosomes promote angiogenesis through the upregulation of *VWF* and *Flt1* in endothelial cells.

In the first experiment, the hUCMSCs exhibited excellent proliferation and fibroblast-like morphology. These results are consistent with the three universal MSC criteria (plastic adhesion, trilineage differentiating capacity, and specific surface antigen expression) recommended by the International Society for Cellular Therapy (Dominici et al., 2006).

With the proposal to convert 1 μg of exosome protein equivalent to 3×10^{10} exosome particles (Webber & Clayton, 2013), the exosomes in this study had a good yield: 0.7 ± 0.029 μg/mL, corresponding to more than 2×10^{10} particles/mL. The number of exosomes collected in 1 mL of the MSCCult MV medium was higher than that in 1 mL of urine (approximately $3–8 \times 10^9$ exosomes) but lower than that in 1 mL of serum ($1–3 \times 10^{12}$) (Li et al., 2014). These results are similar to the exosome yield results of Inas et al. (Helwa et al., 2017).

TEM showed that the size of the exosomes ranged from 30 to 100 nm. As mentioned, exosomes are nanosized EVs derived from cells with a diameter of 30 to 150 nm (Panda et al., 2021). Exosomal proteins have various functional groups, such as tetraspanins, heat shock proteins, membrane transporters, and lipid-binding proteins. Tetraspanins CD9, CD63, and CD81 are the most common exosomal markers. Marker analysis revealed that the purity of the exosome population obtained was fairly high, with a CD63 and CD81 copositivity rate of 84.27%. However, among the CD63- and CD81-positive particles, only 36.18% were also positive for CD9. According to the Minimal Information for Studies of Extracellular Vesicles (MISEV) 2014, the vesicle population obtained should have at least three positive protein markers of EVs, including at least one membrane-bound protein or lipid (Lötvall et al., 2014; Théry & Witwer, 2018). Many studies have only used Western blotting to quantify these markers to demonstrate the presence of exosomes (Gong et al., 2017; Xue et al., 2021). Meanwhile, the MISEV2018 states that CD63 and CD81 are critical markers for determining exosome purity. While CD9 is an associated marker, its positivity rate varies by cell source, and it is even absent in some cell types, such as natural killer cells, B cells, and several types of MSCs (Théry & Witwer, 2018). Our findings outperform those of Rim et al. (Rim & Kim, 2016).

MSCs have shown promising therapeutic efficacy in enhancing angiogenesis. The major mechanism underlying MSC-based therapy is the paracrine function, which secretes a variety of biological factors, including proteins, cytokines, and chemokines (Ratajczak et al., 2012). Recently, MSCs have been exploited to play a role in the endocrine effect by releasing exosomes that contain miRNA (Gong et al., 2017; Heo & Kim, 2022). In this study, the HUVECs cultured in the M200 medium did not exhibit tubular structures on Matrigel. Remarkably, the addition of 20 μg/mL exosomes resulted in tubular structures similar to the HUVECs cultured in the ECGM, a medium with growth factors. This indicates that exosomes derived from hUCMSCs can display bioeffects similar to those of growth factors that support angiogenesis, such as factors in an ECGM. Our results are consistent with those of previous studies (Ratajczak et al., 2012; Yu et al., 2020). However, the addition of 20 μg/mL exosomes in the M200 medium had a rather low efficiency compared with the commercial medium ECGM or ECGM supplemented with 20 μg/mL exosomes. This can be explained by the fact that the features of exosomes derived from MSCs depend on the concentration as well as the culture conditions (Lee et al., 2013).

In the last experiment, the effects of the exosomes on the angiogenic factors of the HUVECs were investigated. Compared with the control (without exosomes), the exosome-treated HUVECs showed an upregulation of HGF, VWF, CD31, Flt1, and Flk1, with the expression of VWF and Flt1 increasing significantly. Flt1, also known as VEGFR-1, is a tyrosine kinase receptor with seven immunoglobulin-like domains in its extracellular region and a tyrosine kinase domain in its intracellular region (Saito, 2021). It was previously identified as a limited angiogenic factor (Shibuya,

2006). Several reports have shown a positive effect of Flt1 on endothelial cell migration and new blood vessel formation (Lee et al., 2011). Moreover, inactivation of Flt1 can lead to marked attenuation of HUVEC migration as well as HUVEC tubule formation on Matrigel (Ji et al., 2018). Flt1 also plays a crucial role in angiogenesis in lung cancer, as it can bind to the surface molecule PIGF, increasing the production of TGF-β1 (Kong et al., 2021).

Von Willebrand factor (VWF) is a large, multimeric glycoprotein found in the peripheral blood stream (Mojzisch & Brehm, 2021). In both in vitro and in vivo studies, VWF deficiency increased VEGFR-2-dependent migration and proliferation and increased capillary reticulum formation. VWF-deficient mice showed an increase in angiogenesis and mature vascular density (Starke et al., 2011). Loss of VWF in the EC also results in uncontrollably increased or probably dysfunctional angiogenesis, which is consistent with clinical observations in some patients with von Willebrand disease (Randi & Laffan, 2017). In particular, the activation of VWF expression promotes a more malignant phenotype in some cancer cell types (Mojzisch & Brehm, 2021). Interestingly, the treatment of HUVECs with exosomes derived from MSCs increased the gene expression of VWF in this study, which could be a promising treatment strategy for some diseases related to blood clotting disorders, such as von Willebrand disease.

Other angiogenic genes, including HGF, Ephrin B2, Ephrin B4, ANG2, ANG1, cadherin, Flk, CD31, MMP2, and TGF-β, were also evaluated in this study. In general, these genes were all expressed with an increased tendency compared with the control group, although the difference was not significant. This result could imply that exosomes derived from MSCs positively affect HUVECs, stimulating their proliferation.

5 Conclusion

In summary, this study demonstrated that the exosomes derived from the hUCMSCs promoted angiogenesis in the endothelial cells (HUVECs). They stimulated angiogenesis through upregulation of the *VWF* and *Flk1* genes. Our findings confirmed that these exosomes directly affect endothelial cells that trigger angiogenesis.

Acknowledgments We thank SCI Biobank for providing mesenchymal stem cells used in this study.

Author's Contributions All authors equally contributed in this work. All authors read and approved the final version of the manuscript for submission.

Availability of Data and Materials Data and materials used and/or analyzed during the current study are available from the corresponding author on reasonable request.

Ethics Approval and Consent to Participate Not applicable.

Consent for Publication Not applicable.

Competing Interests The authors declare that they have no competing interests.

Funding This research is funded by Vietnam National University, Ho Chi Minh City (VNU-HCM) under grant number 562-2020-18-03.

References

Ahmed, L., & Al-Massri, K. (2022). New approaches for enhancement of the efficacy of mesenchymal stem cell-derived exosomes in cardiovascular diseases. *Tissue Engineering and Regenerative Medicine, 19*(6), 1129–1146.

Akan, E., Cetinkaya, B., Kipmen-Korgun, D., Ozmen, A., Koksoy, S., Mendilcioğlu, İ., et al. (2021). Effects of amnion derived mesenchymal stem cells on fibrosis in a 5/6 nephrectomy model in rats. *Biotechnic & Histochemistry, 96*(8), 594–607.

Arslan, F., Lai, R. C., Smeets, M. B., Akeroyd, L., Choo, A., Aguor, E. N., et al. (2013). Mesenchymal stem cell-derived exosomes increase ATP levels, decrease oxidative stress and activate PI3K/Akt pathway to enhance myocardial viability and prevent adverse remodeling after myocardial ischemia/reperfusion injury. *Stem Cell Research, 10*(3), 301–312.

Assis, A. C., Carvalho, J. L., Jacoby, B. A., Ferreira, R. L., Castanheira, P., Diniz, S. O., et al. (2010). Time-dependent migration of systemically delivered bone marrow mesenchymal stem cells to the infarcted heart. *Cell Transplantation, 19*(2), 219–230.

Cao, Y., Ji, C., & Lu, L. (2020). Mesenchymal stem cell therapy for liver fibrosis/cirrhosis. *Annals of Translational Medicine, 8*(8), 562.

Dominici, M., Le Blanc, K., Mueller, I., Slaper-Cortenbach, I., Marini, F., Krause, D., et al. (2006). Minimal criteria for defining multipotent mesenchymal stromal cells. The International Society for Cellular Therapy position statement. *Cytotherapy, 8*(4), 315–317.

Eggenhofer, E., Benseler, V., Kroemer, A., Popp, F. C., Geissler, E. K., Schlitt, H. J., et al. (2012). Mesenchymal stem cells are short-lived and do not migrate beyond the lungs after intravenous infusion. *Frontiers in Immunology, 3*, 297.

Ge, L., Xun, C., Li, W., Jin, S., Liu, Z., Zhuo, Y., et al. (2021). Extracellular vesicles derived from hypoxia-preconditioned olfactory mucosa mesenchymal stem cells enhance angiogenesis via miR-612. *Journal of Nanobiotechnology, 19*(1), 380.

Gong, M., Yu, B., Wang, J., Wang, Y., Liu, M., Paul, C., et al. (2017). Mesenchymal stem cells release exosomes that transfer miRNAs to endothelial cells and promote angiogenesis. *Oncotarget, 8*(28), 45200–45212.

Han, Y., Ren, J., Bai, Y., Pei, X., & Han, Y. (2019). Exosomes from hypoxia-treated human adipose-derived mesenchymal stem cells enhance angiogenesis through VEGF/VEGF-R. *The International Journal of Biochemistry & Cell Biology, 109*, 59–68.

Helwa, I., Cai, J., Drewry, M. D., Zimmerman, A., Dinkins, M. B., Khaled, M. L., et al. (2017). A Comparative study of serum exosome isolation using differential ultracentrifugation and three commercial reagents. *PLoS One, 12*(1), e0170628.

Heo, J. S., & Kim, S. (2022). Human adipose mesenchymal stem cells modulate inflammation and angiogenesis through exosomes. *Scientific Reports, 12*(1), 2776.

Huang, C., Luo, W., Wang, Q., Ye, Y., Fan, J., Lin, L., et al. (2021). Human mesenchymal stem cells promote ischemic repairment and angiogenesis of diabetic foot through exosome miRNA-21-5p. *Stem Cell Research, 52*, 102235.

Ji, S., Xin, H., Li, Y., & Su, E. J. (2018). FMS-like tyrosine kinase 1 (FLT1) is a key regulator of fetoplacental endothelial cell migration and angiogenesis. *Placenta, 70*, 7–14.

Komaki, M., Numata, Y., Morioka, C., Honda, I., Tooi, M., Yokoyama, N., et al. (2017). Exosomes of human placenta-derived mesenchymal stem cells stimulate angiogenesis. *Stem Cell Research & Therapy, 8*(1), 219.

Kong, X., Bu, J., Chen, J., Ni, B., Fu, B., Zhou, F., et al. (2021). PIGF and Flt-1 on the surface of macrophages induces the production of TGF-β1 by polarized tumor-associated macrophages to promote lung cancer angiogenesis. *European Journal of Pharmacology, 912*, 174550.

Lai, R. C., Arslan, F., Lee, M. M., Sze, N. S., Choo, A., Chen, T. S., et al. (2010). Exosome secreted by MSC reduces myocardial ischemia/reperfusion injury. *Stem Cell Research, 4*(3), 214–222.

Lee, H. K., Chauhan, S. K., Kay, E., & Dana, R. (2011). Flt-1 regulates vascular endothelial cell migration via a protein tyrosine kinase-7-dependent pathway. *Blood, 117*(21), 5762–5771.

Lee, J. K., Park, S. R., Jung, B. K., Jeon, Y. K., Lee, Y. S., Kim, M. K., et al. (2013). Exosomes derived from mesenchymal stem cells suppress angiogenesis by down-regulating VEGF expression in breast cancer cells. *PLoS One, 8*(12), e84256.

Li, M., Zeringer, E., Barta, T., Schageman, J., Cheng, A., & Vlassov, A. V. (2014). Analysis of the RNA content of the exosomes derived from blood serum and urine and its potential as biomarkers. *Philosophical Transactions of the Royal Society of London Series B, Biological Sciences, 369*(1652), 20130502.

Li, L., Cheng, D., An, X., Liao, G., Zhong, L., Liu, J., et al. (2021). Mesenchymal stem cells transplantation attenuates hyperuricemic nephropathy in rats. *International Immunopharmacology, 99*, 108000.

Liang, W., Han, B., Hai, Y., Sun, D., & Yin, P. (2021). Mechanism of action of mesenchymal stem cell-derived exosomes in the intervertebral disc degeneration treatment and bone repair and regeneration. *Frontiers in Cell and Developmental Biology, 9*, 833840.

Lötvall, J., Hill, A. F., Hochberg, F., Buzás, E. I., Di Vizio, D., Gardiner, C., et al. (2014). Minimal experimental requirements for definition of extracellular vesicles and their functions: A position statement from the International Society for Extracellular Vesicles. *Journal of Extracellular Vesicles, 3*, 26913.

Masterson, C. H., Tabuchi, A., Hogan, G., Fitzpatrick, G., Kerrigan, S. W., Jerkic, M., et al. (2021). Intra-vital imaging of mesenchymal stromal cell kinetics in the pulmonary vasculature during infection. *Scientific Reports, 11*(1), 5265.

Mojzisch, A., & Brehm, M. A. (2021). The manifold cellular functions of von Willebrand factor. *Cell, 10*(9), 2351.

Panda, B., Sharma, Y., Gupta, S., & Mohanty, S. (2021). Mesenchymal stem cell-derived exosomes as an emerging paradigm for regenerative therapy and Nano-medicine: A comprehensive review. *Life (Basel, Switzerland), 11*(8), 784.

Qiu, X., Liu, J., Zheng, C., Su, Y., Bao, L., Zhu, B., et al. (2020). Exosomes released from educated mesenchymal stem cells accelerate cutaneous wound healing via promoting angiogenesis. *Cell Proliferation, 53*(8), e12830.

Randi, A. M., & Laffan, M. A. (2017). Von Willebrand factor and angiogenesis: Basic and applied issues. *Journal of Thrombosis and Haemostasis, 15*(1), 13–20.

Ratajczak, M. Z., Kucia, M., Jadczyk, T., Greco, N. J., Wojakowski, W., Tendera, M., et al. (2012). Pivotal role of paracrine effects in stem cell therapies in regenerative medicine: Can we translate stem cell-secreted paracrine factors and microvesicles into better therapeutic strategies? *Leukemia, 26*(6), 1166–1173.

Rim, K. T., & Kim, S. J. (2016). Quantitative analysis of exosomes from murine lung cancer cells by flow cytometry. *Journal of Cancer Prevention, 21*(3), 194–200.

Saito, Y. (2021). The role of the PlGF/Flt-1 signaling pathway in the cardiorenal connection. *Journal of Molecular and Cellular Cardiology, 151*, 106–112.

Shabbir, A., Cox, A., Rodriguez-Menocal, L., Salgado, M., & Van Badiavas, E. (2015). Mesenchymal stem cell exosomes induce proliferation and migration of normal and chronic wound fibroblasts, and enhance angiogenesis in vitro. *Stem Cells and Development, 24*(14), 1635–1647.

Shibuya, M. (2006). Vascular endothelial growth factor receptor-1 (VEGFR-1/Flt-1): A dual regulator for angiogenesis. *Angiogenesis, 9*(4), 225–230; discussion 31.

Starke, R. D., Ferraro, F., Paschalaki, K. E., Dryden, N. H., McKinnon, T. A., Sutton, R. E., et al. (2011). Endothelial von Willebrand factor regulates angiogenesis. *Blood, 117*(3), 1071–1080.

Sun, Y., Ju, Y., & Fang, B. (2022). Exosomes from human adipose-derived mesenchymal stromal/stem cells accelerate angiogenesis in wound healing: Implication of the EGR-1/lncRNA-SENCR/DKC1/VEGF-A axis. *Human Cell, 35*(5), 1375–1390.

Takeuchi, R., Katagiri, W., Endo, S., & Kobayashi, T. (2019). Exosomes from conditioned media of bone marrow-derived mesenchymal stem cells promote bone regeneration by enhancing angiogenesis. *PLoS One, 14*(11), e0225472.

Théry, C., & Witwer, K. W. (2018). Minimal information for studies of extracellular vesicles 2018 (MISEV2018): A position statement of the International Society for Extracellular Vesicles and update of the MISEV2014 guidelines. *Journal of Extracellular Vesicles, 7*(1), 1535750.

Valadi, H., Ekström, K., Bossios, A., Sjöstrand, M., Lee, J. J., & Lötvall, J. O. (2007). Exosome-mediated transfer of mRNAs and microRNAs is a novel mechanism of genetic exchange between cells. *Nature Cell Biology, 9*(6), 654–659.

Webber, J., & Clayton, A. (2013). How pure are your vesicles? *Journal of Extracellular Vesicles, 2*. https://doi.org/10.3402/jev.v2i0.19861

Wu, M. C., & Meng, Q. H. (2021). Current understanding of mesenchymal stem cells in liver diseases. *World Journal of Stem Cells, 13*(9), 1349–1359.

Xue, C., Li, X., Ba, L., Zhang, M., Yang, Y., Gao, Y., et al. (2021). MSC-derived exosomes can enhance the angiogenesis of human brain MECs and show therapeutic potential in a mouse model of Parkinson's disease. *Aging and Disease, 12*(5), 1211–1222.

Yan, C., Xv, Y., Lin, Z., Endo, Y., Xue, H., Hu, Y., et al. (2022). Human umbilical cord mesenchymal stem cell-derived exosomes accelerate diabetic wound healing via ameliorating oxidative stress and promoting angiogenesis. *Frontiers in Bioengineering and Biotechnology, 10*, 829868.

Yin, K., Wang, S., & Zhao, R. C. (2019). Exosomes from mesenchymal stem/stromal cells: A new therapeutic paradigm. *Biomarker Research, 7*(1), 8.

Yu, M., Liu, W., Li, J., Lu, J., Lu, H., Jia, W., et al. (2020). Exosomes derived from atorvastatin-pretreated MSC accelerate diabetic wound repair by enhancing angiogenesis via AKT/eNOS pathway. *Stem Cell Research & Therapy, 11*(1), 350.

Zhang, Y., Hao, Z., Wang, P., Xia, Y., Wu, J., Xia, D., et al. (2019). Exosomes from human umbilical cord mesenchymal stem cells enhance fracture healing through HIF-1α-mediated promotion of angiogenesis in a rat model of stabilized fracture. *Cell Proliferation, 52*(2), e12570.

Adv Exp Med Biol - Innovations in Cancer Research and Regenerative Medicine (2023) 4: 47–61
https://doi.org/10.1007/5584_2022_708

Published online: 8 April 2022

Stromal Vascular Fraction and Mesenchymal Stem Cells from Human Adipose Tissue: A Comparison of Immune Modulation and Angiogenic Potential

Tung Dang Xuan Tran, Viet Quoc Pham, Nhan Ngo-The Tran, Hoang Chau Ngo Dang, Nguyet Thi Anh Tran, Ngoc Bich Vu, and Phuc Van Pham

Abstract

Introduction In recent years, both stromal vascular fraction (SVF) from adipose tissue and mesenchymal stem cells (MSC) from adipose tissues were extensively used in both preclinical and clinical treatment for various diseases. Some studies reported differences in treatment efficacy between SVFs and MSCs in animals as well as in humans. Therefore, this study is aimed to evaluate the immune modulation and angiogenic potential of SVFs and MSCs from the same SVF samples to support an explanation when SVFs or MSCs should be used.

Methods The adipose tissue samples from ten female donors with consent forms were collected. SVFs from these samples were isolated according to the published protocols. The

T. D. X. Tran (✉)
NTT Hi-Tech Institute – Nguyen Tat Thanh University, Ho Chi Minh City, Viet Nam

Stem Cells Unit, Van Hanh Hospital, Ho Chi Minh City, Viet Nam
e-mail: drxuantung@gmail.com; tdxtung@ntt.edu.vn

V. Q. Pham and N. B. Vu
Stem Cell Institute, University of Science Ho Chi Minh City, Ho Chi Minh City, Viet Nam

Viet Nam National University Ho Chi Minh City, Ho Chi Minh City, Viet Nam
e-mail: vietpham@sci.edu.vn; ngocvu@sci.edu.vn

N. N.-T. Tran
Viet Nam National University Ho Chi Minh City, Ho Chi Minh City, Viet Nam

Laboratory of Stem Cell Research and Application, University of Science Ho Chi Minh City, Ho Chi Minh City, Viet Nam
e-mail: tntnhan@hcmus.edu.vn

H. C. N. Dang and N. T. A. Tran
Van Hanh Stem Cells Unit, Van Hanh Hospital, Ho Chi Minh City, Viet Nam
e-mail: hoangdcn4390@gmail.com; nguyettran18@gmail.com

P. Van Pham
Stem Cell Institute, University of Science Ho Chi Minh City, Ho Chi Minh City, Viet Nam

Viet Nam National University Ho Chi Minh City, Ho Chi Minh City, Viet Nam

Laboratory of Stem Cell Research and Application, University of Science Ho Chi Minh City, Ho Chi Minh City, Viet Nam
e-mail: pvphuc@hcmus.edu.vn

existence of mesenchymal cells that positive with CD44, CD73, CD90, and CD105 and endothelial progenitor cells that positive with CD31 and CD34 was determined using flow cytometry. Three samples of SVFs with similar percentages of mesenchymal cell portion and endothelial progenitor cell portion were used to isolate MSCs. Obtained MSCs were confirmed as MSCs using the ISCT minimal criteria. To compare the immune modulation of SVF and MSCs, the mixed lymphocyte assay was used. The lymphocyte proliferation, as well as IFN-gamma and TNF-alpha concentrations, were determined. To compare the angiogenic potential, the angiogenesis in quail embryo assay was used. The angiogenesis efficacy was measured based on the vessel areas formed in the embryos after 7 days.

Results The results showed that all SVF samples contained the portions of mesenchymal cells and endothelial progenitor cells. MSCs from SVFs meet all minimal criteria of MSCs that suggested by ISCT. MSCs from SVFs efficiently suppressed the immune cell proliferation compared to the SVFs, especially at ratios of 1:4 (1 MSCs: 4 immune cells). MSCs also inhibited the IFN-gamma and TNF-alpha production more efficiently than SVFs ($p < 0.05$). However, in quail embryo models, SVFs triggered the angiogenesis and neovessel formation better than MSCs with more significant vessel areas after 7 days ($p < 0.05$).

Conclusion This study suggested that SVFs and MSCs have different potentials for immune modulation and angiogenesis. SVFs help the angiogenesis better than MSCs, while MSCs displayed the more significant immune modulation. These results can guide the usage of SVFs or MSCs in disease treatment.

Keywords

Adipose-derived stem cells · Angiogenesis · Immune modulation · Mesenchymal stem cells · Stromal vascular fraction

Abbreviations

ADSC	Adipose-derived stem cell
bFGF	Basic fibroblast growth factor
ECP	Endothelial cell population
HA	Hyaluronic acid
MSC	Mesenchymal stem cell
MSCP	Mesenchymal stem cell population
SVF	Stromal vascular fraction
TNF	Tissue necrosis factor
VEGF	Vascular endothelial growth factor

1 Introduction

Adipose tissue is a stem cell–rich source. To date, stem cells from this source were usually used in two methods, including stromal vascular fraction (SVF) transplantation and adipose-derived mesenchymal stem cell (MSC) transplantation. SVF is a combo of cells extracted from the adipose tissue. It is identified with a variety of cells, including mesenchymal stem cells (Ryu et al., 2013), endothelial progenitor cells (Asahara et al., 2011), endothelial cells (Garipcan et al., 2011), smooth muscle cells (Nunes et al., 2013; Williams et al., 1994), pericytes (Proebstl et al., 2012), and preadipocytes (Lai et al., 2012), and as known as non-expanded stem cells from adipose tissues. MSCs from SVF were defined as mesenchymal stem cells enriched and expanded from the SVFs. In the culture conditions for some passages, the MSCs from SVFs could be successfully isolated by the removal of other cells during by culture and subculture (Van Pham et al., 2013; Pham et al., 2014).

SVFs or MSCs from SVFs are better for applications in regenerative medicine also, though they are controversial issues to date. Indeed, some reports suggested that there was not a difference between SVFs and MSCs in the generation of volume-stable, well-vascularized PU-based constructs (Griessl et al., 2018). Therefore, they suggested that it is better to use SVFs because the culture and expansion of MSCs will be time consuming and cost intensive (Griessl et al., 2018). In another study, Nyberg et al. (2019) compared the usage of MSCs from adipose tissue and SVFs in bone regeneration (Nyberg et al., 2019). In this study, MSCs and SVFs were loaded into the scaffolds and grafted in the murine critical-sized cranial defects. The scaffolds with MSCs had greater (but not statistically significant) bone volume and bone coverage area than scaffolds with SVFs (Nyberg et al., 2019).

Domergue et al. (2016) compared the SVFs and MSCs from adipose tissue in remodeling hypertrophic scars in animal models (Domergue et al., 2016). Authors showed that MSCs reduced the fibrotic scar more efficiently than SVFs (Domergue et al., 2016). In the model of osteoarthritis, Lv et al. (2018) also suggested that MSCs in combination with HA was better than SVF in combination with HA for osteoarthritis treatment (Lv et al., 2018).

Therefore, this study aimed to evaluate and compare the immune modulation and angiogenic potential of SVFs and MSCs isolated from the same SVFs.

2 Methods

2.1 SVF Extraction

Fat tissues from donors were collected at the hospital with the consent forms. All procedures in this study were approved by the ethical committee of the hospital. Ten samples of adipose tissues from ten donors were used in this study. About 150 mL of lipoaspirate was collected from each donor in 3 50 mL sterile syringes. All syringes were kept in a sterile box at 2–8 °C and transferred to the laboratory in 6 h.

At the laboratory, SVFs from all adipose tissues were extracted using the Cell Extraction Kit (Regenmedlab, Ho Chi Minh City, Viet Nam) according to the manufacturer's instructions. Briefly, fats were washed twice with saline to remove blood. Then, it was sucked into the 50 mL syringes. The other syringes contained extraction buffer that were linked to the extractor with collagenase inside. These syringes were connected with syringes containing washed fat. Fat from the syringes was pushed into the other one. This cycle was repeated 15 times with 2-min interval. Finally, the digested fat was transferred to centrifuge 50 mL tubes for centrifuge at 3500 rpm for 15 min to collect SVFs at the bottom of the tubes. SVFs were washed twice with washing buffer to remove the collagenase before they were used in the further assays. The nucleated cells in SVFs were counted by an automatic cell counter (NucleoCounter, ChemoMetec, Denmark).

2.2 SVF Cryopreservation and Thawing

SVFs were cryopreserved and thawed according to the published study (Solodeev et al., 2019). The cryopreservation medium contains 90% fetal bovine serum (Sigma-Aldrich, Louis St. MO) and 10% DMSO (Sigma-Aldrich). SVF cells were frozen in cryopreservation medium at two million SVF cells per 1 mL cryopreservation medium. After aliquoted into cryovials, cryovials were placed on ice and put on −86 °C overnight, then transferred to liquid nitrogen before thawing.

At the thawing step, vials were obtained from the liquid nitrogen, then directly put in the water bath at 37 °C for 2 min. The cell suspension in vials was diluted in phosphate buffer saline supplemented with 10% FBS, then centrifuged at 1500 rpm for 5 min to get a pellet. These cells were counted the viable cells using the automatic cell counter (NucleoCounter).

2.3 Isolation and Expansion of MSCs from SVFs

Three samples of SVFs were used for primary culture to isolate MSCs. SVF cells were cultured in the MSCCult I medium (Regenmedlab, Ho Chi Minh City, Viet Nam) that contains DMEM/F12 supplemented with anti-biotic-mycotic, EGF, basic fibroblast growth factor (bFGF) with 10% human platelet lysate. The cells were plated at 5×10^4 cells/mL in T-75 flasks (SPL, Korea) and incubated at 37 °C with 5% CO_2. After 3 days of incubation, 6 mL of fresh media was added to each flask. After 7 days, the media were replaced with 12 mL of fresh media. The media was subsequently replaced every 3 days until the cells reached 70–80% confluence, where they were subcultured using the De-attachment solution (Regenmedlab, Ho Chi Minh City, Viet Nam).

2.4 MSCs Characterization

MSC candidates were confirmed based on the minimal criteria for MSCs, as suggested by Dominici et al. (2006). All cells have confirmed the expression of CD14, CD34, CD44, CD45, CD73, CD90, CD105, and HLA-DR. All monoclonal antibodies were purchased from BD Biosciences. The cells were stained with the antibodies; then, they were analyzed by FACSCalibur flow cytometer (BD Biosciences, San Jose, CA). Isotype controls were used in all analyses.

To confirm the in vitro differentiation, MSCs were differentiated into adipocytes, osteoblasts, and chondroblasts using the published protocols (Van Pham et al., 2013).

2.5 Evaluation of the Existence of Mesenchymal Cells and Endothelial Progenitor Cells Inside SVFs

To identify the existence of mesenchymal cells, SVFs were stained with cocktails of anti-CD44-FITC, anti-CD73-PE, anti-CD90-PerCP, and anti-CD105-APC. The mesenchymal cells were defined as the cell populations that positive with all these markers. All stained cells were analyzed by FACSCalibur flow cytometer (BD Biosciences). To identify the endothelial cell population, SVF cells were stained with anti-CD31-FITC and anti-CD34-APC. The cell population that positive with both CD31 and CD34 was defined as endothelial cells.

2.6 Mixed Lymphocyte Reaction and CD34 Counting

This assay was performed as the published study (Van Pham et al., 2016a). The allogeneic peripheral blood mononuclear cells (PBMCs) were stimulated with 2.5 μg/ml phytohemagglutinin (PHA) (Gibco). SVF cells and MSCs from SVFs were pretreated with mitomycin C to inhibit the cell proliferation before they were used in this assay for 24 h.

Viable SVF cells or MSCs were mixed with PBMCs at 2 ratios of 1:40 and 1:4 (SVF cells or MSCs versus PBMCs) in the culture medium for 48 h. Unstimulated and PHA-stimulated PBMCs seeded without MSCs/SVFs were used as controls. After 48 h of culture, all floating cells were collected and stained with an anti-CD38-PE monoclonal antibody (BD Bioscience). The level of lymphocyte proliferation was measured by the percentage of CD38 positive cells in the whole population. All experiments were performed in triplicate.

2.7 Cytokine Concentration Measurement

Supernatants were collected and kept frozen at −86 °C for cytokine concentration measurement by ELISA. ELISA kits for IFN-gamma and TNF-alpha were used in this assay. The protocols were used according to the supplier's instructions (Abcam, Cambridge, UK).

2.8 Angiogenesis Assay in Quail Embryos

The eggs of quails were collected directly from The Laboratory Animal Care & Use, Stem Cell Institute, University of Science, VNU-HCM, Viet Nam, to form *ex ovo* quail models. Eggs of quails were collected and hatched in an egg incubator for 60 h at 37 °C of temperature and 65% of humidity. Eggshells were washed and then put the eggs to lie horizontally on the stainless steel test tube rack for 15 min. At the same time, the surface of eggshells was sprayed with Povidine 10% and ethanol 70% to sterilize and clean. All eggs with the test tube rack were transferred to the biosafety cabinet and waited to dry. The top surface of eggs was marked with a pencil. The blade was embedded in a plastic box when the sharp edge was flipped up. Each egg of quail was held on its two heads; the middle of the egg was cut by the blade softly until the eggshell to split into two parts. The ingredients that were inside quail egg, was got into a 6-wells dish. After that, the dishes were moved to the incubator set up 37.5 °C of temperature, 60% of humidity, and fresh air.

Two millions of viable SVF cells or MSCs cells were mixed with 100 uL matrigel and then were put in the quail embryos. There were 5 embryos per group. The PBS mixed with 100 uL matrigel were used as controls for this assay.

Quail embryos were observed and captured on day 1 and day 7. The pictures were analyzed by VesSeg Tool software (version 0.1.4) to calculate and assess the development of the blood vessel in quail embryos.

2.9 Statistical Analysis

The data were analyzed for statistical significance using GraphPad Prism software. Data were presented as mean ± SEM. When applicable, a student's unpaired t-test and one-way ANOVA were used to determine significance; $p < 0.05$ was considered to be statistically significant.

Table 1 The existence of MSCP and EPP detected by flow cytometry in ten samples of SVFs from ten donors

Sample	MSCP (%)	EPP (%)
1	2.34	3.67
2	1.11	5.91
3	0.45	6.31
4	0.59	4.39
5	0.89	3.28
6	1.67	5.11
7	2.02	4.31
8	0.90	3.1
9	1.00	3.09
10	2.17	4.21

3 Results

3.1 SVFs Contain a Small Population of $CD90^{+}CD73^{+}CD105^{+}CD44^{+}$ Cells (Mesenchymal Stem Cells) and a Small Population of $CD31^{+}CD34^{+}$ Cells (Endothelial Progenitors)

The SVFs from 10 donors were collected and analyzed the existence of 2 cell populations inside. The mesenchymal stem cells with markers of CD44 + CD73 + CD90 + CD105+ (mesenchymal stem cell population – MSCP) and endothelial progenitor of CD31 + CD34+ (endothelial progenitor population – EPP) were evaluated by flow cytometry method. The results were shown in Table 1 and Fig. 1.

3.2 Adipose-Derived Mesenchymal Stem Cells Displayed the Standard Phenotype of Mesenchymal Stem Cells

3.2.1 Mesenchymal Stem Cell Particular Marker Expression

Three samples of SVFs (sample 5, sample 8, and sample 9) with similar counts of CD44CD73CD90CD105 and CD31CD34 cell populations were used in further experiments. Each SVFs sample was shared into two parts that were used to expand mesenchymal stem cells and used to cryopreserve. The AT-MSCs were expanded to the fifth passage and used to characterize the MSC phenotype. The results showed that these cells displayed the CD44 (100%), CD73 (97.3 ± 1.001), CD90 (98.18 ± 1.10%), CD105 (96.79 ± 1.03%), but negative with CD14 (0.82 ± 0.34%), CD34 (2.51 ± 0.53%), CD45 (0.84 ± 0.30%), and HLA-DR (1.22 ± 0.63%) (Figs. 2 and 3).

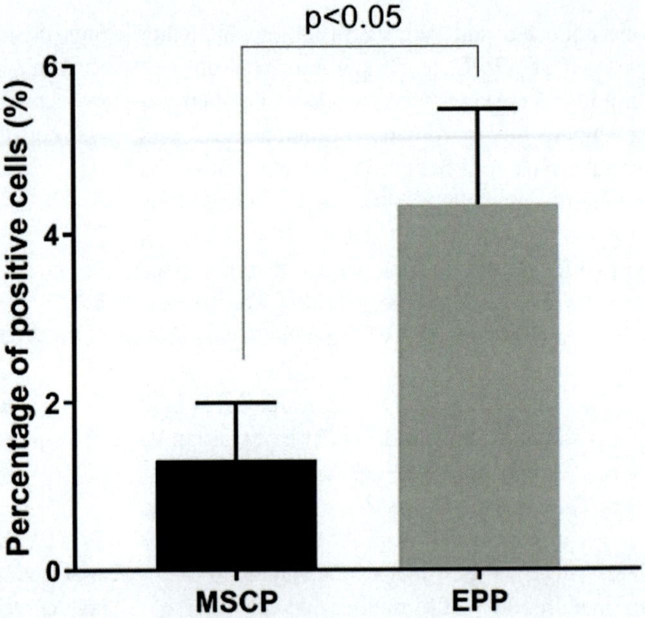

Fig. 1 The existence of small population that positive with CD44CD73CD90CD105 (MSCP) and CD31CD34 (EPP) in the SVFs. The EPP population is significant higher MSCP population in the SVFs ($n = 9$), $p < 0.05$

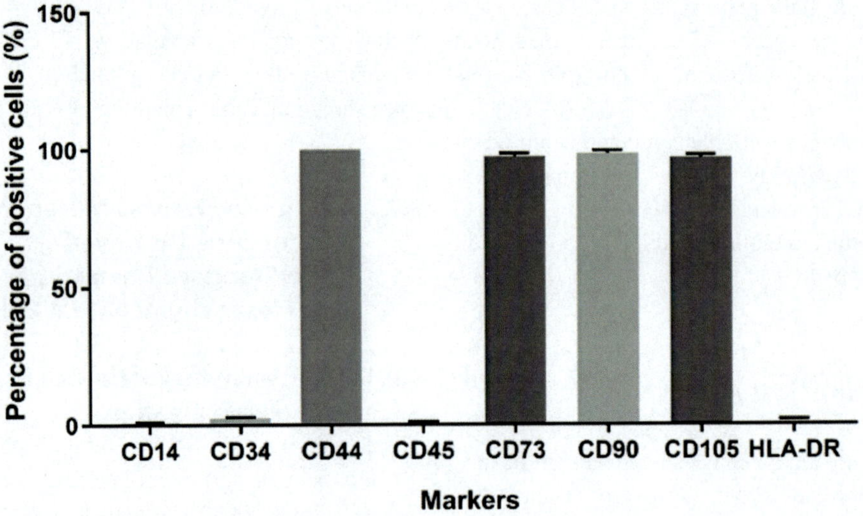

Fig. 2 AT-MSCs expressed the MSC phenotype that was positive with CD44, CD73, CD90, and CD015, but negative with CD14, CD34, CD45, and HLA-DR

3.2.2 The In Vitro Differentiation into Osteoblasts, Chondroblasts, and Adipocytes

MSCs from three samples of SVFs were confirmed in vitro differentiation potential into osteoblasts, chondroblasts, and adipocytes. The results showed that these cells successfully differentiated into osteoblasts, chondroblasts, and adipocytes. After 21 days of being induced in osteogenesis medium, MSCs accumulated the calcium in the matrix that was positive with Alizarin red staining (Fig. 4d–f). They also could be induced into chondroblasts in an inducible medium after 21 days. Induced cells were positive

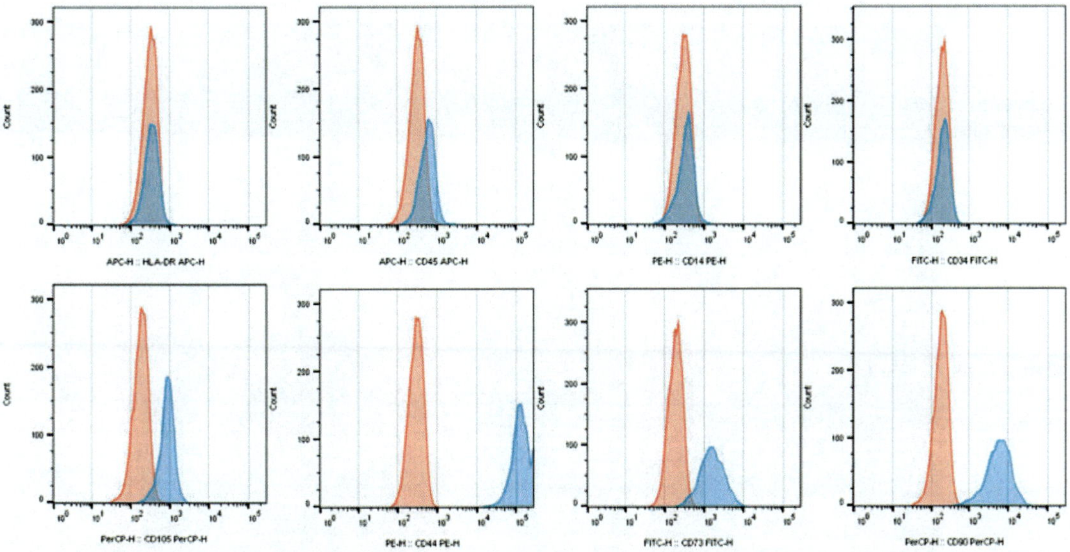

Fig. 3 The expression of some MSC markers was characterized by flow cytometry

with Safranin O staining (Fig. 4g–i). In the adipogenesis medium, MSCs stored the lipid droplets that could be observed after 14 days, and they become bigger after 21 days. These lipid droplets were stained with Oil Red O (Fig. 4j–l and Table 2).

3.3 Lymphocyte Proliferation Is Suppressed by Both SVFs and MSCs

3.3.1 The Percentage of CD38$^+$ Cells Is Evaluated by Flow Cytometry

The lymphocyte proliferation was evaluated by CD38+ counting in the whole cells after mixing PHA stimulated PBMCs with MSCs or SVF cells. The results from three samples (5, 8, 9, and) gave the same trends. In all samples, at the ratios of 1/40, there was not significantly different in the suppression of lymphocyte proliferation between SVF cells and MSCs. However, these differences were significantly different at a ratio of 1/4. Although the suppression efficacy between samples was not similar, the MSCs in all samples stronger suppressed lymphocyte proliferation than SVFs ($p < 0.05$) (Fig. 5).

3.3.2 The Productions of IFN-Gamma and TNF-Alpha of Immune Cells Reduced in Co-culture with SVFs and MSCs

To support the observation about the lymphocyte proliferation of SVFs or MSCs, these supernatants were collected and used for the IFN-gamma and TNF-alpha measurement. The results presented in Fig. 6 showed that there were clear reductions in IFN-gamma and TNF-alpha in supernatant collected from groups of ratios of 1/4 compared to from supernatant of groups of 1/40 and control ($p < 0.05$). Concentrations of two cytokines were significantly lower in the supernatant of MSCs mixed with PBMCs at 1/4 compared to SVF cells mixed with PBMCs at 1/4 ($p < 0.05$).

3.4 SVFs Form the Blood Vessels Better Than MSCs in Quail Embryos

In this experiment, the neovessel formation in quail embryos affected by SVFs or MSCs were evaluated compared to control. The results showed in Figs. 7 and 8. At the same dose of

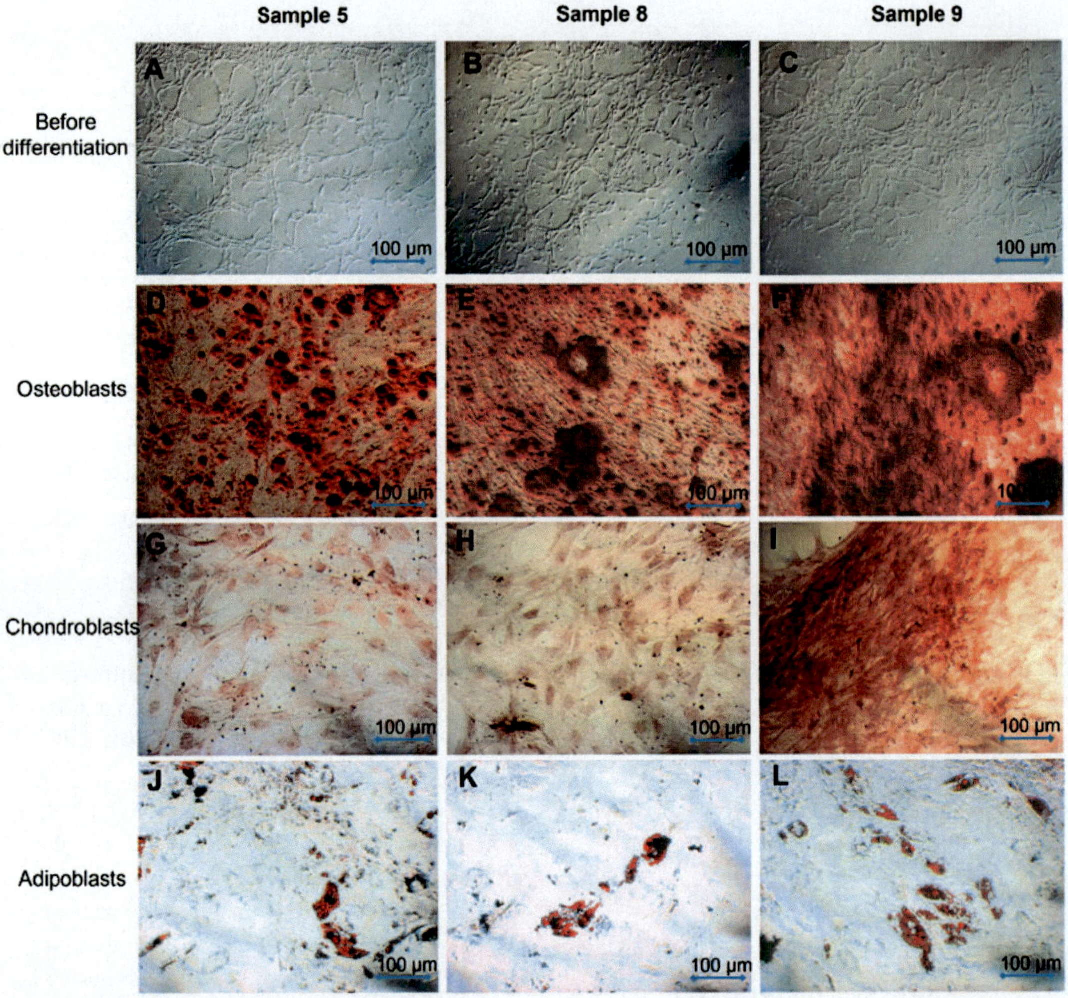

Fig. 4 The AT-MSCs from three samples (samples 5, 8, and 9) were successfully differentiated into adipocytes, osteoblasts, and chondroblasts. Osteoblasts were positive with Alizarin-Red staining that accumulated calcium and magnesium. Chondroblasts were positive with Safranin-O staining. And osteoblasts were positive with Oil Red O staining

Table 2 The expression of MSC markers of AT-MSCs from three samples (sample 5, 8, and 9)

	CD14	CD34	CD44	CD45	CD73	CD90	CD105	HLA-DR
N	3	3	3	3	3	3	3	3
Mean (%)	0.82	2.51	100	0.84	97.30	98.18	96.79	1.22
SD	0.34	0.53	0	0.30	1.00	1.10	1.03	0.63

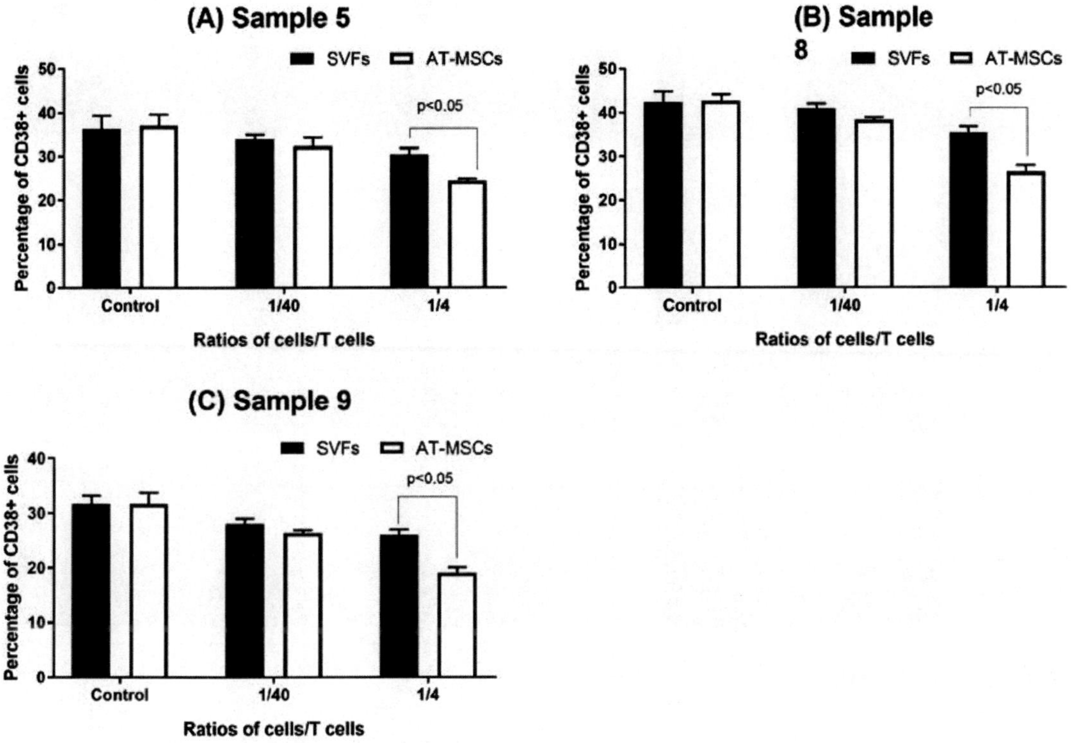

Fig. 5 The lymphocyte proliferation was suppressed by SVFs and AT-MSCs. The results showed that AT-MSCs in all samples efficiently suppressed lymphocyte proliferation. At ratios of 1/40 of SVF cells or AT-MSCs, the difference of suppression efficacy was nonsignificant; however, this difference was clear at ratio of 1/4 of SVFs cells or AT-MSCs and lymphocytes ($p < 0.05$)

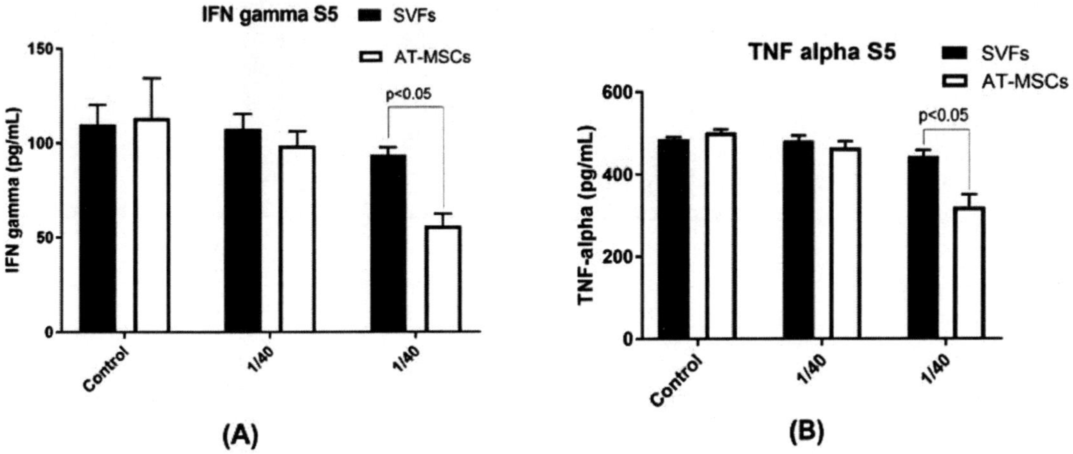

Fig. 6 The supernatant of mixed lymphocyte reaction was used to measure the concentrations of IFN gamma and TNF-alpha. The supernatant of mixed lymphocyte reaction assays of sample S5 was used in this assay

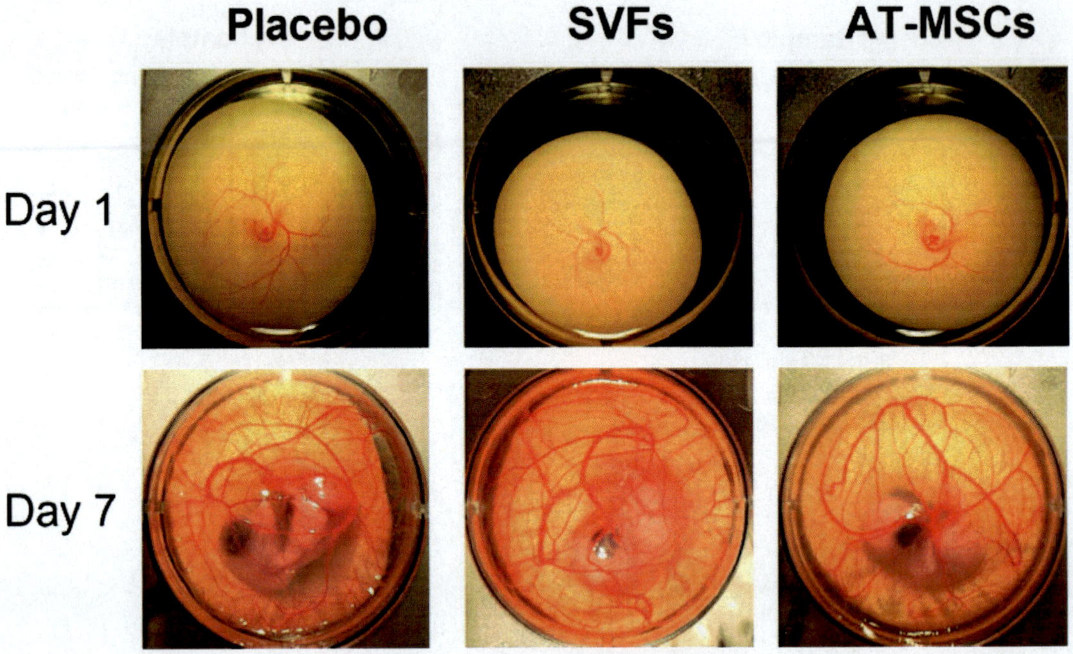

Fig. 7 The angiogenesis in quail embryos effected by SVFs and AT-MSCs transplantation

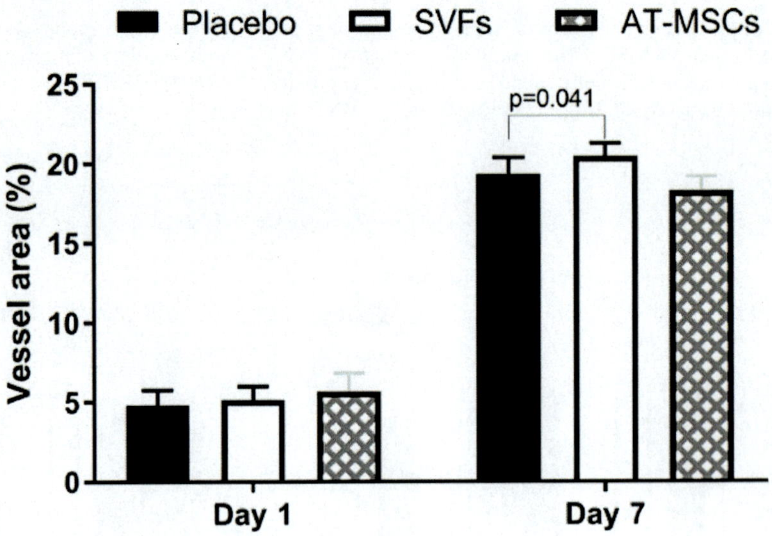

Fig. 8 The vessel area percentage in three groups: placebo, SVFs, and AT-MSCs on day 1 and day 7. The results showed that on day 7, the vessel area of the group of SVFs transplantation is larger than the placebo as well as the group of AT-MSCs

cells and at the same time, after 7 days, in groups of SVF implantation, the vessel areas were significantly higher than in groups of MSC implantation ($p < 0.05$).

4 Discussion

SVFs are combos of various kinds of stem cells, progenitor cells as well as adult cells. Although they were widely in both preclinical trials and clinical treatments. Some studies only isolated the MSC populations from SVFs and expanded by culture for clinical applications. However, there is not any study to compare the immune modulation and angiogenic potential of SVFs and MSCs from the same SVFs to date. Therefore, this study compared immune modulation and angiogenic potentials of SVFs and MSCs from the same SVFs samples. These results from this study will guide how to choose the SVFs or MSCs from SVFs for applications.

The role of immune modulation can relate to the existence of the mesenchymal stem cell population inside the SVFs. While the role of the angiogenic potential of SVFs can relate to the endothelial progenitor population. Therefore in the first experiment, we isolated SVFs from adipose tissues, then using the flow cytometry to determine the existence of two popular populations related to immune modulation and angiogenic potential through their popular markers. The phenotype that are positive with a cocktail of anti-CD44, CD73, CD90, and CD105 antibodies were considered as MSCs. Indeed, according to the ISCT suggestion, MSCs will express these markers (Dominici et al., 2006). Many published studies also used these markers to define the MSCs for adipose tissue–derived mesenchymal stem cells (Camilleri et al., 2016; Petrenko et al., 2020; Secunda et al., 2015). Although only through these markers this subpopulation cannot be confirmed as mesenchymal stem cells, these markers could suggest the existence of MSCs in the SVF samples. The results confirmed that 100% sample (10/10 samples) contained a subpopulation of MSCs from 0.45% to 2.34% of SVF-derived cells. These differences can come from the differences in age and health conditions of donors.

Similarly, we also identified the existence of endothelial cell populations (ECP) in the SVFs samples. We identified the ECP based on the cell population of $CD31^+CD34^+$ cells in SVFs. The results showed that the ECP was richer in SVFs more than MSCP. It accounted for 3.09% to 6.31%. The existence of the ECP with double markers of $CD31^+CD34^+$ in SVFs was determined in the published study by Zimmerlin et al. (2010). In the study of Zimmerlin et al. (2010), they showed that there were 13.7 ± 8.3% ECP in nucleated cells of SVFs. Domergue et al. (2016) identified a cell population with $CD31^+CD34^+$ in the SVFs (Domergue et al., 2016). In the previous study, our group also successfully isolated the ECP population from SVFs (Van Pham et al., 2016b).

In the next experiment, we selected three SVF samples that have similar populations of MSCP and ECP to do further experiments, and to minimize the errors by differences between sample to sample. In this experiment, a part of SVF was used to culture to isolate and enrich the mesenchymal stem cell population. Remained parts of SVFs were frozen for the next experiments.

Using our published protocols (Van Pham et al., 2013; Pham et al., 2014) to isolate and expand mesenchymal stem cells from SVFs, we successfully isolated three samples of MSCs from three samples (sample 5, sample 8, and sample 9). These MSCs satisfied all minimal criteria of mesenchymal stem cells suggested by Dominici et al. (2006). Indeed, they expressed the marker profiles with strong expressions of CD44, CD73, CD90, and CD105, but negative with hematopoietic cell markers included CD14, CD34, CD45, and HLA-DR. They also could be induced into osteoblasts, adipocytes, and chondrocytes. These results agreed with other studies that MSCs from adipose tissues could display the particular phenotype of MSCs (Debnath & Chelluri, 2019; Lee et al., 2004;

Boquest et al., 2006). These MSC populations were used to pair compared to SVFs for immune modulation and angiogenic potentials.

To compare the immune modulation potential of SVF cells and MSCs, the lymphocyte proliferation assays were used. The results were similar in 3 samples from 3 selected donors (sample 5, 8, and 9) that MSCs derived from SVFs were significantly suppressed lymphocyte proliferation more than SVFs cells (nucleated cells from SVFs) at ratios of 1/4 (1 SVFs/MSCs: 4 lymphocytes) ($p < 0.05$). Indeed, in the control group, the CD38+ cells are strongly proliferated by PHA stimulation. The expression of CD38 in a variety of immune cells included T lymphocytes, B lymphocytes, and NK cells (Shubinsky & Schlesinger, 1997). Therefore, in this study, the portion of CD38+ cells in whole cells was used to evaluate the increase of lymphocyte proliferation. Clearly affected by MSCs from adipose tissues, the lymphocyte proliferation was reduced. And this effect created by MSCs was stronger than by SVFs. These results suggested that MSCs from the same SVFs displayed the immune modulation stronger than that SVFs. Although this observation has not been reported in published studies, some articles suggested that MSCs from adipose tissue exhibited immune modulation (Ebrahim et al., 2019; Yu et al., 2019; Shalaby et al., 2016). And this observation was confirmed by the next assay about the concentrations of IFN-gamma and TNF-alpha. These cytokines are produced by lymphocytes. With the effects of immune modulation that their productions could be reduced. At ratio 1:4 of MSCs: lymphocytes, the concentrations of IFN-gamma and TNF-alpha significantly reduced in MSC treatment groups compared to SVFs groups as well as in control groups ($p < 0.05$). The reductions of IFN-gamma and TNF-alpha could be from the lower concentrations of lymphocytes that their proliferation was inhibited by MSCs or the production of these cytokines of lymphocytes were inhibited by MSCs. Indeed, some published studies showed that MSCs from adipose tissue could inhibit the production of inflammatory cytokines such as IL-6, IL-1β, and TNF-α (Ebrahim et al., 2019; Yu et al., 2019). To sum up, MSCs from SVFs displayed the immune modulation was much stronger than the same SVFs. Although SVFs also displayed the immune modulations, these effects were so weak and non-significant compared to control.

In the next experiment, we compared the angiogenic potential of MSCs from SVFs and SVFs. We used the quail embryo models to evaluate the angiogenesis effected by SVFs and MSCs from SVFs. In the same dose of cells, the results confirmed that SVFs triggered the angiogenesis in quail embryos stronger than by MSCs after 7 days. In the recent publication, Bi et al. (2019) compared the effects of SVFs and MSCs from SVFs on wound healing in diabetic mice (Bi et al., 2019). They showed that both SVFs and MSCs from SVFs significantly improved wound healing in hyperglycemic mice; however, it seemed that SVFs promoted wound healing by focusing on angiogenesis and matrix remodeling (Bi et al., 2019). In 2017, Costa et al. showed that in vivo implantation of cell sheets of SVFs in a mouse hind limb ischemia model improved restoration of blood flow (Costa et al., 2017).

These results suggested that SVFs prefer in some treatments related to angiogenesis and blood vessel formation. While MSCs from SVFs prefer in some treatments related to immune modulations. From the literature, we also showed that SVFs transplantation was so successful in both preclinical and clinical trials for hind-limb ischemia or ischemic diabetic feet.

Zhang et al. (2018) reported that grafting of SVF gel displayed a retention rate higher than using standard Colemen fat in nude mice. Especially, there were no necrotic cores in these grafts as Coleman fat grafts. These observations suggested that angiogenesis was efficiently carried out in the SVF gels (Zhang et al., 2018). In the previous study, Sun et al. (2017) used the murine model of wound healing that showed that SVFs could secrete angiogenic factors, including VEGF and bFGF (Sun et al., 2017). In rat models of burn, Karina et al. (2019) concluded that SVFs in combination with platelet-rich plasma stimulated wound healing by supporting angiogenesis (Karina et al., 2019). And the neovessels could be formed in nude mice transplanted with SVF-loaded matrigel (Magalon et al., 2019). The combination of SVF and platelet-rich plasma (PRP) may provide an additive stimulatory effect to support angiogenesis and accelerate the wound healing process; accordingly, this combination is a potential alternative to ADSC treatment.

Clinically, Moon et al. (2019) reported that autologous SVF injection increased TcPO2 from 31.3 ± 7.4 mmHg to 46.4 ± 8.2 mmHg after 12 weeks in ischemic diabetic feet (Moon et al., 2019). In this study, the SVF injection also showed that it improved the cutaneous microvascular blood flow levels from 34.0 ± 21.1 before injection to 76.1 ± 32.5 perfusion unit at 12 weeks after injection (Moon et al., 2019).

MSCs from adipose tissue was used successfully to treat some diseases related to the autoimmune and immune system. In animal models, MSCs from adipose tissue could treat arthritis in mice (Zhang et al., 2017). In arthritis models, MSCs could reduce the inflammatory and T-cell responses (Gonzalez-Rey et al., 2010) or modifying the early adaptive cell responses (Lopez-Santalla et al., 2015) or activation of Treg (Gonzalez-Rey et al., 2010), or inhibited the production of various inflammatory mediators, decreased antigen-specific Th1/Th17 cell expansion, induced production of IL-10 (Zhou et al., 2011). In sheep models with osteoarthritis, Lv et al. (2018) compared the efficacy of transplantation of MSCs or SVFs in combination with hyaluronic acid (HA) in the treatment of osteoarthritis (Lv et al., 2018). Lv et al. (2018) confirmed that MSC + HA injection had better efficacy than SVF + HA in blocking osteoarthritis progression and promoting cartilage regeneration (Lv et al., 2018).

By our evidence and other publications, we suggest that the usage of MSCs from SVF or SVF depends on the pathophysiological mechanism of diseases. It seems that transplantation of SVFs is better for diseases related to ischemia that the treatment needs the angiogenesis, while usage of MSCs from SVFs is better than in treatments of diseases related to immune system conditions including inflammatory or autoimmune diseases. The combinations of both SVFs and MSCs from SVFs also should be considered in some diseases that they relate to both angiogenesis and immune modulation for healing.

5 Conclusion

Adipose tissue is a stem cell–rich source. The stromal vascular fraction from adipose tissue contains a pool of various kinds of cells included mesenchymal stem cells and endothelial progenitor cells. This study showed that both mesenchymal stem cells and endothelial progenitor cells existed in all samples of SVFs (10/10) with different ratios. Although both SVFs or MSCs from SVFs were widely used in both preclinical and clinical treatments in different diseases, differences in the immune modulation and angiogenic potential between them have not been investigated. This study suggested that SVFs strongly exhibited the angiogenesis compared to MSCs, while MSCs from the same SVFs stronger modulate the immune cells than SVFs. These results can guide the usage of SVFs as well as MSCs from SVFs for disease treatment. However, the strictly comparative studies in vivo about differences in the efficacy of SVF and MSC transplantation should be done to verify these findings.

Authors' Contributions All authors contributed to the conceptualization and design of the study, the acquisition, the analysis, and the interpretation of data. PVP were used for drafting the article and revising the article critically for important intellectual content. All authors read and approved the final manuscript.

Availability of Data and Materials Data and materials used and/or analyzed during the current study are available from the corresponding author on reasonable request.

Competing Interests The authors declare that they have no competing interests.

References

Asahara, T., Kawamoto, A., & Masuda, H. (2011). Concise review: Circulating endothelial progenitor cells for vascular medicine. *Stem Cells, 29*, 1650–1655.

Bi, H., Li, H., Zhang, C., et al. (2019). Stromal vascular fraction promotes migration of fibroblasts and angiogenesis through regulation of extracellular matrix in the skin wound healing process. *Stem Cell Research & Therapy, 10*, 302.

Boquest, A. C., Shahdadfar, A., Brinchmann, J. E., & Collas, P. (2006). Isolation of stromal stem cells from human adipose tissue. In *Nuclear reprogramming* (pp. 35–46). Springer.

Camilleri, E. T., Gustafson, M. P., Dudakovic, A., et al. (2016). Identification and validation of multiple cell surface markers of clinical-grade adipose-derived mesenchymal stromal cells as novel release criteria for good manufacturing practice-compliant production. *Stem Cell Research & Therapy, 7*, 107.

Costa, M., Cerqueira, M. T., Santos, T. C., et al. (2017). Cell sheet engineering using the stromal vascular fraction of adipose tissue as a vascularization strategy. *Acta Biomaterialia, 55*, 131–143.

Debnath, T., & Chelluri, L. K. (2019). Standardization and quality assessment for clinical grade mesenchymal stem cells from human adipose tissue. *Hematology, Transfusion and Cell Therapy, 41*, 7–16.

Domergue, S., Bony, C., Maumus, M., et al. (2016). Comparison between stromal vascular fraction and adipose mesenchymal stem cells in remodeling hypertrophic scars. *PLoS One, 11*, e0156161.

Dominici, M., Le Blanc, K., Mueller, I., et al. (2006). Minimal criteria for defining multipotent mesenchymal stromal cells. The International Society for Cellular Therapy position statement. *Cytotherapy, 8*, 315–317.

Ebrahim, N., Mandour, Y. M. H., Farid, A. S., et al. (2019). Adipose tissue-derived mesenchymal stem cell modulates the immune response of allergic rhinitis in a rat model. *International Journal of Molecular Sciences, 20*, 873.

Garipcan, B., Maenz, S., Pham, T., et al. (2011). Image analysis of endothelial microstructure and endothelial cell dimensions of human arteries–A preliminary study. *Advanced Engineering Materials, 13*, B54–B57.

Gonzalez-Rey, E., Gonzalez, M. A., Varela, N., et al. (2010). Human adipose-derived mesenchymal stem cells reduce inflammatory and T cell responses and induce regulatory T cells in vitro in rheumatoid arthritis. *Annals of the Rheumatic Diseases, 69*, 241–248.

Griessl, M., Buchberger, A. M., Regn, S., Kreutzer, K., & Storck, K. (2018). Uncultivated stromal vascular fraction is equivalent to adipose-derived stem and stromal cells on porous polyurethrane scaffolds forming adipose tissue in vivo. *The Laryngoscope, 128*, E206–Ee13.

Karina, S. M. F., Rosadi, I., et al. (2019). Combination of the stromal vascular fraction and platelet-rich plasma accelerates the wound healing process: Pre-clinical study in a Sprague-Dawley rat model. *Stem Cell Investigation, 6*, 18.

Lai, N., Sims, J. K., Jeon, N. L., & Lee, K. (2012). Adipocyte induction of preadipocyte differentiation in a gradient chamber. *Tissue Engineering Part C, Methods, 18*, 958–967.

Lee, R. H., Kim, B., Choi, I., et al. (2004). Characterization and expression analysis of mesenchymal stem cells from human bone marrow and adipose tissue. *Cellular Physiology and Biochemistry, 14*, 311–324.

Lopez-Santalla, M., Mancheño-Corvo, P., Menta, R., et al. (2015). Human adipose-derived mesenchymal stem cells modulate experimental autoimmune arthritis by modifying early adaptive T cell responses. *Stem Cells, 33*, 3493–3503.

Lv, X., He, J., Zhang, X., et al. (2018). Comparative efficacy of autologous stromal vascular fraction and autologous adipose-derived mesenchymal stem cells combined with hyaluronic acid for the treatment of sheep osteoarthritis. *Cell Transplantation, 27*, 1111–1125.

Magalon, J., Velier, M., Simoncini, S., et al. (2019). Molecular profile and proangiogenic activity of the adipose-derived stromal vascular fraction used as an autologous innovative medicinal product in patients with systemic sclerosis. *Annals of the Rheumatic Diseases, 78*, 391–398.

Moon, K. C., Chung, H. Y., Han, S. K., Jeong, S. H., & Dhong, E. S. (2019). Possibility of injecting adipose-derived stromal vascular fraction cells to accelerate microcirculation in ischemic diabetic feet: A pilot study. *International Journal of Stem Cells, 12*, 107–113.

Nunes, S. S., Maijub, J. G., Krishnan, L., et al. (2013). Generation of a functional liver tissue mimic using adipose stromal vascular fraction cell-derived vasculatures. *Scientific Reports, 3*, 2141.

Nyberg, E., Farris, A., O'Sullivan, A., Rodriguez, R., & Grayson, W. (2019). Comparison of stromal vascular fraction and passaged adipose-derived stromal/stem cells as point-of-care agents for bone regeneration. *Tissue Engineering Part A, 25*, 1459–1469.

Petrenko, Y., Vackova, I., Kekulova, K., et al. (2020). A comparative analysis of multipotent mesenchymal stromal cells derived from different sources, with a focus on neuroregenerative potential. *Scientific Reports, 10*, 4290.

Pham, P., Vu, N. B., Phan, N. L.-C., et al. (2014). Good manufacturing practice-compliant isolation and culture of human adipose derived stem cells. *Biomedical Research and Therapy, 4*, 1–9.

Proebstl, D., Voisin, M.-B., Woodfin, A., et al. (2012). Pericytes support neutrophil subendothelial cell crawling and breaching of venular walls in vivo. *Journal of Experimental Medicine, 209*, 1219–1234.

Ryu, Y.-J., Cho, T.-J., Lee, D.-S., Choi, J.-Y., & Cho, J. (2013). Phenotypic characterization and in vivo localization of human adipose-derived mesenchymal stem cells. *Molecules and Cells, 35*, 557–564.

Secunda, R., Vennila, R., Mohanashankar, A. M., Rajasundari, M., Jeswanth, S., & Surendran, R. (2015). Isolation, expansion and characterisation of mesenchymal stem cells from human bone marrow, adipose tissue, umbilical cord blood and matrix: A comparative study. *Cytotechnology, 67*, 793–807.

Shalaby, S. M., Sabbah, N. A., Saber, T., & Abdel Hamid, R. A. (2016). Adipose-derived mesenchymal stem cells modulate the immune response in chronic experimental autoimmune encephalomyelitis model. *IUBMB Life, 68*, 106–115.

Shubinsky, G., & Schlesinger, M. (1997). The CD38 lymphocyte differentiation marker: New insight into its ectoenzymatic activity and its role as a signal transducer. *Immunity, 7*, 315–324.

Solodeev, I., Orgil, M., Bordeynik-Cohen, M., et al. (2019). Cryopreservation of stromal vascular fraction cells reduces their counts but not their stem cell potency. *Plastic and Reconstructive Surgery Global Open, 7*.

Sun, M., He, Y., Zhou, T., Zhang, P., Gao, J., & Lu, F. (2017). Adipose extracellular matrix/stromal vascular fraction gel secretes angiogenic factors and enhances skin wound healing in a murine model. *BioMed Research International, 2017*, 3105780.

Van Pham, P., Bui, K. H.-T., Ngo, D. Q., et al. (2013). Activated platelet-rich plasma improves adipose-derived stem cell transplantation efficiency in injured articular cartilage. *Stem Cell Research & Therapy., 4*, 91.

Van Pham, P., Vu, N. B., & Phan, N. K. (2016a). Umbilical cord-derived stem cells (MODULATISTTM) show strong immunomodulation capacity compared to adipose tissue-derived or bone marrow-derived mesenchymal stem cells. *Biomedical Research and Therapy, 3*, 687–696.

Van Pham, P., Vu, N. B., Nguyen, H. T., & Phan, N. K. (2016b). Isolation of endothelial progenitor cells from human adipose tissue. *Biomedical Research and Therapy, 3*, 1–8.

Williams, S. K., Wang, T. F., Castrillo, R., & Jarrell, B. E. (1994). Liposuction-derived human fat used for vascular graft sodding contains endothelial cells and not mesothelial cells as the major cell type. *Journal of Vascular Surgery, 19*, 916–923.

Yu, S., Cheng, Y., Zhang, L., et al. (2019). Treatment with adipose tissue-derived mesenchymal stem cells exerts anti-diabetic effects, improves long-term complications, and attenuates inflammation in type 2 diabetic rats. *Stem Cell Research & Therapy, 10*, 333.

Zhang, L., Wang, X.-Y., Zhou, P.-J., et al. (2017). Use of immune modulation by human adipose-derived mesenchymal stem cells to treat experimental arthritis in mice. *American Journal of Translational Research, 9*, 2595.

Zhang, Y., Cai, J., Zhou, T., Yao, Y., Dong, Z., & Lu, F. (2018). Improved long-term volume retention of stromal vascular fraction gel grafting with enhanced angiogenesis and adipogenesis. *Plastic and Reconstructive Surgery, 141*, 676e–686e.

Zhou, B., Yuan, J., Zhou, Y., et al. (2011). Administering human adipose-derived mesenchymal stem cells to prevent and treat experimental arthritis. *Clinical Immunology, 141*, 328–337.

Zimmerlin, L., Donnenberg, V. S., Pfeifer, M. E., et al. (2010). Stromal vascular progenitors in adult human adipose tissue. *Cytometry. Part A, 77*, 22–30.

Adv Exp Med Biol - Innovations in Cancer Research and Regenerative Medicine (2023) 4: 63–82
https://doi.org/10.1007/5584_2022_710

Published online: 8 April 2022

Routes of Stem Cell Administration

Sharmila Fagoonee, Shiv Poojan Shukla, Anupam Dhasmana, Alexander Birbrair, Shafiul Haque, and Rinaldo Pellicano

Abstract

Stem cells are very promising for the treatment of a plethora of human diseases. Numerous clinical studies have been conducted to assess the safety and efficacy of various stem cell types. Factors that ensure successful therapeutic outcomes in patients are cell-based parameters such as source, viability, and number, as well as frequency and timing of intervention and disease stage. Stem cell administration routes should be appropriately chosen as these can affect homing and engraftment of the cells and hence reduce therapeutic effects, or compromise safety, resulting in serious adverse events. In this chapter, we will describe the use of stem cells in organ repair and regeneration, in particular, the liver and the available routes of cell delivery in the clinic for end-stage liver diseases. Factors affecting homing and engraftment of stem cells for each administration route will be discussed.

Keywords

Delivery routes · Hepatic diseases · Liver repair and regeneration · Safety and efficacy · Stem cells

S. Fagoonee (✉)
Institute of Biostructure and Bioimaging, National Research Council (CNR), Molecular Biotechnology Center, Turin, Italy
e-mail: sharmila.fagoonee@cnr.it; sharmila.fagoonee@unito.it

S. P. Shukla
Department of Dermatology & Cutaneous Biology, Sydney Kimmel Cancer Center Thomas Jefferson University, Philadelphia, PA, USA
e-mail: shiv.poojan@jefferson.edu

A. Dhasmana
Department of Immunology and Microbiology and South Texas Center of Excellence in Cancer Research, School of Medicine, The University of Texas Rio Grande Valley, McAllen, TX, USA

Department of Biosciences and Cancer Research Institute, Himalayan Institute of Medical Sciences, Swami Rama Himalayan University, Dehradun, India
e-mail: anudhas007@gmail.com

A. Birbrair
Department of Pathology, Federal University of Minas Gerais, Belo Horizonte, MG, Brazil

Department of Radiology, Columbia University Medical Center, New York, NY, USA
e-mail: birbrair@icb.ufmg.br

S. Haque
Research and Scientific Studies Unit, College of Nursing and Allied Health Sciences, Jazan University, Jazan, Saudi Arabia

Bursa Uludağ University Faculty of Medicine, Nilüfer, Bursa, Turkey
e-mail: shafiul.haque@hotmail.com

R. Pellicano
Unit of Gastroenterology, Molinette Hospital, Turin, Italy
e-mail: rinaldo_pellican@hotmail.com

Abbreviations

ALB:	Albumin
AST	Aspartate Transaminase
BAL	Bioartificial liver
BM-HSCs	Bone marrow–derived HSCs
BM-MNCs	Bone marrow–derived mononuclear stem cells
BM-MSCs	Bone marrow–derived stem cells
BM-SCs	Bone marrow–derived stem cells
ESCs	Embryonic stem cells
EVs	Extracellular vesicles
GPSCs	Germline cell–derived pluripotent stem cells
GvHD	Graft-versus-host disease
HBV	Hepatitis B virus
HBV-DLC	HBV-related decompensated liver cirrhosis
HLSC	Human liver stem cells
HOCs	Hepatic oval cells
HSCs	Hematopoietic stem cells
IL	Interleukin
iPSC	Induced pluripotent stem cells
MELD	Model for end-stage liver disease
MNCs	Mononuclear stem cells
MSCs	Mesenchymal/stromal stem cells
PT	Prothrombin
PVE	Portal vein embolization
TBIL	Total bilirubin
TGF	Transforming Growth Factor
UC-MSCs	UC-derived MSCs
UC-SCs	Cord blood–derived stem cells

1 Introduction

Stem cells, thanks to their inherent capacity for self-renewal and differentiation into different cell types, are very promising for the treatment of a plethora of human diseases. An array of stem cell types has been described as clinically useful for the resolution of tissue damage caused by genetic defects, deregulated metabolism, or injury. Pluripotent stem cells such as induced pluripotent stem cells (iPSCs), embryonic stem cells (ESCs), and germline cell–derived pluripotent stem cells (GPSCs, or pluripotent spermatogonial stem cells) have the potential to adopt different cell fates as demonstrated by various in vitro and preclinical studies (Liu & Zheng, 2019; Pellicano et al., 2021; Mehta et al., 2010). Pluripotent stem cells are being employed in clinical trials but practical issues regarding immunogenicity and heterogeneity limit their use (recently extensively discussed by S. Yamanaka) (Yamanaka, 2020). Moreover, even the scarce possibility of residual undifferentiated stem cells persisting after induced differentiation or genetic alterations occurring during in vitro culture or activation of gene expression networks shared by stem cells and cancer cells predisposing the former to tumorigenic fates once injected in vivo has solicited cautiousness in the clinical progress of these cells (Yamanaka, 2020; Lee et al., 2013). Despite these issues, Mandai et al. showed that autologous transplantation of retinal pigment epithelial cell sheets made with iPSCs derived from skin fibroblasts of one patient with age-related macular degeneration was feasible (Mandai et al., 2017). The 1-year post-surgery follow-up showed that the graft remained intact and that there were no signs of local or systemic malignancy. While most clinical studies registered in the National Institute of Health (NIH) registry (www.clinicaltrials.gov) regard mainly generating iPSCs from patient tissues to study disease mechanism or for master cell bank creation, few of these are employing iPSCs as intention-to-treat clinical studies. An interventional Phase I/II study is recruiting 20 patients with age-related macular degeneration to perform autologous transplantation of iPSC-derived retinal pigment epithelium grown on a biodegradable poly lactic-co-glycolic acid scaffold (NCT04339764). The primary endpoint is safety of the transplantation of retinal pigment epithelial scaffold in the subretinal space, and the secondary outcome is to evaluate visual acuity change as well as report any adverse events. Long-term follow-up is needed to assure patients and clinicians about the safety of pluripotent stem cell–based treatment. Mesenchymal stromal/

stem cells (MSCs), on the other hand, have gained tremendous interest because of their hypo-immunogenicity, immunosuppressive, and paracrine properties, for their ease of isolation from adult tissues and multipotency, and for their tolerable safety profile, as well as therapeutic evidence in a number of clinical studies. In fact, MSCs are currently being investigated and used in several clinical settings, and have been approved as treatment in some countries, for instance, for orthopedic surgery for joint degenerative and inflammatory diseases (Lopa et al., 2019; Pham et al., 2021; Vu et al., 2021). The use of allogeneic MSCs in certain disease settings have shown some promises but in vivo studies in animal models have pointed out how delivery routes exert differential effects on the maintenance of the MSCs in an immuno-privileged state (Gu et al., 2015).

Despite the promising preclinical results obtained for the treatment of various pathologies in animal models of human diseases and the initial enthusiasm regarding human stem cell transplantation, stem cells administration in patients has been limited by several setbacks. In case of allogeneic stem cells transplantation, for instance, immune rejection or graft-versus-host disease, requiring life-long treatment with immunosuppressant to avoid clearance of injected cells by the immune system, has been an important concern. Thus, stem cells have moved at a slow pace into the clinic, with fine control required at each step, from the stem cell production process to the in vivo transplantation for feasibility and safety assessment. Parameters that need careful evaluation before stem cells transplantation include cell dosage, frequency of cell administration, and timing of intervention. This leads to another crucial decision-making, that is, the route of stem cell delivery. Stem cell administration via different routes are possible in patients, and some comparative studies have been done associating delivery route with clinical outcome, discussed in this chapter, in particular regarding liver diseases.

2 Factors Affecting Stem Cell Transplantation Efficacy

Successful homing and engraftment of injected stem cells are crucial in achieving therapeutic efficacy. This depends on several factors, which need to be assessed carefully before performing stem cell treatment in patients affected by life-threatening diseases, and discussed hereafter.

2.1 Source of Stem Cells

The correct choice of stem cells for transplantation is essential for achieving successful therapeutic outcome in the clinics. Autologous stem cell sources can bypass the problems related to immune rejection. Thus, efforts are being made to generate transplantable stem cells from patient-derived tissues. Adult stem cells, which show major promises with respect to pluripotent stem cells in patients due to the above-mentioned concerns, are under close investigation in several disease types. To date, MSCs are the most employed stem cells in clinical studies. MSCs show pleiotropic effects and exert their action by either being present at the site of injury/inflammation or distally (paracrine action). MSCs can in fact release paracrine factors (cytokines, chemokines, and extracellular vesicles) which can induce tissue repair (Ahangar et al., 2020).

MSCs can be obtained from bone marrow, as well as in a noninvasive way from different adult tissues, such as the placenta, umbilical cord, amniotic fluid, and adipose tissue. Despite their similar behavior in culture and presence of MSC surface markers, these cells show different degrees of differentiation potential. For instance, a significant difference in proliferative capacity was seen for MSC derived from umbilical cord blood, with respect to MSCs from other sources (Alfaifi et al., 2018). Differences among MSCs derived from adipose tissue and bone marrow were also noted, with the former being more potent in terms of immunomodulatory effects due to the higher capacity of cytokine secretion (Interleukin (IL)-6 and Transforming Growth Factor (TGF)-β1) (Melief et al., 2013). Thus, in-depth studies on the properties of stem cells need to be performed to obtain the optimal therapeutic effects in patients.

2.2 Cell Vitality, Dosage, Frequency of Administration, and Timing of Intervention

Cryopreservation may affect some adult stem cell properties (Bahsoun et al., 2019). It was shown that cryopreservation did not change bone marrow–derived MSC morphology, surface marker expression, differentiation capacity, and proliferation potential but viability, colony-forming ability, attachment and migration, genomic stability, and paracrine function could be variably affected by the freezing–thawing process (Bahsoun et al., 2019). Thus, accurate quantification and characterization of vital cells are an important requisite, as only therapeutically viable cells will be able to engraft and proceed with tissue repair in vivo.

Stem cell dosage is another important issue. Analysis of number of MSCs used in clinical trials registered in the Clinicaltrials.gov repository was recently performed by Kabat et al. (2020). The authors found that doses were indicated either as total number of cells/patients or as number of cells/kg. The cell dosage was dependent on route of administration, with intravenous route requiring higher number of MSCs (median of 100 million MSCs/patient/dose), probably because a significant number of cells entrap in the lungs and thus do not home to target tissues. Regarding this delivery route, it was found that the mean dosage of 100–150 million cells/patient was the minimal effective dosage, while lower or higher doses did not provide any significant therapeutic outcome. Thus, early stage clinical trials are necessary to establish the minimal effective dose before proceeding with larger trials.

It is now accepted that one dose of stem cells is not enough to give sufficient therapeutic effects, especially in chronic diseases that take years to develop and involve multistep processes. Administering repeated doses results in cumulative increase in the number of transplanted cells that can persist longer in vivo (Wysoczynski et al., 2018). Clinical studies have been undertaken to investigate the effects of both dosage and frequency of stem cell administration. For instance, a Phase I/IIa clinical trial employed three different doses of adipose tissue–derived autologous MSCs (ten million, 20 million, and 50 million) injected intra-articularly at study beginning (Time 0) and 3 weeks after in patients with knee osteoarthritis (NCT01809769). It was thereafter reported that three injections at the highest dose of 50 million cells was safe and the most therapeutically effective, showing improvement in pain, function, and cartilage volume of the knee joint (Song et al., 2018).

The optimal time window for stem cell treatment in each disease has to be determined in preclinical models prior to clinical intervention. For instance, in the case of ischemic stroke, timing is crucial. Pharmacologic intervention with tissue-type plasminogen activator (tPA) provides neurorestorative effects at 4.5 h intravenously and 24 h endovascularly but not at later time points (He et al., 2020). In experimental models, it was found that intravenous administration of bone marrow–derived MSCs at 3 and 24 h post-injury, but not at 7 days, significantly decreased the lesion volume and improved motor deficits (Wang et al., 2014). Investigation on the correct timing of intervention with stem cells for ischemic stroke in clinical studies has brought variable outcomes, and is still ongoing for other diseases as well (He et al., 2020; Zhang et al., 2020).

2.3 Cell Visualization Methods

Cell visualization is another hurdle in this journey of stem cell transplantation. Stem cell labelling is possible for preclinical studies, hence allowing tracking of infused cells from the site of delivery to the organ of interest, in order to be able to associate any therapeutic effect to the cells.

Quantifying the biodistribution and clearance of cells in the long term help in optimizing the cell dose, frequency of injections, as well as stem cell transplantation route (Schmuck et al., 2016). Currently available cell tracking tools, with high spatial and temporal resolution, have dictated cell-labelling approaches used in vivo. In rodents, for instance, it is possible to use magnetic-, fluorescence- or bioluminescence-based labelling methods due to the availability of sophisticated instruments such as magnetic resonance imagers, fluorescence microscopes, and bioluminescence imagers. Most of these imaging modalities allow tracking of injected stem cells, but each strategy has its limitations (Leahy et al., 2016). Magnetic resonance imaging, using for instance, with the use of the Food and Drug Administration (FDA)–approved superparamagnetic iron oxide (SPIO) nanoparticles (which are biocompatible and show high relaxivity), provides good sensitivity and tissue penetration (Wáng & Idée, 2017). However, use of the correct dosage of iron oxide nanoparticles is essential, as high concentrations of these substances may generate oxidative stress and toxicity, thus negatively impacting on cell survival (Ohki et al., 2020). On the other hand, fluorescence-based imaging takes advantage from the vast array of fluorescent dyes developed to suit in vivo cell tracking and cell fate mapping requirements. In animal studies, fluorescent dyes (PKH26, CFSE, DiR) are mainly used for stem cell tracing. However, dye leakage or macrophage-engulfed labelled cells and tissue auto-fluorescence may give misleading results. Near-infrared fluorescent dyes, such as Quantum Dots, which provide a narrow emission spectrum to bypass tissue auto-fluorescence, and that allow real-time evaluation and longitudinal analyses, have resulted in successful tracking of stem cells in vivo. This field is under constant evolution, and clinically useful stem cell visualization strategies are being improved. Thus, to date, preclinical studies come closest at estimating the cell dosage, frequency of administration, timing of intervention, as well as route of cell delivery that can provide therapeutic benefits, in order to be able to set the baseline for clinical studies.

3 Stem Cell Administration Routes

Multiple routes of stem cell delivery have been studied in preclinical models, in order to achieve optimal homing to and engraftment in the diseased organ. Several studies have shown that the delivery route has a deep impact on stem cell biodistribution and mechanism of action, and hence the clinical outcome. Stem cells can be delivered locally into the target tissue or by the systemic route (Table 1) (Kean et al., 2013; Bonaventura et al., 2021; Lv et al., 2021;

Table 1 Stem cell administration routes commonly employed in the clinic

Organ	Cell administration route	References
Brain	Intracerebroventricular Intravascular Intranasal Lumbar puncture Intrathecal Stereotactic brain surgery	Bonaventura et al. (2021), Lv et al. (2021)
Heart	Transendocardial Direct intramyocardial Intravenous Intracoronary	Golpanian et al. (2016), Kanelidis et al. (2017)
Liver	Intravenous Intrahepatic Intrasplenic	Sanchez-Diaz et al. (2021), Yang et al. (2021)
Kidney	Intravenous Intra-arterial stent placement	Wong (2021)
Joints	Intra-articular Intravenous	Lopez-Santalla et al. (2020)

Organ-wise delivery routes are shown

Golpanian et al., 2016; Kanelidis et al., 2017; Sanchez-Diaz et al., 2021; Yang et al., 2021; Wong, 2021; Lopez-Santalla et al., 2020). For local delivery, stem cells can be embedded in cell carrier materials to increase residence time of viable cells in the tissue in order to enhance their engraftment (Antunes et al., 2014; O'Cearbhaill et al., 2014). For instance, scaffold-based stem cell therapy can lead to marked improvement in the structure and function of tissues by providing controlled release of therapeutic biomolecules at sites of injury (Gu et al., 2021). Systemic delivery, whilst the most employed both in experimental models and in patients, have several drawbacks, such as entrapment of injected cells in the lungs as first-pass organ. This reduces the availability of stem cells to target organs, and diminishes the therapeutic effects. Allogeneic stem cells may also be rapidly cleared by the immune system, for instance, by natural killer–mediated killing (Petrus-Reurer et al., 2021). However, some types of stem cells, like the MSCs, through the secretion of paracrine factors and extracellular vesicles, can act even from a distance to confer therapeutic effects on the target organ. Thus, the choice of stem cells, as well as improved delivery protocol is imperative for their clinical use. Modification of these cells to release higher concentration of therapeutic molecules can also be envisaged (as extensively described in (Naldini, 2019; Ratajczak & Suszynska, 2016)).

4 Stem Cell Delivery Routes for Liver Diseases

4.1 The Liver

The liver, as the largest internal organ of the body, is a central and vital hub through which several processes such as detoxification, regulation of metabolism, and control of homeostasis take place. The prevalence of liver diseases is increasing globally (Collaborators, 2020). The liver is constantly exposed to viruses and drug-induced damages, and has evolved several mechanisms to endogenously repair itself. One main mechanism is by self-regeneration whereby hepatic cell cycle to replenish the damaged cells (Van Haele et al., 2019; Burra et al., 2021). However, chronic insults to the liver can buffer organ repair so that liver functionality is gradually affected (Fig. 1) (Rosso et al., 2020). Orthotopic

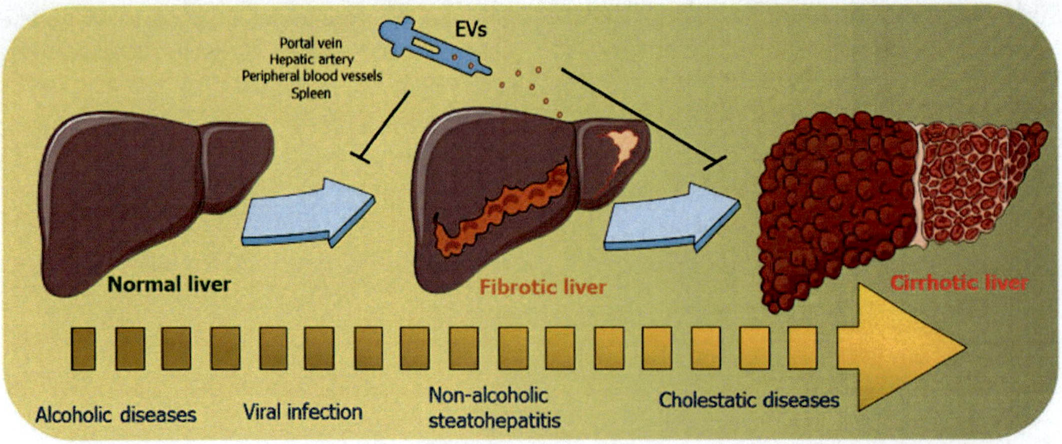

Fig. 1 Disease progression in the liver exposed to chronic insults. Excessive alcohol consumption, viral infection, dysmetabolism, genetic causes, or drugs can lead to the development of liver scarring. Liver fibrosis is the first stage of liver scarring which can progress into liver cirrhosis. Stem cell–derived extracellular vesicles (EVs), directed to the liver and delivered through different routes, including portal vein, hepatic artery, splenic artery or systemic routes, could dampen liver fibrosis and cirrhosis development

liver transplantation is widely used to treat liver disease upon organ failure (Burra et al., 2021; Iansante et al., 2018). The fact that this procedure is complex and donor organs are scarce have prompted the search for alternative therapeutic options to treat end-stage liver diseases (Cernigliaro et al., 2020). Thus, exogenous systems have been implemented to support hepatic functions in case of liver failure. These include extracorporeal liver support devices that purify blood by removing albumin-bound toxic substances or water-soluble substances, such as ammonia, creatinine, or urea by dialysis (García Martínez & Bendjelid, 2018). Several clinical studies have been conducted with these devices with the primary outcome being survival (Lee et al., 2016). Bioartificial liver (BAL) support systems that incorporate primary hepatocytes into a bioreactor have also been used in phase I studies or controlled clinical trials (He et al., 2019). As these support systems still require some improvements, alternatives, such as cell-based treatments, have gained tremendous interest in the field of liver regenerative medicine. Transplantation of liver cells to promote regeneration has been reported in the clinical setting. Albeit successful, it is difficult to obtain a routine supply of good-quality hepatocytes. Thus, much hope has turned toward stem cells, such as liver progenitor and multipotent stem cells from several sources, which represent clinically usable cell sources, discussed below (Cernigliaro et al., 2020; Alwahsh et al., 2018).

4.2 Types of Stem Cells in Clinical Use for the Treatment of Liver Diseases

Stem cells are highly sensitive to various factors and their procurement is not very easy, and all the procedure must follow medical ethics. Before their transplantation in patients with liver diseases, stem cells must show the capability to differentiate into functional hepatocytes and should have low immunogenicity. These cells must be viable and small enough to pass in the liver sinus quickly, as well as should have good proliferative index without being tumorigenic. Several types of stem and progenitor cells have been employed for the treatment of liver diseases in preclinical models, such as the ESCs, hepatic oval cells, iPSCs, MSCs, small hepatocyte-like progenitor cells, and immortalized hepatocytes, with clinical transability as the major goal (Fig. 2). However, the pluripotent stem cells

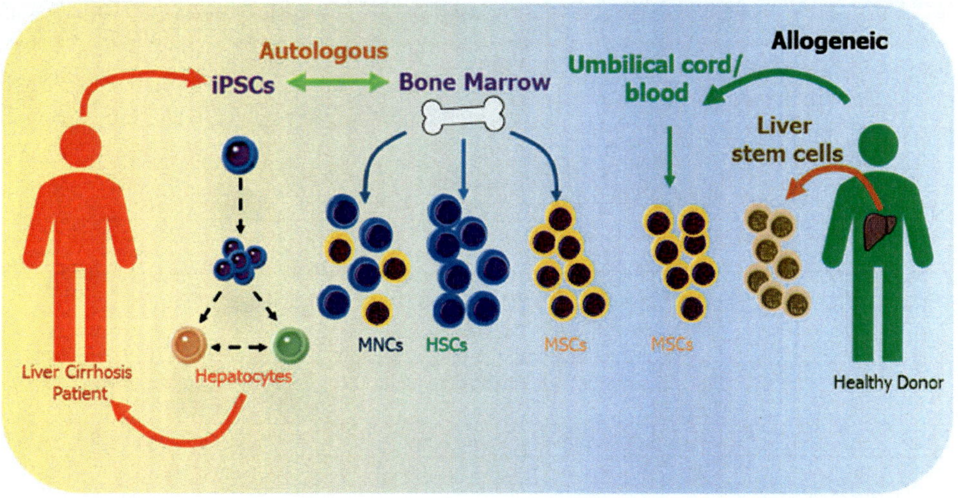

Fig. 2 Stem cell types employed for the treatment of liver diseases. Mesenchymal stem cells (MSCs) obtained from different sources, such as umbilical cord or bone marrow, liver stem cells, induced pluripotent stem cells (iPSCs)–derived hepatocytes, hematopoietic stem cells (HSCs), and mononuclear stem cells (MNCs) are the most promising for the resolution of liver diseases in the clinic

face the above-mentioned hurdles, thus limiting the types of stem cells used in clinical studies. The following stem cell types have better perspectives in the field of clinical hepatology.

4.3 Hepatic Oval Cells

Hepatic oval cells (HOCs) are adult hepatic stem cells with multilineage differentiation potential. Most endogenous hepatocytes tend to lose functionality after serious liver tissue damage mature hepatocytes proliferation gradually ceasing. For the betterment of liver function, a variety of cytokines has been shown to induce differentiation of HOCs into functional mature hepatocytes and biliary epithelial cells; thus, it can be expected that HOC transplantation can cure various acute and subacute liver diseases. A few studies found activated and proliferated HOCs in chronic liver diseases and hepatocellular carcinoma (Chen et al., 2007; de Lima et al., 2008). Due to difficulties encountered regarding their purification, identification, and tumorigenic behavior, the application of HOC in the clinic, albeit very promising, still needs to be implemented.

4.4 Mesenchymal Stem Cells

MSCs are the most widely used multipotent adult stem cells, and are regarded as the ideal cell source for cell therapy and regenerative medicine. Several studies have shown that in vitro proliferation and differentiation of MSCs into hepatocytes can be promoted by several growth factors; thereafter, these cells were efficiently used in liver regeneration and have become an important extrahepatic source of hepatocytes (Lysy et al., 2008). MSCs are promising as cell-based therapy for the treatment of liver diseases due to their proliferative ability, differentiation, and immunomodulatory capacities. MSCs have become a new research focus in regenerative medicine because of their high degree of self-replication and multidirectional differentiation potential, as well as their uniquely low immunogenicity, immunoregulation, and anti-inflammatory effects (Lysy et al., 2008). As mentioned above, MSCs can also be derived from a variety of tissues and organs, including postpartum tissues and unwanted surgical specimens, hence offering the opportunity to fulfill the demands of the rising numbers of clinical trials. The United States National Institutes of Health stem cell registry recorded 5,888 stem cell–related clinical trials starting January 2017, among which 325 have used cord blood–derived stem cells (UC-SCs), and 186 postpartum tissue. Clinical trials have been performed using human UC-derived MSCs (UC-MSCs) through various administration routes.

Stem cells with mesenchymal properties derived from the liver have also been employed in clinical studies for hepatic function restoration and have shown positive outcomes (Table 2) (Mohamadnejad et al., 2013, 2016; Spada et al.,

Table 2 Preclinical studies helpful in deciding the optimal administration routes for liver diseases

Route	Experimental subject	Findings	References
Intravenous	Mice	ConA-induced hepatitis, GFP-labelled MSCs were found only in the lungs but not in the liver	Higashimoto et al. (2013)
Intravenous	Pigs	MSCs accumulation in the lungs, but also distributing throughout the body and other organs	Moll et al. (2019), Fischer et al. (2009), Schrepfer et al. (2007), Gao et al. (2001), Lee et al. (2009)
Intravenous		Stem cells accumulate in the lung	Eggenhofer et al. (2012)
Intraportal	Acute liver failure pigs	Stem cells restored hepatic function	Cao et al. (2012)
Peripheral vein and Intraportal route	Acute liver failure pigs	Delivery of stem cells through the peripheral vein. No rescue of ALF pigs; delivery by intraportal injection. ALF pigs survived for over 6 months after	Wei and Lv (2013)

Some routes of stem cell delivery in the most commonly used animal models (mice and pigs) are shown with summary of findings

2020; Smets et al., 2019; Bruno et al., 2021; Khan et al., 2008; Zhang, 2017). MSCs, due to the above-listed properties, can be easily incorporated into therapeutic regimens. A number of comparative analyses was performed on MSCs derived from different tissues, and some source-dependent inherent differences were found. Based on the molecular profiles, MSCs derived from bone marrow and adipose tissue, were found to have higher differentiation potential and better immunomodulatory effects, and could thus represent optimal stem cell sources for tissue regeneration (Heo et al., 2016). Interestingly, MSCs also showed propensity to differentiate into hepatocyte-like cells showing hepatic morphology and function in vitro, hence offering promises for liver regeneration and repair too (Schwartz et al., 2002). MSCs have been employed in several clinical studies for end-stage liver disease; however, the outcome of these studies has not yet been published.

Mesenchymal-like stem cells derived from the human adult liver (human liver stem cells or HLSC) have also been studied in several mouse models before their use in patients (Herrera et al., 2006, 2013; Famulari et al., 2020). The safety of HLSC administration in inherited neonatal onset hyperammonemia was recently assessed in a Phase I clinical study on three pediatric patients suffering from argininosuccinic aciduria (one patient) and methylmalonic academia (2 patients) (Spada et al., 2019). Patients received two doses (at 1-week interval) of HLSC by percutaneous intrahepatic injections. It was found that HLSC injections in the first months of life of these patients was safe. Moreover, analysis of clinical and biochemical parameters showed that following HLSC treatment, ammonia concentration was stable despite increase in dietary protein intake, hence showing some efficacy. Stem cells from adult livers (Heterologous Human Adult Liver-derived Progenitor Cells or HepaStem) were also employed by Smets et al. in a safety study in pediatric patients with urea cycle disorders or Crigler–Najjar syndrome (Smets et al., 2019). Cell infusion in these patients safe, albeit some adverse events were noted. Importantly, increased ureagenesis with significant improvement of de novo urea formation was observed in the urea cycle disorders patients following HepaStem infusion. Thus, adult liver–derived stem cells are indeed promising for the treatment of severe human liver diseases.

5 Administration Routes of Stem Cells for Hepatic Diseases

In cases of liver cirrhosis, liver failure, or other liver disorders, various delivery routes have been used to target the liver. The commonly used routes for stem cell transplantation include peripheral vein, portal vein and hepatic artery. Mostly intravenous or systemic injection routes have been used, followed by intrahepatic administration as well as by intrasplenic delivery (Fig. 3). The delivery route for stem cells seems to be crucial for therapeutic efficiency. Systemic delivery of cells may cause rapid loss of cells within the capillaries, especially in the lungs, through the first-pass effect. Therefore, the choice of administration route should follow some principles: (1) there should be much available space to ensure sufficient engraftment of transplanted cells, (2) there should be a good blood supply, (3) an adequate nutritional hepatic factors from the portal vein is required. Thus, the transplantation route based on anatomical considerations is acceptable. Depending on the graft location, transplantation routes can be divided into orthotopic and heterotopic transplantation (Hu et al., 2021). Orthotopic transplantation involves intrahepatic and splenic routes while heterotopic one regards transplantation under the renal capsule, for instance (Hu et al., 2021).

6 Intravenous Injection

The most commonly employed route for MSC transplantation is the intravenous route, due to its ease and least-invasive behavior, which therefore allows the distribution of stem cells throughout the body. Repeated infusions of stem cells through the peripheral vein of the arm are thus

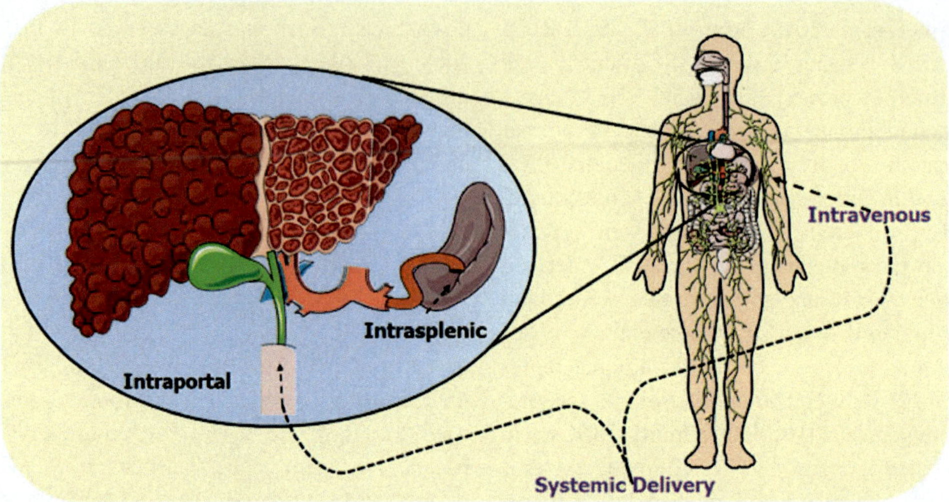

Fig. 3 Main routes of stem cell delivery to the diseased liver. Stem cells can be delivered locally (via the portal vein or through the spleen) into the target tissue or by the systemic route (peripheral vessels). The latter is the most employed in the clinic

feasible compared to other routes of administration. However, this delivery route, in almost all cases led to the biodistribution of MSCs mostly in lungs, but also in spleen, liver, bone marrow, thymus, kidney, and to skin, but the cell fate was different in different studies and models used, and was independent of the nature of the stem cells (autologous/syngeneic, allogeneic, or even xenogeneic) (Barbash et al., 2003). In a study, it was observed that most of the cells (12 ~ 19% in spleen and 4 ~ 5% in lymph nodes) localized in lymphoid tissues following tail vein injection compared to portal vein injection, which brought 5 ~ 7% and 1 ~ 2% of the cells to the spleen and lymph nodes, respectively (Kurtz, 2008). Portal vein injection had a benefit over tail vein injection for engraftment in the liver, where cells could be detected after 1 and 7 days. Very few cells were detectable in the lungs after portal vein injection, in contrast to tail vein injection. After 7 and 21 days, cells were almost undetectable in all tissues. Hitherto, there are no clinical trials comparing the effectiveness of intravenous injection and other routes, thus putting preclinical studies at the forefront in deciding the optimal administration routes for liver diseases, as shown in Table 2.

Like intra-arterial delivery of cell therapies, intravenous administration is most often associated with mechanisms involving secondary signaling effector cell systems and interactions with the host immune system (Walker et al., 2010, 2011; Lee et al., 2009). Intravenous administrations are most frequently adopted in commercial cell therapy design and development, and the well-known, systemically IV-infused Remestemcel-L (Prochymal), for instance, has been developed to primarily treat graft-vs.-host disease (GvHD) (Locatelli et al., 2017). Another work regards the use of multistem, a multipotent adult progenitor cell, which shows resemblance to MSCs, to treat ischemic stroke (Mays & Savitz, 2018). On the other hand, intravenous administration generates concerns regarding formation of emboli or thrombi. However, limited number of cells may reach target tissues resulting in a more transient persistence of cells, and a dilution of paracrine factors reaching target tissues following intravenous infusion. Finally, despite the fact that MSCs can modulate immune responses, intravenous delivery may rapidly lead to the removal of clinical MSCs by innate host immune cells (Moll et al., 2016, 2019; de Witte et al., 2018).

Several clinical trials have assessed the therapeutic potential of MSCs derived from bone marrow or umbilical cord blood in liver cirrhosis, providing, however, conflicting results. There are some cases in which administrative route was intravenous and the results as discussed hereafter.

6.1 Therapeutic Applications of Umbilical Cord–Derived Stem Cells in Liver Cirrhosis Patients

A clinical trial (ChiCTR-ONC-12002103) for decompensated liver cirrhosis has been performed in 103 patients with 50 subjects receiving 4.0–4.5 × 10^8 hUC-MSC through the intravenous route and 53 patients serving as control. In this study, a decrease in aspartate transaminase (AST), increase in albumin (ALB), total bilirubin (TBIL), prothrombin (PT), improvement in model for end-stage liver disease (MELD) and Child–Pugh scores were observed (Fang et al., 2018). A study has also been performed (Trial number NCT01218464) in 43 cases; a dose of 0.5 × 10^6/kg hUC-MSCs were infused intravenously in 24 patients versus 19 controls. After a follow-up of 72 weeks, hUC-MSC injection led to an improvement in MELD score and increase in ALB and PT (Shi et al., 2012). Zhang et al. found a reduction in ascites, increase in ALB, TBIL, and an improvement in MELD score (Trial number NCT01220492) during a follow-up period of 48 weeks in 45 cases with 30 patients receiving a dose of 0.5 × 10^6/kg hUC-MSCs and 15 serving as control (Zhang et al., 2012). For the ischemic-type biliary lesion disorder (NCT02223897), infusion of 1 × 10^6/kg hUC-MSCs (in week 1, 2, 4, 8, 12, and 16) in 12 patients resulted in a decrease in BIL, ALP, γ-GT, and improvement in graft survival (Zhang et al., 2017). In another study (NCT01690247) on liver transplant patients, 14 patients were treated with a dose of 1 × 10^6 UC-MSC/kg body weight, and a decrease in ALT, AST, TBIL, acute rejection, and an improvement in liver allograft histology were observed (Shi et al., 2017). The efficacy of hUC-MSC as clinical treatment of 60 patients for HBV-related decompensated liver cirrhosis (HBV-DLC) was observed by Yu et.al. for a 3-month regimen of UC-MSC at a dose of 0.5–1.0 × 10^6 cells/kg/month versus a control group of HBV-DLC patients treated with standard medical treatment (Yu et al., 2016). The group treated with combined UC-MSC treatment exhibited a significant rise in globulin, cholinesterase, and alkaline phosphatase and reduced Child–Pugh scores during the follow-up period. Safety and efficacy of hUC-MSC transplantation combined with plasma exchange for acute-on-chronic liver failure caused by hepatitis B virus (HBV) was studied in 120 patients (NCT01724398) after providing conventional treatment for 48 weeks and slow infusion of hUC-MSC through peripheral vein for 30 min (1 × 10^5/kg, once a week, four times), though the outcome is still unknown.

MSCs are also derived from bone marrow, though only 0.001–0.01% of all bone marrow cells are stem cells, which is significantly low. Moreover, their capacity for proliferation and differentiation decreases noticeably with age (Pouikli et al., 2021; Parekkadan & Milwid, 2010). The therapeutic applications of bone marrow derived–stem cells (BM-MSCs) in liver cirrhosis patients have been studied in last few years (Rajaram et al., 2017; Esmaeilzadeh et al., 2019). An increase in serum ALB, decrease in CPS and MELD score, bilirubin, and ascites were observed in a study on 40 patients, when given single median dose through intrahepatic and intrasplenic route (Liu et al., 2015). In another study, BM-MSCs were transplanted through peripheral vein route with single median dose of 30–40 × 10^6 in a group of four patients and with a single median dose of 31.73 × 10^6 in eight patients (Mohamadnejad et al., 2007; Kharaziha et al., 2009). The outcome of this study showed improved liver function and increased serum ALB, serum creatinine, PT, and bilirubin accompanying a decreased MELD score. However, bone marrow aspiration can be a painful procedure, which raises ethical concerns, thus warranting the search for alternative MSC sources.

6.2 Therapeutic Applications of Bone Marrow–Derived Stem Cells (BM-SCs) in Liver Cirrhosis Patients

Several clinical trials showed efficient liver regeneration with reduced fibrosis, improved liver function, and enhanced survival rates after successful transfusion of autologous bone marrow–derived mononuclear stem cells (BM-MNCs) (Pilat et al., 2013; Huebert & Rakela, 2014). The therapeutic utility of BM-MNCs has been largely evaluated in liver cirrhosis patients associated with hepatitis B, hepatitis C, alcoholic liver diseases, and decompensated and biliary duct liver cirrhosis. In a study, Lyra et al. injected BM-SCs through hepatic artery route in a group of 10 patients, using a single median dose of 1×10^8 and observed increased serum ALB and decreased bilirubin (Lyra et al., 2007). Rajkumar et al. studied a group of 22 patients treated with a single median dose of 240–1,027 $\times 10^6$ through intravenous route, and found increased ALB and decreased bilirubin and ascites (Rajkumar et al., 2007). Some studies with varied sample size (5–10) have shown increased serum ALB, decreased CPS, decreased ascites, and decreased PT time when BM-SCs were administered through the peripheral vein. Stem cells can secrete a variety of cytokines in the damaged liver, inhibit liver cell apoptosis, promote endogenous liver stem cell differentiation, stimulate endogenous hepatocyte proliferation, and promote liver angiogenesis. Importantly, they regulate the immune system in order to reduce the inflammatory response to repair the liver tissue. Therapeutic applications of bone marrow–derived HSCs (BM-HSCs) in liver cirrhosis patients have also been studied. In liver cirrhosis patients, transplantation of BM-HSCs at a dose range of 1×10^6 to 2×10^8 either through the portal vein, peripheral vein, or hepatic artery has been considered safe and tolerable (Mohamadnejad et al., 2007; Stutchfield et al., 2010). These stem cells can differentiate into hepatocyte-like cells, improve the function of damaged liver, and reduce the degree of liver fibrosis (Yannaki et al., 2005).

7 Portal Vein/Intrahepatic Stem Cell Delivery

Portal vein is the most often used route for stem cell transplantation due to multiple vascular accesses. However, accessing this route by transjugular or transhepatic percutaneous approaches poses some difficulties in the presence of ascites and has major bleeding risk, coagulopathy, and diathesis. Liver acts as best portal vein infusion and intrahepatic injection organ due to its unique structure and abundance in portal vein blood supply. The increased proliferation and differentiation induced by hepatocyte growth factors and regenerative factors are due to the direct contact of liver mesenchymal cells with transplanted cells, which thus create an optimal environment for grafted cell survival.

Few clinical trials have used multiple infusion routes simultaneously. In an Egyptian clinical trial, MSCs were infused through the portal vein or intrasplenic route for the treatment of end-stage liver failure, and after 6 months results were compared (Amer et al., 2011). It was shown that portal vein route was more effective than intrasplenic route, as revealed by the fatigue impact scale and MELD score. Apart from portal vein injection, the hepatic artery can also be used for intrahepatic stem cell delivery. In clinical trials for MSC transplantation, hepatic artery delivery route is employed more frequently than the portal vein injection route. In a preclinical study, by the increase in survival time and decrease in liver injury observed, it was evident that MSC delivery through intraportal route was superior in inducing liver injury repair in swine with acute liver failure than hepatic intra-arterial injection, peripheral intravenous injection, and in situ intrahepatic injection (Sang et al., 2016). Intra-arterial delivery of MSCs allows for infusion of cells within the local vascular system of the target organ without the pitfalls of intravenous administration, especially the trapping of cells within the lung microvasculature, hence ensuring that more cells reach the target tissue (Kean et al., 2013; Watanabe & Yavagal, 2016; Fischer et al., 2009). In a clinical study, decompensated liver cirrhosis patients receiving HSC transplantation through hepatic artery showed significant decrease ($P < 0.01$) in the mean MELD score, and after 6 months of follow-up, the patients showed improvement in multiple diagnostics and biochemical parameters (Kean et al., 2013). After 6 months to 1 year post-HSC transplantation, angiogram analysis of the patients showed significant improvement in the branching of hepatic arteries and veins. Transplantation of human hepatic progenitor cells in five patients suffering from decompensated liver cirrhosis led to a marked clinical recovery with decline in the MELD score, but without significant variation in T-cell subpopulation (CD3, CD4, CD8, and CD4/CD8 ratio), natural killer cells, and in serum cytokine levels between pre- and post-transplantation, hence demonstrating the safety of $EpCAM^+$ hepatic progenitor cell transplantation for the treatment of liver cirrhosis (Habeeb et al., 2015).

Autologous BM-MSCs were injected into end-stage liver disease patients through peripheral or portal vein and the procedure was found to be safe and well accepted along with improved liver function and clinical features (Mohamadnejad et al., 2007; Kharaziha et al., 2009). On the other hand, in malignant liver lesion patients, an increase in hepatic regeneration was observed after portal vein embolization (PVE) and administration of CD133+ BM-MSCs when compared to PVE alone (Fürst et al., 2007). Interestingly, cirrhotic patients showed improved Child–Pugh score and ALB levels following autologous BM cell infusion (Terai et al., 2006). In another study, after intraportal administration of mobilized $CD34^+$ BM-MSCs in acute liver failure patients, an improvement in biochemical and histopathology was achieved (Gasbarrini et al., 2007). The safety of granulocyte colony–stimulating factor administration was evaluated in liver failure patients by collecting and reinfusing $CD34^+$ cells through the intrahepatic routes. In more than 50% of the patients, an improved hepatic function was observed after a long-term follow-up (Gordon et al., 2006; Levicar et al., 2008). In the context of MSC transplantation for liver disease treatments, intraportal injection is the best possible route due to speedier engraftment and prevention of off-target accumulation. Importantly, before undertaking any route for treatment, health conditions of patients and the possible risks involved must be carefully evaluated. In studies with ALF pigs where human placenta–derived MSCs were infused through the portal vein, decreased liver inflammation, reduced hepatic denaturation and necrosis, as well as enhanced liver regeneration were observed (Cao et al., 2012, 2014). In addition, Li et al. found increased regeneration of the damaged liver and reversion to normal liver structure and function in young pigs (Li et al., 2012).

8 Splenic Route of Stem Cell Transplantation

Spleen is the optimal anatomical location for cell transplantation and is widely used since the splenic sinus has huge space, a certain mesh structure, and rich in blood supply (Sato et al., 2005). The cells transplanted through the splenic route were shown to have a long survival time in small animals. Cell transplantation into the splenic parenchyma offers an optimal environment for hepatocyte growth (Mito et al., 1979). Few studies reported that blood flow could dilute the transplanted cells, which can be ingested by phagocytes, therefore affecting stem cell engraftment and therapeutic activity. Cell injection in the spleen can efficiently bypass this problem. However, the spleen can contain less than 3% of normal liver cells, and can induce portal hypertension. Moreover, spleen transplantation is a very complex procedure. Altogether, these factors restrict stem cell transplantation through the splenic route in clinical studies. Compared to infusion of hepatocytes through the splenic artery, intrasplenic approach has been found to be a superior approach (Khan et al., 2010). Transplantation of fetal hepatocytes through the splenic artery in a patient with last-stage liver cirrhosis resulted in improvement in clinical conditions (Nagata et al., 2003; Siefert et al., 2019). The translocation of intrasplenically infused cells has also been demonstrated (Gridelli et al., 2012). Only few clinical trials have used multiple infusion routes simultaneously, and in one clinical trial, BM-MSCs were used to treat end-stage liver failure and were injected through the portal vein and intrasplenic route. In a study, after 6 months of follow-up, the authors reported that stem cell delivery through the portal vein was more effective than intrasplenic one for the treatment of liver failure, based on fatigue impact scale and MELD score (Amer et al., 2011). Some complications can be induced by intrasplenic injection: It was found that 70% of patients with intrasplenic injection had fever after cell transplantation compared to only 30% of those after intrahepatic injection. This shows that stem cell delivery routes may induce differential side effects in the severe liver disease patients.

9 Current Challenges

Despite the tremendous multidisciplinary efforts made for including stem cells in the therapeutic regimens for the treatment of severe liver diseases, several challenges still need to be faced. One of these regards the standardization of stem cell production under good manufacturing practices (GMP) guidelines. It is still not clear, for example, whether the GMP manufacturing processes may affect stem cell properties in vivo. The positive results obtained in animal studies may not fully translate into benefits for humans with similar pathologies. Liver regeneration in animals and humans are different in terms of signaling pathways and cellular sources that control this process (Alwahsh et al., 2018).

In patients with hematopoietic diseases, the efficacy of stem cell intervention can be determined quite easily by analyzing the presence of stem cell–specific markers on circulating blood cells. It is also essential to consider any histological alteration caused by transplanted stem cells in order to trace the infused cells in the patients, or to assess any fibrotic changes that may be induced upon cell transplantation. However, it is not always possible to obtain biopsies from solid organs from patients. Thus, there is an intense search for specific markers, especially biomarkers, such as those enclosed within circulating extracellular vesicles in the form of proteins or RNA, in order to assess the outcome of stem cell treatment. Biochemical parameters also support in monitoring improvement in organ function.

Determining which stem cell source best fits the organ under therapy is still under scrutiny. The liver is made up of different cells types including hepatocytes, cholangiocytes, liver progenitor cells, hepatic stellate cells, Kupffer cells, and endothelial cells (Alwahsh et al., 2018). Upon injury, all liver cell types present as the site of damage are affected. A still unanswered question is whether the stem cells can replenish all these hepatic cell types when delivered in patients.

10 Conclusions

Stem cell therapy is promising for the treatment of human liver diseases, and stem cell local delivery has shown better therapeutic efficacy in clinical trials, thus pointing out to the importance of stem cell delivery routes. Several important issues still have to be addressed to meet primary efficacy end-points and before stem cells can be routinely applied in the clinics for hepatic diseases. It is important to determine at which disease stage stem cell–based intervention should be undertaken to improve organ function. Potential off-target effects of cell-based therapies should be analyzed in long-term studies. At present, despite the tremendous progress in the field of stem cell therapy, successful stem-based therapies in the clinic are still too few. Stem cell therapy represents an appealing but challenging goal in translational medicine. Due to the variables described in this chapter, and the source of stem cells as well as receiving organ microenvironment quality, it is important to direct stem cell therapy in the clinic toward personalized rather than general procedures.

Author Contribution

SF: Conceptualization, Writing – original draft, Writing – review & editing, Supervision.
SPS: Writing – original draft, Writing – review & editing.
AD: Writing – original draft, Writing – review & editing.
SH: Visualization, Writing – review & editing.
AB: Visualization, Writing – review & editing.
RP: Visualization, Writing – review & editing, Supervision.

Conflict of Interest The authors declare no conflict of interest.

References

Ahangar, P., Mills, S. J., & Cowin, A. J. (2020). Mesenchymal stem cell secretome as an emerging cell-free alternative for improving wound repair. *International Journal of Molecular Sciences, 21*(19). https://doi.org/10.3390/ijms21197038

Alfaifi, M., Eom, Y. W., Newsome, P. N., & Baik, S. K. (2018). Mesenchymal stromal cell therapy for liver diseases. *Journal of Hepatology, 68*(6), 1272–1285. https://doi.org/10.1016/j.jhep.2018.01.030

Alwahsh, S. M., Rashidi, H., & Hay, D. C. (2018). Liver cell therapy: Is this the end of the beginning? *Cellular and Molecular Life Sciences, 75*(8), 1307–1324. https://doi.org/10.1007/s00018-017-2713-8

Amer, M. E., El-Sayed, S. Z., El-Kheir, W. A., Gabr, H., Gomaa, A. A., El-Noomani, N., et al. (2011). Clinical and laboratory evaluation of patients with end-stage liver cell failure injected with bone marrow-derived hepatocyte-like cells. *European Journal of Gastroenterology & Hepatology, 23*(10), 936–941. https://doi.org/10.1097/MEG.0b013e3283488b00

Antunes, M. A., Abreu, S. C., Cruz, F. F., Teixeira, A. C., Lopes-Pacheco, M., Bandeira, E., et al. (2014). Effects of different mesenchymal stromal cell sources and delivery routes in experimental emphysema. *Respiratory Research, 15*, 118. https://doi.org/10.1186/s12931-014-0118-x

Bahsoun, S., Coopman, K., & Akam, E. C. (2019). The impact of cryopreservation on bone marrow-derived mesenchymal stem cells: A systematic review. *Journal of Translational Medicine, 17*(1), 397. https://doi.org/10.1186/s12967-019-02136-7

Barbash, I. M., Chouraqui, P., Baron, J., Feinberg, M. S., Etzion, S., Tessone, A., et al. (2003). Systemic delivery of bone marrow-derived mesenchymal stem cells to the infarcted myocardium: Feasibility, cell migration, and body distribution. *Circulation, 108*(7), 863–868. https://doi.org/10.1161/01.CIR.0000084828.50310.6A

Bonaventura, G., Munafò, A., Bellanca, C. M., La Cognata, V., Iemmolo, R., Attaguile, G. A., et al. (2021). Stem cells: Innovative therapeutic options for neurodegenerative diseases? *Cell, 10*(8), 1992. https://doi.org/10.3390/cells10081992

Bruno, S., Herrera Sanchez, M. B., Chiabotto, G., Fonsato, V., Navarro-Tableros, V., Pasquino, C., et al. (2021). Human liver stem cells: A liver-derived mesenchymal stromal cell-like population with pro-regenerative properties. *Frontiers in Cell and Development Biology, 9*, 644088. https://doi.org/10.3389/fcell.2021.644088

Burra, P., Bizzaro, D., Forza, G., Feltrin, A., Volpe, B., Ronzan, A., et al. (2021). Severe acute alcoholic hepatitis: Can we offer early liver transplantation? *Minerva Gastroenterology, 67*(1), 23–25. https://doi.org/10.23736/S2724-5985.20.02778-6

Cao, H., Yang, J., Yu, J., Pan, Q., Li, J., Zhou, P., et al. (2012). Therapeutic potential of transplanted placental mesenchymal stem cells in treating Chinese miniature pigs with acute liver failure. *BMC Medicine, 10*, 56. https://doi.org/10.1186/1741-7015-10-56

Cao, H., Ma, J., Yang, J., Su, X., Chen, D., Yu, J., et al. (2014). A metabonomics study of Chinese miniature pigs with acute liver failure treated with transplantation of placental mesenchymal stem cells. *Metabolomics, 10*(4), 651–662. https://doi.org/10.1007/s11306-013-0603-0

Cernigliaro, V., Peluso, R., Zedda, B., Silengo, L., Tolosano, E., Pellicano, R., et al. (2020). Evolving cell-based and cell-free clinical strategies for treating severe human liver diseases. *Cell, 9*(2). https://doi.org/10.3390/cells9020386

Chen, Q. R., Xiang, J., Liao, B., Liu, Q. B., Che, L. H., Xue, L., et al. (2007). Evolutive characters of oval cells in experimental hepatocarcinogenesis. *Ai Zheng, 26*(7), 719–723.

Collaborators, G. C. (2020). The global, regional, and national burden of cirrhosis by cause in 195 countries and territories, 1990-2017: A systematic analysis for the global burden of disease study 2017. *The Lancet Gastroenterology & Hepatology, 5*(3), 245–266. https://doi.org/10.1016/S2468-1253(19)30349-8

de Lima, V. M., Oliveira, C. P., Alves, V. A., Chammas, M. C., Oliveira, E. P., Stefano, J. T., et al. (2008). A rodent model of NASH with cirrhosis, oval cell proliferation and hepatocellular carcinoma. *Journal of*

Hepatology, 49(6), 1055–1061. https://doi.org/10.1016/j.jhep.2008.07.024

de Witte, S. F. H., Luk, F., Sierra Parraga, J. M., Gargesha, M., Merino, A., Korevaar, S. S., et al. (2018). Immunomodulation by therapeutic mesenchymal stromal cells (MSC) is triggered through phagocytosis of MSC by Monocytic cells. *Stem Cells, 36*(4), 602–615. https://doi.org/10.1002/stem.2779

Eggenhofer, E., Benseler, V., Kroemer, A., Popp, F. C., Geissler, E. K., Schlitt, H. J., et al. (2012). Mesenchymal stem cells are short-lived and do not migrate beyond the lungs after intravenous infusion. *Frontiers in Immunology, 3*, 297. https://doi.org/10.3389/fimmu.2012.00297

Esmaeilzadeh, A., Ommati, H., Kooshyar, M. M., Jarahi, L., Akhavan Rezayat, K., Saberi, S., et al. (2019). Autologous bone marrow stem cell transplantation in liver cirrhosis after correcting nutritional anomalies, A controlled clinical study. *Cell Journal, 21*(3), 268–273. https://doi.org/10.22074/cellj.2019.6108

Famulari, E. S., Navarro-Tableros, V., Herrera Sanchez, M. B., Bortolussi, G., Gai, M., Conti, L., et al. (2020). Human liver stem cells express UGT1A1 and improve phenotype of immunocompromised Crigler Najjar syndrome type I mice. *Scientific Reports, 10*(1), 887. https://doi.org/10.1038/s41598-020-57820-2

Fang, X., Liu, L., Dong, J., Zhang, J., Song, H., Song, Y., et al. (2018). A study about immunomodulatory effect and efficacy and prognosis of human umbilical cord mesenchymal stem cells in patients with chronic hepatitis B-induced decompensated liver cirrhosis. *Journal of Gastroenterology and Hepatology, 33*(4), 774–780. https://doi.org/10.1111/jgh.14081

Fischer, U. M., Harting, M. T., Jimenez, F., Monzon-Posadas, W. O., Xue, H., Savitz, S. I., et al. (2009). Pulmonary passage is a major obstacle for intravenous stem cell delivery: The pulmonary first-pass effect. *Stem Cells and Development, 18*(5), 683–692. https://doi.org/10.1089/scd.2008.0253

Fürst, G., Schulte Am Esch, J., Poll, L. W., Hosch, S. B., Fritz, L. B., Klein, M., et al. (2007). Portal vein embolization and autologous CD133+ bone marrow stem cells for liver regeneration: Initial experience. *Radiology, 243*(1), 171–179. https://doi.org/10.1148/radiol.2431060625

Gao, J., Dennis, J. E., Muzic, R. F., Lundberg, M., & Caplan, A. I. (2001). The dynamic in vivo distribution of bone marrow-derived mesenchymal stem cells after infusion. *Cells Tissues Organs, 169*(1), 12–20. https://doi.org/10.1159/000047856

García Martínez, J. J., & Bendjelid, K. (2018). Artificial liver support systems: What is new over the last decade? *Annals of Intensive Care, 8*(1), 109. https://doi.org/10.1186/s13613-018-0453-z

Gasbarrini, A., Rapaccini, G. L., Rutella, S., Zocco, M. A., Tittoto, P., Leone, G., et al. (2007). Rescue therapy by portal infusion of autologous stem cells in a case of drug-induced hepatitis. *Digestive and Liver Disease, 39*(9), 878–882. https://doi.org/10.1016/j.dld.2006.06.037

Golpanian, S., Schulman, I. H., Ebert, R. F., Heldman, A. W., DiFede, D. L., Yang, P. C., et al. (2016). Concise review: Review and perspective of cell dosage and routes of administration from preclinical and clinical studies of stem cell therapy for heart disease. *Stem Cells Translational Medicine, 5*(2), 186–191. https://doi.org/10.5966/sctm.2015-0101

Gordon, M. Y., Levicar, N., Pai, M., Bachellier, P., Dimarakis, I., Al-Allaf, F., et al. (2006). Characterization and clinical application of human CD34+ stem/progenitor cell populations mobilized into the blood by granulocyte colony-stimulating factor. *Stem Cells, 24*(7), 1822–1830. https://doi.org/10.1634/stemcells.2005-0629

Gridelli, B., Vizzini, G., Pietrosi, G., Luca, A., Spada, M., Gruttadauria, S., et al. (2012). Efficient human fetal liver cell isolation protocol based on vascular perfusion for liver cell-based therapy and case report on cell transplantation. *Liver Transplantation, 18*(2), 226–237. https://doi.org/10.1002/lt.22322

Gu, L. H., Zhang, T. T., Li, Y., Yan, H. J., Qi, H., & Li, F. R. (2015). Immunogenicity of allogeneic mesenchymal stem cells transplanted via different routes in diabetic rats. *Cellular & Molecular Immunology, 12*(4), 444–455. https://doi.org/10.1038/cmi.2014.70

Gu, C., Feng, J., Waqas, A., Deng, Y., Zhang, Y., Chen, W., et al. (2021). Technological advances of 3D scaffold-based stem cell/exosome therapy in tissues and organs. *Frontiers in Cell and Development Biology, 9*, 709204. https://doi.org/10.3389/fcell.2021.709204

Habeeb, M. A., Vishwakarma, S. K., Bardia, A., & Khan, A. A. (2015). Hepatic stem cells: A viable approach for the treatment of liver cirrhosis. *World Journal of Stem Cells, 7*(5), 859–865. https://doi.org/10.4252/wjsc.v7.i5.859

He, Y. T., Qi, Y. N., Zhang, B. Q., Li, J. B., & Bao, J. (2019). Bioartificial liver support systems for acute liver failure: A systematic review and meta-analysis of the clinical and preclinical literature. *World Journal of Gastroenterology, 25*(27), 3634–3648. https://doi.org/10.3748/wjg.v25.i27.3634

He, J. Q., Sussman, E. S., & Steinberg, G. K. (2020). Revisiting stem cell-based clinical trials for ischemic stroke. *Frontiers in Aging Neuroscience, 12*, 575990. https://doi.org/10.3389/fnagi.2020.575990

Heo, J. S., Choi, Y., Kim, H. S., & Kim, H. O. (2016). Comparison of molecular profiles of human mesenchymal stem cells derived from bone marrow, umbilical cord blood, placenta and adipose tissue. *International Journal of Molecular Medicine, 37*(1), 115–125. https://doi.org/10.3892/ijmm.2015.2413

Herrera, M. B., Bruno, S., Buttiglieri, S., Tetta, C., Gatti, S., Deregibus, M. C., et al. (2006). Isolation and characterization of a stem cell population from adult human liver. *Stem Cells, 24*(12), 2840–2850. https://doi.org/10.1634/stemcells.2006-0114

Herrera, M. B., Fonsato, V., Bruno, S., Grange, C., Gilbo, N., Romagnoli, R., et al. (2013). Human liver stem cells improve liver injury in a model of fulminant liver failure. *Hepatology, 57*(1), 311–319. https://doi.org/10.1002/hep.25986

Higashimoto, M., Sakai, Y., Takamura, M., Usui, S., Nasti, A., Yoshida, K., et al. (2013). Adipose tissue derived stromal stem cell therapy in murine ConA-derived hepatitis is dependent on myeloid-lineage and CD4+ T-cell suppression. *European Journal of Immunology, 43*(11), 2956–2968. https://doi.org/10.1002/eji.201343531

Hu, C., Yu, J., Cao, H., & Li, J. (2021). Cell transplantation therapy for liver failure. In L. Li (Ed.), *Artificial liver*. Springer.

Huebert, R. C., & Rakela, J. (2014). Cellular therapy for liver disease. *Mayo Clinic Proceedings, 89*(3), 414–424. https://doi.org/10.1016/j.mayocp.2013.10.023

Iansante, V., Mitry, R. R., Filippi, C., Fitzpatrick, E., & Dhawan, A. (2018). Human hepatocyte transplantation for liver disease: Current status and future perspectives. *Pediatric Research, 83*(1–2), 232–240. https://doi.org/10.1038/pr.2017.284

Kabat, M., Bobkov, I., Kumar, S., & Grumet, M. (2020). Trends in mesenchymal stem cell clinical trials 2004-2018: Is efficacy optimal in a narrow dose range? *Stem Cells Translational Medicine, 9*(1), 17–27. https://doi.org/10.1002/sctm.19-0202

Kanelidis, A. J., Premer, C., Lopez, J., Balkan, W., & Hare, J. M. (2017). Route of delivery modulates the efficacy of mesenchymal stem cell therapy for myocardial infarction: A meta-analysis of preclinical studies and clinical trials. *Circulation Research, 120*(7), 1139–1150. https://doi.org/10.1161/CIRCRESAHA.116.309819

Kean, T. J., Lin, P., Caplan, A. I., & Dennis, J. E. (2013). MSCs: Delivery routes and engraftment, cell-targeting strategies, and immune modulation. *Stem Cells International, 2013*, 732742. https://doi.org/10.1155/2013/732742

Khan, A. A., Parveen, N., Mahaboob, V. S., Rajendraprasad, A., Ravindraprakash, H. R., Venkateswarlu, J., et al. (2008). Safety and efficacy of autologous bone marrow stem cell transplantation through hepatic artery for the treatment of chronic liver failure: A preliminary study. *Transplantation Proceedings, 40*(4), 1140–1144. https://doi.org/10.1016/j.transproceed.2008.03.111

Khan, A. A., Shaik, M. V., Parveen, N., Rajendraprasad, A., Aleem, M. A., Habeeb, M. A., et al. (2010). Human fetal liver-derived stem cell transplantation as supportive modality in the management of end-stage decompensated liver cirrhosis. *Cell Transplantation, 19*(4), 409–418. https://doi.org/10.3727/096368910X498241

Kharaziha, P., Hellström, P. M., Noorinayer, B., Farzaneh, F., Aghajani, K., Jafari, F., et al. (2009). Improvement of liver function in liver cirrhosis patients after autologous mesenchymal stem cell injection: A phase I-II clinical trial. *European Journal of Gastroenterology & Hepatology, 21*(10), 1199–1205. https://doi.org/10.1097/MEG.0b013e32832a1f6c

Kurtz, A. (2008). Mesenchymal stem cell delivery routes and fate. *International Journal of Stem Cells, 1*(1), 1–7. https://doi.org/10.15283/ijsc.2008.1.1.1

Leahy, M., Thompson, K., Zafar, H., Alexandrov, S., Foley, M., O'Flatharta, C., et al. (2016). Functional imaging for regenerative medicine. *Stem Cell Research & Therapy, 7*(1), 57. https://doi.org/10.1186/s13287-016-0315-2

Lee, R. H., Pulin, A. A., Seo, M. J., Kota, D. J., Ylostalo, J., Larson, B. L., et al. (2009). Intravenous hMSCs improve myocardial infarction in mice because cells embolized in lung are activated to secrete the anti-inflammatory protein TSG-6. *Cell Stem Cell, 5*(1), 54–63. https://doi.org/10.1016/j.stem.2009.05.003

Lee, A. S., Tang, C., Rao, M. S., Weissman, I. L., & Wu, J. C. (2013). Tumorigenicity as a clinical hurdle for pluripotent stem cell therapies. *Nature Medicine, 19*(8), 998–1004. https://doi.org/10.1038/nm.3267

Lee, K. C., Stadlbauer, V., & Jalan, R. (2016). Extracorporeal liver support devices for listed patients. *Liver Transplantation, 22*(6), 839–848. https://doi.org/10.1002/lt.24396

Levicar, N., Pai, M., Habib, N. A., Tait, P., Jiao, L. R., Marley, S. B., et al. (2008). Long-term clinical results of autologous infusion of mobilized adult bone marrow derived CD34+ cells in patients with chronic liver disease. *Cell Proliferation, 41*(Suppl 1), 115–125. https://doi.org/10.1111/j.1365-2184.2008.00491.x

Li, J., Zhang, L., Xin, J., Jiang, L., Zhang, T., Jin, L., et al. (2012). Immediate intraportal transplantation of human bone marrow mesenchymal stem cells prevents death from fulminant hepatic failure in pigs. *Hepatology, 56*(3), 1044–1052. https://doi.org/10.1002/hep.25722

Liu, L. P., & Zheng, Y. W. (2019). Predicting differentiation potential of human pluripotent stem cells: Possibilities and challenges. *World Journal of Stem Cells, 11*(7), 375–382. https://doi.org/10.4252/wjsc.v11.i7.375

Liu, W. H., Song, F. Q., Ren, L. N., Guo, W. Q., Wang, T., Feng, Y. X., et al. (2015). The multiple functional roles of mesenchymal stem cells in participating in treating liver diseases. *Journal of Cellular and Molecular Medicine, 19*(3), 511–520. https://doi.org/10.1111/jcmm.12482

Locatelli, F., Algeri, M., Trevisan, V., & Bertaina, A. (2017). Remestemcel-L for the treatment of graft versus host disease. *Expert Review of Clinical Immunology, 13*(1), 43–56. https://doi.org/10.1080/1744666X.2016.1208086

Lopa, S., Colombini, A., Moretti, M., & de Girolamo, L. (2019). Injective mesenchymal stem cell-based treatments for knee osteoarthritis: From mechanisms of action to current clinical evidences. *Knee Surgery, Sports Traumatology, Arthroscopy, 27*(6), 2003–2020. https://doi.org/10.1007/s00167-018-5118-9

Lopez-Santalla, M., Fernandez-Perez, R., & Garin, M. I. (2020). Mesenchymal stem/stromal cells for rheumatoid arthritis treatment: An update on clinical applications. *Cell, 9*(8), 1852. https://doi.org/10.3390/cells9081852

Lv, Z. Y., Li, Y., & Liu, J. (2021). Progress in clinical trials of stem cell therapy for cerebral palsy. *Neural Regeneration Research, 16*(7), 1377–1382. https://doi.org/10.4103/1673-5374.300979

Lyra, A. C., Soares, M. B., da Silva, L. F., Fortes, M. F., Silva, A. G., Mota, A. C., et al. (2007). Feasibility and safety of autologous bone marrow mononuclear cell transplantation in patients with advanced chronic liver disease. *World Journal of Gastroenterology, 13*(7), 1067–1073. https://doi.org/10.3748/wjg.v13.i7.1067

Lysy, P. A., Campard, D., Smets, F., Najimi, M., & Sokal, E. M. (2008). Stem cells for liver tissue repair: Current knowledge and perspectives. *World Journal of Gastroenterology, 14*(6), 864–875. https://doi.org/10.3748/wjg.14.864

Mandai, M., Watanabe, A., Kurimoto, Y., Hirami, Y., Morinaga, C., Daimon, T., et al. (2017). Autologous induced stem-cell-derived retinal cells for macular degeneration. *The New England Journal of Medicine, 376*(11), 1038–1046. https://doi.org/10.1056/NEJMoa1608368

Mays, R. W., & Savitz, S. I. (2018). Intravenous cellular therapies for acute ischemic stroke. *Stroke, 49*(5), 1058–1065. https://doi.org/10.1161/STROKEAHA.118.018287

Mehta, A., Mathew, S., Viswanathan, C., & Sen Majumdar, A. (2010). Intrinsic properties and external factors determine the differentiation bias of human embryonic stem cell lines. *Cell Biology International, 34*(10), 1021–1031. https://doi.org/10.1042/CBI20100283

Melief, S. M., Zwaginga, J. J., Fibbe, W. E., & Roelofs, H. (2013). Adipose tissue-derived multipotent stromal cells have a higher immunomodulatory capacity than their bone marrow-derived counterparts. *Stem Cells Translational Medicine, 2*(6), 455–463. https://doi.org/10.5966/sctm.2012-0184

Mito, M., Ebata, H., Kusano, M., Onishi, T., Saito, T., & Sakamoto, S. (1979). Morphology and function of isolated hepatocytes transplanted into rat spleen. *Transplantation, 28*(6), 499–505. https://doi.org/10.1097/00007890-197912000-00013

Mohamadnejad, M., Alimoghaddam, K., Mohyeddin-Bonab, M., Bagheri, M., Bashtar, M., Ghanaati, H., et al. (2007). Phase 1 trial of autologous bone marrow mesenchymal stem cell transplantation in patients with decompensated liver cirrhosis. *Archives of Iranian Medicine, 10*(4), 459–466. Erratum in: Arch Iran Med. 2008 Jan;11(1):135. PMID: 17903050

Mohamadnejad, M., Alimoghaddam, K., Bagheri, M., Ashrafi, M., Abdollahzadeh, L., Akhlaghpoor, S., et al. (2013). Randomized placebo-controlled trial of mesenchymal stem cell transplantation in decompensated cirrhosis. *Liver International, 33*(10), 1490–1496. https://doi.org/10.1111/liv.12228

Mohamadnejad, M., Vosough, M., Moossavi, S., Nikfam, S., Mardpour, S., Akhlaghpoor, S., et al. (2016). Intraportal infusion of bone marrow mononuclear or CD133+ cells in patients with decompensated cirrhosis: A double-blind randomized controlled trial. *Stem Cells Translational Medicine, 5*(1), 87–94. https://doi.org/10.5966/sctm.2015-0004

Moll, G., Geißler, S., Catar, R., Ignatowicz, L., Hoogduijn, M. J., Strunk, D., et al. (2016). Cryopreserved or fresh mesenchymal stromal cells: Only a matter of taste or key to unleash the full clinical potential of MSC therapy? *Advances in Experimental Medicine and Biology, 951*, 77–98. https://doi.org/10.1007/978-3-319-45457-3_7

Moll, G., Ankrum, J. A., Kamhieh-Milz, J., Bieback, K., Ringdén, O., Volk, H. D., et al. (2019). Intravascular mesenchymal stromal/stem cell therapy product diversification: Time for new clinical guidelines. *Trends in Molecular Medicine, 25*(2), 149–163. https://doi.org/10.1016/j.molmed.2018.12.006

Nagata, H., Ito, M., Shirota, C., Edge, A., McCowan, T. C., & Fox, I. J. (2003). Route of hepatocyte delivery affects hepatocyte engraftment in the spleen. *Transplantation, 76*(4), 732–734. https://doi.org/10.1097/01.TP.0000081560.16039.67

Naldini, L. (2019). Genetic engineering of hematopoiesis: Current stage of clinical translation and future perspectives. *EMBO Molecular Medicine, 11*(3). https://doi.org/10.15252/emmm.201809958

O'Cearbhaill, E. D., Ng, K. S., & Karp, J. M. (2014). Emerging medical devices for minimally invasive cell therapy. *Mayo Clinic Proceedings, 89*(2), 259–273. https://doi.org/10.1016/j.mayocp.2013.10.020

Ohki, A., Saito, S., & Fukuchi, K. (2020). Magnetic resonance imaging of umbilical cord stem cells labeled with superparamagnetic iron oxide nanoparticles: Effects of labelling and transplantation parameters. *Scientific Reports, 10*(1), 13684. https://doi.org/10.1038/s41598-020-70291-9

Parekkadan, B., & Milwid, J. M. (2010). Mesenchymal stem cells as therapeutics. *Annual Review of Biomedical Engineering, 12*, 87–117. https://doi.org/10.1146/annurev-bioeng-070909-105309

Pellicano, R., Caviglia, G. P., Ribaldone, D., Altruda, F., & Fagoonee, S. (2021). Induced pluripotent stem cells from spermatogonial stem cells: Potential applications. In *Cell sources for iPSCs* (Vol. 7, pp. 15–30). Elsevier.

Petrus-Reurer, S., Romano, M., Howlett, S., Jones, J. L., Lombardi, G., & Saeb-Parsy, K. (2021). Immunological considerations and challenges for regenerative cellular therapies. *Communications Biology, 4*(1), 798. https://doi.org/10.1038/s42003-021-02237-4

Pham, V. Q., Tran, N. N., Vu, B. T., Le, H. T., Vu, N. B., & Pham, P. V. (2021). Angiogenic potential of hypoxia preconditioned human adipose and umbilical cord-derived mesenchymal stem cells: A comparative study. *Minerva Biotechnology and Biomolecular Research, 33*(3), 146–156. https://doi.org/10.23736/S2724-542X.21.02749-X

Pilat, N., Unger, L., & Berlakovich, G. A. (2013). Implication for bone marrow derived stem cells in hepatocyte regeneration after orthotopic liver transplantation. *International of J Hepatology, 2013*, 310612. https://doi.org/10.1155/2013/310612

Pouikli, A., Parekh, S., Maleszewska, M., Nikopoulou, C., Baghdadi, M., Tripodi, I., et al. (2021). Chromatin remodeling due to degradation of citrate carrier impairs osteogenesis of aged mesenchymal stem cells. *Nature Aging, 1*(9), 810–825. https://doi.org/10.1038/s43587-021-00105-8

Rajaram, R., Subramani, B., Abdullah, B. J. J., & Mahadeva, S. (2017). Mesenchymal stem cell therapy for advanced liver cirrhosis: A case report. *JGH Open, 1*(4), 153–155. https://doi.org/10.1002/jgh3.12027

Rajkumar, J., Baskar, S., Senthil Nagarajan, R., Murugan, P., Terai, S., Sakaida, I., et al. (2007). Autologous bone marrow stem cell infusion (AMBI) therapy for chronic liver diseases. *Journal of Stem Cells Regenarative Medicine, 3*(1), 26–37.

Ratajczak, M. Z., & Suszynska, M. (2016). Emerging strategies to enhance homing and engraftment of hematopoietic stem cells. *Stem Cell Reviews and Reports, 12*(1), 121–128. https://doi.org/10.1007/s12015-015-9625-5

Rosso, C., Caviglia, G. P., Younes, R., Ribaldone, D. G., Fagoonee, S., Pellicano, R., et al. (2020). Molecular mechanisms of hepatic fibrosis in chronic liver diseases. *Minerva Biotecnologica, 32*(3), 121–127. https://doi.org/10.23736/S1120-4826.20.02619-1

Sanchez-Diaz, M., Quiñones-Vico, M. I., Sanabria de la Torre, R., Montero-Vílchez, T., Sierra-Sánchez, A., Molina-Leyva, A., et al. (2021). Biodistribution of mesenchymal stromal cells after administration in animal models and humans: A systematic review. *Journal of Clinical Medicine, 10*(13). https://doi.org/10.3390/jcm10132925

Sang, J. F., Shi, X. L., Han, B., Huang, T., Huang, X., Ren, H. Z., et al. (2016). Intraportal mesenchymal stem cell transplantation prevents acute liver failure through promoting cell proliferation and inhibiting apoptosis. *Hepatobiliary & Pancreatic Diseases International, 15*(6), 602–611. https://doi.org/10.1016/s1499-3872(16)60141-8

Sato, Y., Araki, H., Kato, J., Nakamura, K., Kawano, Y., Kobune, M., et al. (2005). Human mesenchymal stem cells xenografted directly to rat liver are differentiated into human hepatocytes without fusion. *Blood, 106*(2), 756–763. https://doi.org/10.1182/blood-2005-02-0572

Schmuck, E. G., Koch, J. M., Centanni, J. M., Hacker, T. A., Braun, R. K., Eldridge, M., et al. (2016). Biodistribution and clearance of human mesenchymal stem cells by quantitative three-dimensional Cryo-imaging after intravenous infusion in a rat lung injury model. *Stem Cells Translational Medicine, 5*(12), 1668–1675. https://doi.org/10.5966/sctm.2015-0379

Schrepfer, S., Deuse, T., Reichenspurner, H., Fischbein, M. P., Robbins, R. C., & Pelletier, M. P. (2007). Stem cell transplantation: The lung barrier. *Transplantion Proceedings, 39*(2), 573–576. https://doi.org/10.1016/j.transproceed.2006.12.019

Schwartz, R. E., Reyes, M., Koodie, L., Jiang, Y., Blackstad, M., Lund, T., et al. (2002). Multipotent adult progenitor cells from bone marrow differentiate into functional hepatocyte-like cells. *The Journal of Clinical Investigation, 109*(10), 1291–1302. https://doi.org/10.1172/JCI15182

Shi, M., Zhang, Z., Xu, R., Lin, H., Fu, J., Zou, Z., et al. (2012). Human mesenchymal stem cell transfusion is safe and improves liver function in acute-on-chronic liver failure patients. *Stem Cells Translational Medicine, 1*(10), 725–731. https://doi.org/10.5966/sctm.2012-0034

Shi, M., Liu, Z., Wang, Y., Xu, R., Sun, Y., Zhang, M., et al. (2017). A pilot study of mesenchymal stem cell therapy for acute liver allograft rejection. *Stem Cells Translational Medicine, 6*(12), 2053–2061. https://doi.org/10.1002/sctm.17-0134

Siefert, J., Hillebrandt, K. H., Moosburner, S., Podrabsky, P., Geisel, D., Denecke, T., et al. (2019). Hepatocyte transplantation to the liver via the splenic artery in a juvenile large animal model. *Cell Transplantation, 28*(1 suppl), 14S–24S. https://doi.org/10.1177/0963689719885091

Smets, F., Dobbelaere, D., McKiernan, P., Dionisi-Vici, C., Broue, P., Jacquemin, E., et al. (2019). Phase I/II trial of liver-derived mesenchymal stem cells in pediatric liver-based metabolic disorders: A prospective, open label, multicenter, partially randomized, safety study of one cycle of heterologous human adult liver-derived progenitor cells (HepaStem) in urea cycle disorders and Crigler-Najjar syndrome patients. *Transplantation, 103*(9), 1903–1915. https://doi.org/10.1097/TP.0000000000002605

Song, Y., Du, H., Dai, C., Zhang, L., Li, S., Hunter, D. J., et al. (2018). Human adipose-derived mesenchymal stem cells for osteoarthritis: A pilot study with long-term follow-up and repeated injections. *Regenerative Medicine, 13*(3), 295–307. https://doi.org/10.2217/rme-2017-0152

Spada, M., Porta, F., Righi, D., Gazzera, C., Tandoi, F., Ferrero, I., et al. (2019). Intrahepatic Administration of Human Liver Stem Cells in infants with inherited neonatal-onset Hyperammonemia: A phase I study. *Stem Cell Reviews and Reports*. https://doi.org/10.1007/s12015-019-09925-z

Spada, M., Porta, F., Righi, D., Gazzera, C., Tandoi, F., Ferrero, I., et al. (2020). Intrahepatic administration of human liver stem cells in infants with inherited neonatal-onset hyperammonemia: A phase I study. *Stem Cell Reviews and Reports, 16*(1), 186–197. https://doi.org/10.1007/s12015-019-09925-z

Stutchfield, B. M., Forbes, S. J., & Wigmore, S. J. (2010). Prospects for stem cell transplantation in the treatment of hepatic disease. *Liver Transplantation, 16*(7), 827–836. https://doi.org/10.1002/lt.22083

Terai, S., Ishikawa, T., Omori, K., Aoyama, K., Marumoto, Y., Urata, Y., et al. (2006). Improved liver function in patients with liver cirrhosis after autologous bone marrow cell infusion therapy. *Stem Cells, 24*(10), 2292–2298. https://doi.org/10.1634/stemcells.2005-0542

Van Haele, M., Snoeck, J., & Roskams, T. (2019). Human liver regeneration: An etiology dependent process.

International Journal of Molecular Sciences, 20(9). https://doi.org/10.3390/ijms20092332

Vu, N. B., Nguyen, H. T., Palumbo, R., Pellicano, R., Fagoonee, S., & Pham, P. V. (2021). Stem cell-derived exosomes for wound healing: Current status and promising directions. *Minerva Medica, 112*(3), 384–400. https://doi.org/10.23736/S0026-4806.20.07205-5

Walker, P. A., Shah, S. K., Jimenez, F., Gerber, M. H., Xue, H., Cutrone, R., et al. (2010). Intravenous multipotent adult progenitor cell therapy for traumatic brain injury: Preserving the blood brain barrier via an interaction with splenocytes. *Experimental Neurology, 225*(2), 341–352. https://doi.org/10.1016/j.expneurol.2010.07.005

Walker, P. A., Letourneau, P. A., Bedi, S., Shah, S. K., Jimenez, F., & Cox, C. S. (2011). Progenitor cells as remote "bioreactors": Neuroprotection via modulation of the systemic inflammatory response. *World Journal of Stem Cells, 3*(2), 9–18. https://doi.org/10.4252/wjsc.v3.i2.9

Wáng, Y. X., & Idée, J. M. (2017). A comprehensive literatures update of clinical researches of superparamagnetic resonance iron oxide nanoparticles for magnetic resonance imaging. *Quantitative Imaging in Medicine and Surgery, 7*(1), 88–122. https://doi.org/10.21037/qims.2017.02.09

Wang, L. Q., Lin, Z. Z., Zhang, H. X., Shao, B., Xiao, L., Jiang, H. G., et al. (2014). Timing and dose regimens of marrow mesenchymal stem cell transplantation affect the outcomes and neuroinflammatory response after ischemic stroke. *CNS Neuroscience & Therapeutics, 20*(4), 317–326. https://doi.org/10.1111/cns.12216

Watanabe, M., & Yavagal, D. R. (2016). Intra-arterial delivery of mesenchymal stem cells. *Brain Circulation, 2*(3), 114–117. https://doi.org/10.4103/2394-8108.192522

Wei, T., & Lv, Y. (2013). Immediate intraportal transplantation of human bone marrow mesenchymal stem cells prevents death from fulminant hepatic failure in pigs. *Hepatology, 58*(1), 451–452. https://doi.org/10.1002/hep.26143

Wong, C. Y. (2021). Current advances of stem cell-based therapy for kidney diseases. *World Journal of Stem Cells, 13*(7), 914–933. https://doi.org/10.4252/wjsc.v13.i7.914

Wysoczynski, M., Khan, A., & Bolli, R. (2018). New paradigms in cell therapy: Repeated dosing, intravenous delivery, immunomodulatory actions, and new cell types. *Circulation Research, 123*(2), 138–158. https://doi.org/10.1161/CIRCRESAHA.118.313251

Yamanaka, S. (2020). Pluripotent stem cell-based cell therapy-promise and challenges. *Cell Stem Cell, 27*(4), 523–531. https://doi.org/10.1016/j.stem.2020.09.014

Yang, Y., Zhao, Y., Zhang, L., Zhang, F., & Li, L. (2021). The application of mesenchymal stem cells in the treatment of liver diseases: Mechanism, efficacy, and safety issues. *Frontiers in Medicine, 8*, 655268. https://doi.org/10.3389/fmed.2021.655268

Yannaki, E., Athanasiou, E., Xagorari, A., Constantinou, V., Batsis, I., Kaloyannidis, P., et al. (2005). G-CSF-primed hematopoietic stem cells or G-CSF per se accelerate recovery and improve survival after liver injury, predominantly by promoting endogenous repair programs. *Experimental Hematology, 33*(1), 108–119. https://doi.org/10.1016/j.exphem.2004.09.005

Yu, S. J., Chen, L. M., Lyu, S., Li, Y. Y., Yang, B., Geng, H., et al. (2016). Safety and efficacy of human umbilical cord derived-mesenchymal stem cell transplantation for treating patients with HBV-related decompensated cirrhosis. *Zhonghua Gan Zang Bing Za Zhi, 24*(1), 51–55. https://doi.org/10.3760/cma.j.issn.1007-3418.2016.01.010

Zhang, D. (2017). A clinical study of bone mesenchymal stem cells for the treatment of hepatic fibrosis induced by hepatolenticular degeneration. *Genetics and Molecular Research, 16*(1). https://doi.org/10.4238/gmr16019352

Zhang, Z., Lin, H., Shi, M., Xu, R., Fu, J., Lv, J., et al. (2012). Human umbilical cord mesenchymal stem cells improve liver function and ascites in decompensated liver cirrhosis patients. *Journal of Gastroenterology and Hepatology, 27*(Suppl 2), 112–120. https://doi.org/10.1111/j.1440-1746.2011.07024.x

Zhang, Y. C., Liu, W., Fu, B. S., Wang, G. Y., Li, H. B., Yi, H. M., et al. (2017). Therapeutic potentials of umbilical cord-derived mesenchymal stromal cells for ischemic-type biliary lesions following liver transplantation. *Cytotherapy, 19*(2), 194–199. https://doi.org/10.1016/j.jcyt.2016.11.005

Zhang, S., Yang, Y., Fan, L., Zhang, F., & Li, L. (2020). The clinical application of mesenchymal stem cells in liver disease: The current situation and potential future. *Annals of Translational Medicine, 8*(8), 565. https://doi.org/10.21037/atm.2020.03.218

Adv Exp Med Biol - Innovations in Cancer Research and Regenerative Medicine (2023) 4: 83–100
https://doi.org/10.1007/5584_2022_709

Published online: 8 April 2022

Optimal Delivery Route of Mesenchymal Stem Cells for Cardiac Repair: The Path to Good Clinical Practice

Dragica Miloradovic, Dragana Miloradovic, Biljana Ljujic, and Marina Gazdic Jankovic

Abstract

Research has shown that mesenchymal stem cells (MSCs) could be a promising therapy for treating progressive heart disease. However, translation into clinics efficiently and successfully has proven to be much more complicated. Many questions remain for optimizing treatment. Application method influences destiny of MSCs and afterwards impacts results of procedure, yet there is no general agreement about most suitable method of MSC delivery in the clinical setting. Herein, we explain principle of most-frequent MSCs delivery techniques in cardiology. This chapter summarizes crucial translational obstacles of clinical employment of MSCs for cardiac repair when analysed trough a prism of latest research centred on different techniques of MSCs application.

Keywords

Cardiac repair · Clinical practice · Delivery route · Mesenchymal stem cells

D. Miloradovic, D. Miloradovic, B. Ljujic, and M. G. Jankovic (✉)
Faculty of Medical Sciences, Department of Genetics, University of Kragujevac, Kragujevac, Serbia
e-mail: marinagazdic87@gmail.com

Abbreviations

AICI	Antegrade intracoronary infusion
AICI	antegrade intracoronary infusion
AMI	acute myocardial infarction
AT-MSCs	adipose tissue–derived mesenchymal stem cells
bFGF	basic fibroblast growth factor
BM-MSCs	bone marrow–derived mesenchymal stem/stromal cells
CABG	coronary artery bypass grafting
CCR2	chemokine receptor type-2 surface receptors
CROs	contract research organizations
CVDs	Cardiovascular diseases;
CXCR2	C–X–C chemokine receptor type 2
CXCR-4	C–X–C chemokine receptor type 4
DI	direct implantation
EPCs	endogenous endothelial regenerative progenitor cells
EPCs	endogenous endothelial regenerative progenitor cells GCP-good clinical practice
ET-1	endothelin-1
GM-CSF	Granulocyte-macrophage colony-stimulating factor
HF	heart failure
HGF	hepatocyte growth factor
HO-1	heme oxygenase-1
IA	intra-arterial infusion

IC	intracoronary
IC	ischemic cardiomyopathy
IDO	indoleamine 2,3-dioxygenase
IGF	insulin-like growth factor
IHD	ischemic heart disease
IL-10	interleukin-10
IM	intramyocardial
iNOS	inducible nitric oxide synthase
IV	intravenous
LIF	leukaemia inhibitory factor
LVAD	left ventricular assist device
LVEF	left ventricle ejection fraction
LVF	left ventricular function
MCP-1	monocyte chemoattractant protein 1
MHC	the major histocompatibility complex
MMP-9	Matrix metallopeptidase 9
MSCs	mesenchymal stem cells
NICM	non-ischemic cardiomyopathy
PCI	percutaneous transluminal coronary intervention
PDGF	Platelet-derived growth factor
PGE2	prostaglandin E_2
RCVI	retrograde coronary venous infusion
SDF1	stromal cell–derived factor 1
TE	transendocardial
TESI	transendocardial stem cell injection
TGF-β	transforming growth factor-β
UC-MSC	Umbilical cord–derived mesenchymal stem cells
VEGF	Vascular endothelial growth factor

1 Introduction

Cardiovascular disease (CVD) encompasses a vast range of diseases including diseases of the myocardium, the vasculature, the heart's electrical circuit, and congenital heart disease (Roger et al., 2012). Most of the people die due to brain and heart vascular occlusions that end fatally in their late 1960s (https://www.who.int/health-topics/cardiovascular-diseases/#tab=tab_1). A combination of risk factors, such as hypertension, diabetes, hyperlipidaemia, cigarette smoking, overweight, sedentary lifestyle, excessive and frequent alcohol consumption are the major cause of CVDs. Heart attacks are acute events and arise mostly because of the presence of atherosclerotic plaque that block normal bloodstream to the heart. Notwithstanding the improvement of pharmacological and surgical administration together with rehabilitation access, CVDs continue to manifest the growing mortality and disability worldwide. World Health Organization (WHO) claims that one decade from now CVDs will cause death of more than 20 million people, projecting that this will remain a serious public health issue (https://www.who.int/health-topics/cardiovascular-diseases/#tab=tab_1). Differentiation capacity and immunomodulatory ability make MSCs not only an excellent tool for research, but also result in treatment of various pathologies together with CVDs (Gazdic et al., 2017). However, translation into clinics according to the good clinical practice has proven considerably more difficult. Delivery strategy and low cell survival remain main obstacles when it comes to MSCs heart transplantation. Herein, we debate justification of universal MSCs delivery techniques in cardiology. This chapter summarizes crucial translational obstacles to clinical applications of MSCs for cardiac repair based on the latest research focusing on various techniques of MSCs delivery.

2 Mesenchymal Stem Cells: The Promising Tool for Cardiac Repair

Mesenchymal stem cells (MSCs) are adult, undifferentiated multipotent stem cells as well as encouraging tool in regenerative medicine. There are three norms that define multipotent MSCs: adherence to culture flasks; expression of cluster of differentiation CD105, CD73 and CD90; and absence of CD45, CD34, CD11b, CD14, CD79a, CD31 and MHC class II molecules. Also, MSCs have potential to transform into various types of tissues (Dominici et al., 2006). MSCs represent attractive candidates for therapy of HF due to simplicity of their isolation and because they can be easily expanded (Volarevic et al., 2018). Bone marrow, adipose tissue, umbilical cord and dental pulp have shown as the best sources of MSCs, yet it still remains unclear which source of MSCs is optimal (Markovic et al., 2018; Harrell et al., 2019). BM-MSCs have many traits that allow their therapeutic usage: simple acquisition, fast proliferation in vitro, minimal immunological rejection, protracted standing after transplantation in the host, preservation of capacity to differentiate subsequently to repetitive passaging but also simplicity of transplantation (Chamberlain et al., 2007; Xie et al., 2012). Anyhow, the procurement of BM-MSCs is not so simple due to the complicating process of harvesting and due to the fact that size and longevity of BM-MSCs fairly diminishes by ageing (Mueller & Glowacki, 2001; Stenderup et al., 2003). In order to avoid such drawbacks, alternate sources for MSCs isolation are suggested (Lu et al., 2006; Zuk et al., 2002). Umbilical cord blood and adipose tissue have been examined as a possible source for isolation and therapeutic utilization of MSCs thanks to its larger proliferative potential, high cell yields and simplicity of collection (Kim et al., 2010; Davies & Walker, 2017). BM-MSCs, UCB-MSCs and AT-MSCs all possess akin morphological as well as operational traits (Volarevic et al., 2011).

According to data reviewed earlier, it is easy to comprehend why are MSCs such a great instrument for prevention and treatment of cardiac complications (Volarevic et al., 2017). MSCs can migrate in the damaged tissue after cell transplantation, directly to inflamed zones caused by ischemia (White et al., 2016; Ye & Zhang, 2017; Shin et al., 2018; Rice et al., 2006). In particular, during this chemotaxis process, MSCs distinguish SDF1 and MCP-1 – by interplay accompanying the CXCR-4 and 1 and CXCR2-C-X-C (Kang et al., 2012) – which in turn results in discriminating migration and subsequently the intravenous application. Importantly, MSCs can be allogeneically transplanted without immune reaction of the recipient, even if there is no MHC histocompatibility matching between patients (Ryan et al., 2005). Although MSCs differentiate into functional cells expressing cardiomyogenic phenotype (Kawada et al., 2004; Gopinath et al., 2010), MSCs exhibit regenerative potential that overcomes their ability to differentiate into myocardial cells (Kang et al., 2012). MSCs can impact vascular growth, inflaming, oxidative stress, fibrosis, cell apoptosis and necrosis, resulting in therapeutically benefit in heart disease (Anderson et al., 2016; Lee et al., 2012; Arslan et al., 2013; Teng et al., 2015; Salomon et al., 2013; Wen et al., 2012). Several research groups emphasized on an important role for VEGF, IGF, HGF, ET-1, and bFGF for MSC-dependent suppression of cardiomyocyte death. MSC-derived VEGF, PDGF as well as MMP-9, all together have a role in cardiac remodelling and regeneration (Samper et al., 2013; Ankrum & Karp, 2010). Mechanisms by which MSCs promote angiogenesis also include arranging of pericytes as well as endothelial supporting cells all along the angiogenesis (Ghajar et al., 2010; Tigges et al., 2013), spur of the EPCs when ischemia occurs (Detante et al., 2017) and immune-regulation of the micro-environment (Le Blanc & Mougiakakos, 2012; Heldring et al., 2015). By direct cell-to-cell connection as well as autocrine manner moderated via IDO PGE2, iNOS, HO-1, IL-10, and TGF-β, MSCs have capacity to adjust initiation, movement or differentiation of particular immune system cells (Gazdic et al., 2015). Moreover, it was demonstrated that MSCs enhance heart regeneration via cell-independent influence that leads to proliferation and differentiation of host myocardial precursor cells in functional cardiac cells (Hatzistergos et al., 2010a).

Despite the benefits of MSCs therapy in cardiac repair (Lee et al., 2012; Arslan et al., 2013; Teng et al., 2015; Salomon et al., 2013; Wen et al., 2012; Liu et al., 2017; Ma et al., 2018; Bian et al., 2014; Wollert et al., 2017; Wang et al., 2012; Mazhari & Hare, 2007; Gnecchi et al., 2008; Zuba-Surma et al., 2015), remedial outcome of MSCs application is restricted due to poor survival within injured site after transplantation (Toma et al., 2002). Besides, injured myocardium represents hostile environment for MSCs survival and engraftment due to oxidative stress, inflammation, lack of nutrients and aggravation of the patient's health (Wu et al., 2011; Monsel et al., 2014; van Rhijn-Brouwer et al., 2018). In spite of all advantages of MSCs therapy, there are

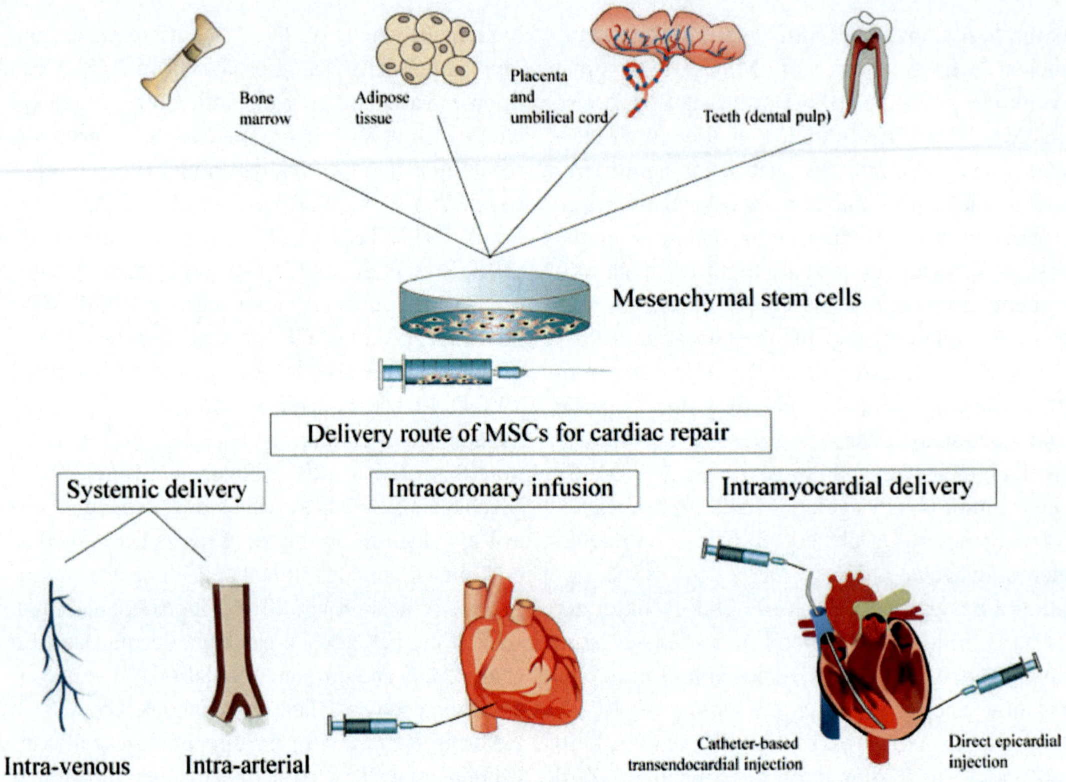

Fig. 1 Delivery route of MSCs for cardiac repair. Schematic depicting the main tissue sources of MSCs. In this setting, MSCs delivery has been achieved predominantly by systemic delivery (intravenous or intra-arterial), intracoronary infusion or direct intramyocardial injection, either transendocardially or epicardially. *Abbreviation*: MSC, mesenchymal stem cells

important matters to consider as to pro arrhythmic and tumorigenic potential of MSCs, but also as to differentiation in unsuited sort of cells (Jeong et al., 2011; Breitbach et al., 2007). Accordingly, to make MSCs therapy successful it is crucial to overcome these obstacles (Fig. 1).

3 Routes of Therapeutic MSCs Delivery

Many examinations tried to reveal the difference between various types of MSCs delivery (Liu et al., 2016a; Lim et al., 2018; Prockop & Olson, 2007; Liu et al., 2016b; Hare et al., 2009; Detante et al., 2017; Galipeau & Sensébé, 2018; Nijboer et al., 2018). The ideal delivery method would have to complete certain norms such as (i) prior transplantation, donor cells should preserve capacity for efficient and direct transfer to injured heart areas; ii) the stable survival and endurance in the heart of grafted donor cells; (iii) the safety of the procedure for the heart of host; (iv) the good tolerance of the procedure by patients with low-risk complication rate; (v) easily accessible in hospitals with no need for supplementary equipment; (vi) the cost of treatment aspire not overly expensive; and mostly, (vii) cells delivered in this manner would bring to many health condition welfares of the patients, such as a decrease of mortality and morbidity. Furthermore, method should be germane for most sorts of cells and clinical scenarios including acute myocardial infarction (AMI) conditions in patients who are recovering post AMI, to chronic cardiac diseases. Herein major techniques for MSCs administration will be displayed, such intravenous (IV), intra-arterial (IA) infusion, intracoronary (IC) together with intramyocardial (IM) with epicardial and transendocardial administration (Dib et al., 2011).

3.1 Intra-Arterial Infusion

Intra-arterial (IA) infusion might seem to be mainly effective technique for some types of indications, but also risky in other types of indications. IA infusion enables delivery of MSCs in coronary vasculature, therefore minimizes perils of both direct implantation (DI) and intravenous (IV). In that way the capture of cells in lung can be avoided and therefore cells can reach target sites (Watanabe & Yavagal, 2016). Studies showed that intracoronary (IC) administration of MSCs and other BM cell populations is secure and induce enhancement of AMI patients' health condition AMI (Chen et al., 2004; Leistner et al., 2011). SafeCell Heart study showed considerable improvements in left ventricle ejection fraction (LVEF) with IC MSC delivery (Lalu et al., 2018). Furthermore, the results also imply that reduced dosages of MSCs are much more efficiently administered via the IA route at 24 h (Yavagal et al., 2014). However, not only delivery route, diameter and quantity of cells, but also infusion pace need to be taken into consideration, particularly when it comes to administration of MSCs in vasculature of heart or brain (Janowski et al., 2013). Thus, the effectiveness of IA MSCs delivery is still in examination and it's employment remains to be elucidated.

3.2 Intravenous Administration

IV administration of MSCs is one of the most frequently utilized technique because of the relative simplicity and limited risk (Moll et al., 2019). Also, it is the most-suitable and least-invasive route, used more frequently in AMI due the predominance of physiological homing signals, which permit the cells to migrate towards the injured myocardium (Kocher et al., 2007). After administration, MSCs migrate via systemic circulation, graft within injured myocardium, and directed by regional signals start to differentiate (Boomsma et al., 2007; Carr et al., 2008). BM-MSCs are mostly drawn to and stored within ischemic area, and consequently may induce cardiac repair (Barbash et al., 2003). In particular, MSCs engrafted in the myocardium differentiate towards cardiomyocytes and vascular endothelial cells, having the effect of myogenesis and angiogenesis which leads to decreased myocardial infarct dimension and subsequently to enhanced heart action of AMI rats (Nagaya et al., 2004). Allogenic BM-MSC- and UC-MSC-administered IV proved as unscathed and efficacious for patients that have ICM, NICM, as well as HF, while observing LVEF, operational state and well-being (Luger et al., 2017; Bartolucci et al., 2017; Butler et al., 2017). IV MSC administration subsequently to heart reperfusion of patients with acute MI enhanced LVF, reversed remodelling and changed electrophysiologic traits by systemic anti-inflammatory effects (Luger et al., 2017; Price et al., 2006; Chullikana et al., 2015; Hare et al., 2009).

IV delivery is known to be the most useful manner of cell delivery, especially because it demands external venous approach. However, in order to reach injured myocardium, intravenously transplanted cells must pass through the lungs, with the possibility of being entrapped in the lungs, liver, and spleen (Carr et al., 2008; Barbash et al., 2003; Gao et al., 2001). Accordingly, the number of MSCs committed to refurbish ischaemic area could drastically reduce (Saito et al., 2002). In order to let the administered MSCs to reach the infarcted artery, this route of cell delivery demands numerous infusions (Strauer, 1979; Gregg & Fisher, 1963). Ergo, systemic delivery is likely to cause vascular occlusion (Walczak et al., 2008). Also, it is found that 0.5×10^6 cells per kg/body weight could lead to MI, alike blood vessels of healthy patients. This is serious issue with the IV delivery of MSCs, due to a necessity of high dose of cells that can also induce a systemic immune response (Hoogduijn et al., 2013). Taking all of the aforementioned into consideration, it can be said that

IV delivery is more appropriate for early stages of heart insult and its employment must be further examined.

3.3 Intracoronary Administration

Intracoronary (IC) administration of cells is conducted by regular over-the-wire balloon angioplasty catheter (Nagaya et al., 2004). Consequently, the angioplasty catheter is planted in the coronary artery and the balloon is distended at a low pressure so that blood flow is blocked during cell infusion on distal part of catheter (Qi et al., 2008). This method has been firstly employed by Strauer and co-workers who transplanted bone marrow cells (BMCs) following percutaneous transluminal coronary intervention (PCI) for AMI (Strauer et al., 2001). IC transfer of autologous BMCs within 24 h of optimal reperfusion therapy fails to improve global LV function after MI, but might successfully have an impact on infarct remodelling (Janssens et al., 2006). Single dose of intracoronary BMCs provided an acceleration of LVEF recovery after AMI compared with control group, however there were no long-term advantages on cardiac function after BMCs therapy (Meyer et al., 2006). Similarly, a study of MI porcine model showed that IC-transplanted MSCs lead to enhanced engraftment (Freyman et al., 2006). This method of MSCs delivery hasn't proved any safety concerns except the distal microvascular occlusion, especially in case of applying high dose of MSCs (Hong et al., 2014).

Just like every delivery route, there are pros and cons regarding IC MSCs delivery. Interventional cardiologists are familiar with this technique especially with the option to deliver MSCs while conducting PCI in order to treat MI patients. Distinguishing from IM, the IC administration technique may obtain increased homogeneous cell allocation in designated areas, avoiding cell aggregates generation as well as elevated immune system's response (Copland, 2011). Having said that, the possible blood deficiency at the time of coronaries obstruction and the absence of vascularity within ischemic tissue can negatively influence cell delivery. Despite the aforementioned, a small percentage of transplanted MSCs stay in infracted area, therefore, suggesting therapeutic potential of IC infusion. There are two ways of delivering cells via IC method: antegrade intracoronary infusion (AICI) and retrograde intracoronary venous infusion (RCVI).

3.3.1 Antegrade Intracoronary Infusion

AICI is a method based on usage of transcutaneous catheterization that is implanted in the artery (usually femoral artery). Catheter is led by fluoroscopic screening until it reaches the target coronary artery (Campbell & Suzuki, 2012). AICI approach enables homogeneous distribution of cells at the target area. Myoblast transplantation by catheter into people with heart failure led to advancement in parameters such as New York Heart Association (NYHA) classification, The Minnesota Living with Heart Failure Questionnaire (MLHFQ), dynamism of ventriculus, together with proving of reversal ventriculus remodelling. Singular MSC injection at 20 ml/min declined lethality and improved cardiac circulation with no declining in effectiveness of delivery (Assmus et al., 2013). The process of implantation of balloon angioplasty catheter should be implanted carefully within area of infarcted artery otherwise an injury and an embolism may occur (Dib et al., 2011; Vulliet et al., 2004). Administration of smaller dimension of MSCs and in more rapid pace is a must due to possibility of micro vascular plugging that occurs due to cell aggregation.

3.3.2 Retrograde Coronary Venous Infusion

RCVI technique enables close approach to cardiac destination of ischemia and infarct. Using the approach via the internal jugular or femoral vein, a central line is inserted into the coronary sinus alongside the infusion catheter implanted through a wire. Concerning lower risk in regard to embolism and avoiding the wash away of distributed cells in the bloodstream the detached coronary sinus is occluded with a balloon (Hou et al., 2005a; Dib et al., 2010; Pohl et al., 2004; Raake et al., 2004; von Degenfeld et al., 2003). Formigli and co-workers conducted research on a rat model and found higher cell distribution within front left ventricle's wall. Improved grafting of cells lasted or more than 2 weeks and during that period left ventricle remodelling was diminished, but heart activity was improved (Formigli et al., 2007). RCVI was beneficial in animal models of MI and IHD (Huang et al., 2013; Yokoyama et al., 2006). Also, outcomes from medical examination of people suffering from cardiac insufficiency as well as chronic refractory angina showed that this method is secure and leads to improved cardiac function (Patel et al., 2015; Tuma et al., 2011). Therefore, RCVI is a choice in situations of austere subtotalling stenosis of coronary arteries of severe aortic stenosis. Also, this method can be performed easily by cardiac surgeons in hospitals with cine-angiography systems. Nevertheless, there is a high risk for patients with vulnerable coronary sinus due to possibility of rupture (Llano et al., 2009). Compared to AICI, RCVI less often causes embolus of coronaries and empowers distribution of cell to ischemic sites. However, the level of MSCs homing and their effect on heart renovation remains uncertain (Gathier et al., 2018).

3.4 Intramyocardial Injection

Intramyocardial (IM) is mostly invading transfer means since it demands thoracotomy or sternotomy, followed by donor cells injection with needle directly into myocardium of the patients, under the direct visualisation of infarcted myocardium. Also, if perforation happens, during the injection, it can be handled by sutures. The process is based on injection of cells in the periphery of ischemic myocardium. This delivery method is safe and even less likely to provoke embolism and induce arrhythmias (Ang et al., 2008; Fukushima et al., 2007; Charwat et al., 2008). Numerous preclinical and clinical trials examined the effect of IM cell delivery in heart, but here, only few will be highlighted.

Porcine BM-MSCs injected into the cardiomyopathic hamsters provided significant ventricular function improvement and increased systolic wall thickening 3 weeks after a second IM delivery (Shabbir et al., 2009). MSCs transplantation reduced myocardial apoptosis and fibrosis by ~50% in HGF-, LIF-, and GM-CSF-dependent manner (Shabbir et al., 2009). In line with these outcomes, Zisa et al. found that IM-delivered BM-MSCs improved LVEF by 30%, 4 weeks after cell therapy in the hamsters (Zisa et al., 2009). MSC-derived VEGF was considered to be the main factor that improved cardiac repair. IM injection of BMCs as well as MSCs was secure and significantly improved LV functioning and remodelling in people suffering from coronary insufficiency (Shabbir et al., 2009; Zisa et al., 2009; Tian et al., 2014; Mathiasen et al., 2015). The PROMETHEUS trial revealed that IM injections of MSCs in heart of patients with coronary ischemia going through coronary artery bypass grafting (CABG) enhanced not only regional myocardial but also global LV function (Karantalis et al., 2014). It is supposed that transplanted MSCs manifest beneficial impacts mainly at the puncture site through discharge of anti-fibrotic matrix metalloproteases and soluble factors that stimulate neovascularization. In accordance, application of cytokines and BMSCs also have an advantageous effect specifically on LVEF (Hamshere et al., 2015; Choudhury et al., 2017). Recent study showed that IM delivery of exosomes followed by MSC transplantation in people with MI led to better heart function, diminished scar diameter and enhanced novel vascular formation (Huang et al., 2019).

Despite of many advantages of this method, many questions remain. Needle injection can cause mechanical injury and, in that way, induce the donor cell damage directly or secondarily by causing inflammation (Fukushima et al., 2008; Suzuki et al., 2004). Considering previously mentioned struggles, it is crucial to refine IM shot pressuring together with, bulk of cell suspensions, and the sort of syringe in order to enhance cell grafting and remedial impact of MSCs.

There are three routes for IM injection: transepicardial, transendocardial and transcoronary.

3.4.1 Transendocardial Injection

Transendocardial stem cell injection (TESI) enables MSCs delivery straight in the myocardium by a catheter (guided by fluoroscopic direction or electroanatomic mapping) in sinister ventricle cavum, while MSCs are being administered (Qi et al., 2008). Swine model of MI showed the capability of MSCs delivered by TESI to have beneficial and improving effect on heart activity and scaring reduction (Hatzistergos et al., 2010b; Amado et al., 2005; Schuleri et al., 2008). There are studies that proved the safety of transendocardial delivery of MSCs or BMCs in ICM chronic ICM and LV dysfunction (Heldman et al., 2014) and non-ischemic dilated cardiomyopathy (NIDCM) (Mushtaq et al., 2014). Moreover, delivery of autologous BMCs by the helical needle TE catheter increased EF in people with OMI (de la Fuente et al., 2007). Hare and co-workers compared therapeutic effects of allogeneic and autologous BM-MSCs applied by TESI in people suffering from IHD and revealed functional improvement in patients (Hare et al., 2012). TESI is an attractive delivery technique that gives the option to avoid the open-heart surgery guided by 2D or 3D system and enables estimation of myocardial viability. Minimal invasiveness and capacity to precisely infuse MSCs within infarcted area, provide great cell retention and enhancement of the cardiac function (Hou et al., 2005b; Gyöngyösi et al., 2008; Bervar et al., 2017; Sherman et al., 2006). TESI appears to be the most favourable route of delivery due to higher retention rate of cells delivered in this manner and improvement of LVEF in AMI preclinical and clinical trials (Kanelidis et al., 2017). Despite all the benefits, TESI requires highly experienced and professional staff, and has a risk of perforation of myocardium and cardiac tamponade subsequently, together with arrhythmias.

3.4.2 Epicardial Injection

The epicardial injection enables cells administration into the specific area directly visualized or endoscopically observed, contrarily to endocardial and transvascular IM injection. Cells are delivered by 27-gauge pre-bent needle (needle is inserted into the myocardium, and then repeatedly suctioned in order to check whether the needle is in the ventricle cavity) (Menasché et al., 2008). Afterwards, retraction of the needle is done and cells are repeatedly injected around the affected area. Epicardial injection improved NYHA, MLHFQ, ventricular viability, and evidence of reverse ventricular remodelling (Dib et al., 2009). Besides, clinical trials displayed this manner of cell delivery as advantageous for patients undergoing coronary surgery (Viswanathan et al., 2010). Even though potentially safe and efficient, epicardial injection has its bad sides such as leakage of the cells, limited cell delivery, inability to control the dosage of delivered cells and, finally, reduced cell engraftment (Passier et al., 2008). To overcome such deficiencies, future studies should reconsider the usage of apparatus for injecting with regulable period, tension as well as puncture mark (Dib et al., 2011) along with the reinstatement of diluent, cells transplantation on decomposable polymer scaffold and other biologically acceptable patches (Perea-Gil et al., 2016; Wang et al., 2017). One of new attempts made to overcome low cell survival and engraftment of cells delivered epicardially is epicardial placement. It is stated that epicardial placing of MSCs as a 'cell-sheet' considerably enhanced chance of cells to survive and augmented healing ramification to IM injection rats with acute MI and HF (Narita et al., 2013; Tano et al., 2014). Furthermore, epicardial placement has minimal chances of causing embolization and arrhythmias (Fukushima et al., 2007; Narita et al., 2013). Also, the effectiveness of an alternate methods to epicardial placement has been displayed along with fibrin glue or predesigned tissue engineered contracts (Atluri et al., 2014; Mewhort et al., 2016). Self-organizing peptide hydrogels might accomplish better epicardial placement of MSCs on the heart. One of the most examined types of hydrogels is PuraMatrix® (PM; 3-D Matrix, Ltd.) (Zhang, 2003; Yokoi et al., 2005). Also, a study tried to advance the epicardial placement to epicardial 'coating' along immediately formed hydrogel-MSC system, which turned out to be advantageous for cell delivery and heart activity in HF patients (Ichihara et al., 2018).

3.5 Advantages and Disadvantages of Strategies for Delivering Mesenchymal Stem Cells to the Damaged Heart

In spite of various attempts that are made to discover optimal delivery manner (Liu et al., 2016a, b; Lim et al., 2018; Prockop & Olson, 2007; Hare et al., 2009; Detante et al., 2017; Galipeau & Sensébé, 2018; Nijboer et al., 2018), no perfect solution has been elucidated yet (Table 1). As mentioned above many criteria should be accomplished so that a method could be named as safe and beneficial. As least invasive, easy to conduct, inexpensive and with low risk of complications, IV administration is a method that could be used as an alternative (Moll et al., 2019; Kocher et al., 2007; Boomsma et al., 2007; Carr

Table 1 Effects of different delivery routes of mesenchymal stem cells on cardiac function

Delivery route	Disease	Results	Model	References
IA	AMI	Improved left ventricular function	Human	Chen et al. (2004) Leistner et al. (2011)
IV	HF, NICM AMI	Reversed remodelling and changed electro physiologic traits Improved left ventricular function Improved functional status and quality of life	Human	Bartolucci et al. (2017) Butler et al. (2017) Price et al. (2006) Hare et al. (2009) Chullikana et al. (2015)
IV	AMI ICM, NICM, HF	Successful engraftment and migration Decreased infarct size improved cardiac function	Mouse Rat	Boomsma et al. (2007) Carr et al. (2008) Nagaya et al. (2004) Luger et al. (2017)
IC	AMI	Infarct size reduction Increased cardiac index Improved systolic function No long-term advantage on cardiac function	Human	Strauer et al. (2001) Janssens et al. (2006) Meyer et al. (2006)
AICI	CHF	Improvement of LV function	Human	Assmus et al. (2013)
RCVI	HF Chronic refractory angina	Improved cardiac function	Human	Patel et al. (2015) Tuma et al. (2011)
IM	IHF Coronary ischemia NICM ICM	Improved LVEF Reduction in NYHA class	Human	Mathiasen et al. (2015) Karantalis et al. (2014) Hamshere et al. (2015) Choudhury et al. (2017)
TESI	OMI NIDCM AMI	Improved diastolic function, LVEF and electromechanical parameters	Human	De la Fuente et al. (2007) Bervar et al. (2017) Krause et al. (2009).
TESI	AMI	Infarct size reduction Improved LVEF	Porcine	Kanelidis et al. (2017)
Epicardial injection	AMI MI HF	Improved perfusion Improved NYHA Evidence of reverse ventricular remodelling	Human	Viswanathan et al. (2010) Dib et al. (2009)

Abbreviations: *AMI* – acute myocardial infarction, *IV* – intravenous, *IA* – intra-arterial, *IC* – intracoronary, *IM* – intramyocardial, *AICI* – antegrade intracoronary infusion, *RCVI* – retrograde coronary venous infusion, *TESI* – transendocardial stem cell injection, *LVEF* – left ventricle ejection fraction, *ICM* – ischemic cardiomyopathy, *NICM* – non-ischemic cardiomyopathy, *HF* – heart failure, *IHD* – ischemic heart disease, *OMI* – old myocardial infarction, *IHF* – ischaemic heart failure, *NIDCM* – non-ischemic dilated cardiomyopathy, *CHF* – chronic heart failure et al., 2008; Barbash et al., 2003). Yet, IV administration has some drawbacks, such as entrapment in other organs and relatively fast elimination of MSCs by innate immune system cells. In order to achieve desired effect, numerous infusions are necessary which in turns elevates the risk of thromboembolism (Carr et al., 2008; Barbash et al., 2003; Gao et al., 2001; Saito et al., 2002; Strauer, 1979; Gregg & Fisher, 1963; Walczak et al., 2008; Hoogduijn et al., 2013). This manner of MSCs delivery should be reconsidered especially in patients with AMI and chronic IHD because of the lack of necessary local signals in such cardiac pathologies. IA infusion enables transfer of a larger number of MSCs exactly in desired vascular area, avoiding the halt in other organs (Watanabe & Yavagal, 2016; Kean et al., 2013). Just like IV, IA delivery is the least-invasive technique compared to any other delivery method. Still, the embolism is potential risk of IA delivery, so various factors (delivery route, diameter, quantity of cells and infusion pace) should be reconsidered prior to its clinical appliance (Janowski et al., 2013). This method must be further examined in order to be applied in clinical practice.

When it comes to comparing IC administration to other methods, certain advantages emerge. Unlike IM, IC administration is less invasive and an intervention a cardiologist can perform simply due to standard equipment present in the internal cardiology clinics. IC administration enables superior MSCs delivery and approximately homogeneous distribution to injured myocardium zones when compared to IM (Freyman et al., 2006; Copland, 2011). However, MSCs turns out to be prone to causing microvascular occlusion (Hong et al., 2014) and can't survive long when delivered intracoronary. Likewise, there are not such small rodents that can serve as a proper model for investigation of concrete dynamics and mechanisms of MSCs when delivered IC, demanding further and more detailed research. AICI is attractive method but can cause vascular damage, embolism and cell aggregation if not conducted carefully and with inexperienced staff (Dib et al., 2011; Vulliet et al., 2004). Even if improved in such manner, AICI, just like other methods, should be examined more thoroughly so that its application can be considered justified. When it comes to RCVI, it is an option in scenarios of austere subtotal stenosis of coronary arteries or severe aortic stenosis. Also, comparing to AICI, RCVI is less likely to provoke coronary embolism and better distribution of cell to ischaemic zones and poor arterial supply. However, RCVI can lead to rupture of sensitive coronary sinus of treated patients (Llano et al., 2009). Despite outcomes of studies that displayed RCVI as a beneficial, exact extent of MSCs homing and its impact on cardiac repair are not completely clarified (Dib et al., 2010; Gathier et al., 2018).

IM enables MSCs delivery even in zones with poor vascularity, distinguished from IC. Moreover, arrhythmias and embolism risk are much more diminished when compared to all the previously mentioned systems of cell delivery (Ang et al., 2008; Fukushima et al., 2007; Charwat et al., 2008). When compared to IC, IM injection proved as a method with improved blood supply to the cardiac tissue (Freyman et al., 2006). Stem cells turning into islet-like clusters, and ventricular arrhythmias when compared to IV and IC, poor retention rate, big invasiveness and potential risk of inflammation caused by needle injection require IM injection to be further examined (Freyman et al., 2006; Fukushima et al., 2008; Suzuki et al., 2004).

TESI, conducted percutaneously, is less invasive distinguishing from epicardial injection but more invasive distinguishing from IC (Krause et al., 2009; Losordo et al., 2007). TESI is a pretty advantageous technique due to its ability to deliver MSCs straight to the myocardium, and with much bigger cell retention compared to other techniques. Also, there are reports that confirmed the safety of TESI in terms of arrhythmias; however, cardiac tamponade or rupture may occur (Freyman et al., 2006). An electroanatomical mapping system of the left ventricle gives precise insight of specific areas and viability of myocardium, and therefore enables direct delivery of MSCs in injured zones with higher retention rate (Hou et al., 2005b; Gyöngyösi et al., 2008; Bervar et al., 2017; Sherman et al., 2006; Kanelidis et al., 2017). However, electroanatomical mapping system has high costs which many clinics can't afford, requires a lot of time and experienced medical staff. Epicardial injection, on the other hand, is a method that is used in most preclinical and clinical research and gives similar retention rate to TESI (Dib et al., 2009; Perin & López, 2006; Psaltis et al., 2010). Intramyocardial or epicardial necrotic changes can be treated only by epicardial access. Major issue of epicardial injection is that it must be performed simultaneously with CABG or left LVAD (Ota et al., 2008), which is suitable for patients who undergo coronary surgery (Karantalis et al., 2014; Viswanathan et al., 2010). Accordingly, high invasiveness and high costs are major disadvantages when it comes to epicardial injection. Additionally, outflow of the cells, restricted cell distribution, unmanageable dose of delivered cells and diminished cell engraftment (Passier et al., 2008) make this technique questionable. Yet, improvements in regard to epicardial injection such as epicardial placement in order to diminish risks of embolism and arrhythmias (Fukushima et al., 2007; Narita et al., 2013) could overcome mentioned drawbacks, even though future and more detailed research is necessary.

3.6 Mesenchymal Stem Cell Perspective: Good Clinical Practice for Cardiac Repair

In addition to clinical challenges associated with optimal delivery routes such as low retention rate, poor survival and limited homing to the aimed tissue, the obstacles for MSC therapies in cardiology also include challenges resulting from the sourcing and manufacturing of MSCs. MSCs are heterogeneous population of cells whose therapeutic effect depends on the characteristics of the donor (health condition, congenital traits, sex, and age), tissue of origin, isolation technique, and culture procedures (growth medium formulation, oxygen concentration, degree of cell confluence, 2D/3D culture, ageing in vitro) (Levy et al., 2020). Moreover, freezing/thawing protocols can affect the viability and function of MSCs. Accordingly, it is not difficult to understand why MSCs administration in various clinical trials gives incongruent and unsatisfactory outcomes in regards to regeneration and immunomodulation. Thus, optimizing the handling of MSCs and standardizing therapeutic potency of the MSC product are crucial before beginning clinical application. Intensive research and thorough elaboration of each one of these factors are the only solutions to really overcome obstacles of MSCs delivery in heart.

While the potency of the MSC product and the delivery route are critical parameters for the efficacy of MSC therapies in clinical trials, transplant recipient factors must also be taken into consideration on the path to the good clinical practice for cardiac repair. Variations in the host disease stage/severity and immune responses against MSCs after administration can affect their therapeutic outcome for cardiac diseases (Wang et al., 2014). Other important host factors that can influence MSC function in vivo are inflammation status, level of chemokine secretion, and tissue micro-environment such as hypoxia and extracellular matrix (Shi et al., 2018; Kyriakou et al., 2008). Thus, evaluating the status of recipients – especially the disease stage – might help to optimize the dosing and improve the clinical predictions of therapeutic response to MSC therapies for cardiac diseases (Levy et al., 2020).

4 Conclusion

MSCs could be a promising therapy for treating progressive heart disease due to their multilineage capacity for differentiation, stimulating effect on angiogenesis and immuno-modulating traits. Medical examinations displayed safety and benefits of regional as well as systemic application of MSCs for injured cardiac tissue caused by various pathologies. The manner of cell administration has crucial effect on the destiny of MSCs, and therefore affects results of the treatment; however, there is no general agreement of best yet approach for MSC delivery in the clinical setting. To address this concern, further examinations should solve these obstacles so that regenerative potential of MSCs can be completely exploited in the treatment of heart diseases.

Acknowledgements This study was supported by Serbian Ministry of Sciences (project number ON 175103) and Faculty of Medical Sciences University of Kragujevac (JP 05/20).

Conflicts of Interest The authors acknowledge lack of any conflict of interests.

References

Amado, L. C., Saliaris, A. P., Schuleri, K. H., et al. (2005). Cardiac repair with intramyocardial injection of allogeneic mesenchymal stem cells after myocardial infarction. *Proceedings of the National Academy of Sciences of the United States of America, 102*(32), 11474–11479.

Anderson, J. D., Johansson, H. J., Graham, C. S., et al. (2016). Comprehensive proteomic analysis of mesenchymal stem cell exosomes reveals modulation of angiogenesis via nuclear factor-KappaB signaling. *Stem Cells, 34*(3), 601–613.

Ang, K. L., Chin, D., Leyva, F., et al. (2008). Randomized, controlled trial of intramuscular or intracoronary injection of autologous bone marrow cells into scarred myocardium during CABG versus CABG alone. *Nature Clinical Practice. Cardiovascular Medicine, 5*(10), 663–670.

Ankrum, J., & Karp, J. M. (2010). Mesenchymal stem cell therapy: Two steps forward, one step back. *Trends in Molecular Medicine, 16*(5), 203–209.

Arslan, F., Lai, R. C., Smeets, M. B., et al. (2013). Mesenchymal stem cell-derived exosomes increase ATP levels, decrease oxidative stress and activate PI3K/Akt pathway to enhance myocardial viability and prevent adverse remodeling after myocardial ischemia/reperfusion injury. *Stem Cell Research, 10*(3), 301–312.

Assmus, B., Walter, D. H., Seeger, F. H., et al. (2013). Effect of shock wave-facilitated intracoronary cell therapy on LVEF in patients with chronic heart failure: The CELLWAVE randomized clinical trial [published correction appears in JAMA. 2013 may 15;309(19):1994]. *Journal of the American Medical Association, 309*(15), 1622–1631.

Atluri, P., Miller, J. S., Emery, R. J., et al. (2014). Tissue-engineered, hydrogel-based endothelial progenitor cell therapy robustly revascularizes ischemic myocardium and preserves ventricular function. *The Journal of Thoracic and Cardiovascular Surgery, 148*(3), 1090–1098.

Barbash, I. M., Chouraqui, P., Baron, J., et al. (2003). Systemic delivery of bone marrow-derived mesenchymal stem cells to the infarcted myocardium: Feasibility, cell migration, and body distribution. *Circulation, 108*(7), 863–868.

Bartolucci, J., Verdugo, F. J., González, P. L., et al. (2017). Safety and efficacy of the intravenous infusion of umbilical cord mesenchymal stem cells in patients with heart failure: A phase 1/2 randomized controlled trial (RIMECARD trial [randomized clinical trial of intravenous infusion umbilical cord mesenchymal stem cells on Cardiopathy]). *Circulation Research, 121*(10), 1192–1204.

Bervar, M., Kozelj, M., Poglajen, G., et al. (2017). Effects of Transendocardial CD34+ cell transplantation on diastolic parameters in patients with nonischemic dilated cardiomyopathy. *Stem Cells Translational Medicine, 6*(6), 1515–1521.

Bian, S., Zhang, L., Duan, L., et al. (2014). Extracellular vesicles derived from human bone marrow mesenchymal stem cells promote angiogenesis in a rat myocardial infarction model. *Journal of Molecular Medicine (Berlin, Germany), 92*(4), 387–397.

Boomsma, R. A., Swaminathan, P. D., & Geenen, D. L. (2007). Intravenously injected mesenchymal stem cells home to viable myocardium after coronary occlusion and preserve systolic function without altering infarct size. *International Journal of Cardiology, 122*(1), 17–28.

Breitbach, M., Bostani, T., Roell, W., et al. (2007). Potential risks of bone marrow cell transplantation into infarcted hearts. *Blood, 110*(4), 1362–1369.

Butler, J., Epstein, S. E., Greene, S. J., et al. (2017). Intravenous allogeneic mesenchymal stem cells for nonischemic cardiomyopathy: Safety and efficacy results of a phase II-A randomized trial. *Circulation Research, 120*(2), 332–340.

Campbell, N. G., & Suzuki, K. (2012). Cell delivery routes for stem cell therapy to the heart: Current and future approaches. *Journal of Cardiovascular Translational Research, 5*(5), 713–726.

Carr, C. A., Stuckey, D. J., Tatton, L., et al. (2008). Bone marrow-derived stromal cells home to and remain in the infarcted rat heart but fail to improve function: An in vivo cine-MRI study. *American Journal of Physiology. Heart and Circulatory Physiology, 295*(2), H533–H542.

Chamberlain, G., Fox, J., Ashton, B., & Middleton, J. (2007). Concise review: Mesenchymal stem cells: Their phenotype, differentiation capacity, immunological features, and potential for homing. *Stem Cells, 25*(11), 2739–2749.

Charwat, S., Gyöngyösi, M., Lang, I., et al. (2008). Role of adult bone marrow stem cells in the repair of ischemic myocardium: Current state of the art. *Experimental Hematology, 36*(6), 672–680.

Chen, S. L., Fang, W. W., Ye, F., et al. (2004). Effect on left ventricular function of intracoronary transplantation of autologous bone marrow mesenchymal stem cell in patients with acute myocardial infarction. *The American Journal of Cardiology, 94*(1), 92–95.

Choudhury, T., Mozid, A., Hamshere, S., et al. (2017). An exploratory randomized control study of combination cytokine and adult autologous bone marrow progenitor cell administration in patients with ischaemic cardiomyopathy: The REGENERATE-IHD clinical trial. *European Journal of Heart Failure, 19*(1), 138–147.

Chullikana, A., Majumdar, A. S., Gottipamula, S., et al. (2015). Randomized, double-blind, phase I/II study of intravenous allogeneic mesenchymal stromal cells in acute myocardial infarction. *Cytotherapy, 17*(3), 250–261.

Copland, I. B. (2011). Mesenchymal stromal cells for cardiovascular disease. *Journal of Cardiovascular Disease Research, 2*(1), 3–13.

Davies, J. E., & Walker, J. T. (2017). Keating a concise review: Wharton's jelly: The rich, but enigmatic, source of mesenchymal stromal cells. *Stem Cells Translational Medicine, 6*, 1620–1630.

de la Fuente, L. M., Stertzer, S. H., Argentieri, J., et al. (2007). Transendocardial autologous bone marrow in chronic myocardial infarction using a helical needle catheter: 1-year follow-up in an open-label, nonrandomized, single-center pilot study (the TABMMI study). *American Heart Journal, 154*(1), 79.e1–79.e797.

Detante, O., Rome, C., & Papassin, J. (2017). How to use stem cells for repair in stroke patients. *Revue Neurologique (Paris), 173*(9), 572–576.

Dib, N., Dinsmore, J., Lababidi, Z., et al. (2009). One-year follow-up of feasibility and safety of the first U.S., randomized, controlled study using 3-dimensional guided catheter-based delivery of autologous skeletal myoblasts for ischemic cardiomyopathy (CAuSMIC study). *JACC. Cardiovascular Interventions, 2*(1), 9–16.

Dib, N., Menasche, P., Bartunek, J. J., et al. (2010). Recommendations for successful training on methods of delivery of biologics for cardiac regeneration: A report of the International Society for Cardiovascular Translational Research. *JACC. Cardiovascular Interventions, 3*(3), 265–275.

Dib, N., Khawaja, H., Varner, S., McCarthy, M., & Campbell, A. (2011). Cell therapy for cardiovascular disease: A comparison of methods of delivery. *Journal of Cardiovascular Translational Research, 4*(2), 177–181.

Dominici, M., Le Blanc, K., Mueller, I., et al. (2006). Minimal criteria for defining multipotent mesenchymal stromal cells. The International Society for Cellular Therapy position statement. *Cytotherapy, 8*(4), 315–317.

Formigli, L., Perna, A. M., Meacci, E., et al. (2007). Paracrine effects of transplanted myoblasts and relaxin on post-infarction heart remodelling. *Journal of Cellular and Molecular Medicine, 11*(5), 1087–1100.

Freyman, T., Polin, G., Osman, H., Crary, J., Lu, M., Cheng, L., Palasis, M., & Wilensky, R. L. (2006). A quantitative, randomized study evaluating three methods of mesenchymal stem cell delivery following myocardial infarction. *European Heart Journal, 27*(9), 1114–1122.

Fukushima, S., Varela-Carver, A., Coppen, S. R., et al. (2007). Direct intramyocardial but not intracoronary injection of bone marrow cells induces ventricular arrhythmias in a rat chronic ischemic heart failure model. *Circulation, 115*(17), 2254–2261.

Fukushima, S., Coppen, S. R., Lee, J., et al. (2008). Choice of cell-delivery route for skeletal myoblast transplantation for treating post-infarction chronic heart failure in rat. *PLoS One, 3*(8), e3071.

Galipeau, J., & Sensébé, L. (2018). Mesenchymal stromal cells: Clinical challenges and therapeutic opportunities. *Cell Stem Cell, 22*(6), 824–833.

Gao, J., Dennis, J. E., Muzic, R. F., Lundberg, M., & Caplan, A. I. (2001). The dynamic in vivo distribution of bone marrow-derived mesenchymal stem cells after infusion. *Cells, Tissues, Organs, 169*(1), 12–20.

Gathier, W. A., van Ginkel, D. J., van der Naald, M., van Slochteren, F. J., Doevendans, P. A., & Chamuleau, S. A. J. (2018 Jun). Retrograde coronary venous infusion as a delivery strategy in regenerative cardiac therapy: An overview of preclinical and clinical data. *Journal of Cardiovascular Translational Research, 11*(3), 173–181.

Gazdic, M., Volarevic, V., Arsenijevic, N., & Stojkovic, M. (2015). Mesenchymal stem cells: A friend or foe in immune-mediated diseases. *Stem Cell Reviews and Reports, 11*(2), 280–287.

Gazdic, M., Arsenijevic, A., Markovic, B. S., et al. (2017). Mesenchymal stem cell-dependent modulation of liver diseases. *International Journal of Biological Sciences, 13*(9), 1109–1117.

Ghajar, C. M., Kachgal, S., Kniazeva, E., et al. (2010). Mesenchymal cells stimulate capillary morphogenesis via distinct proteolytic mechanisms. *Experimental Cell Research, 316*(5), 813–825.

Gnecchi, M., Zhang, Z., Ni, A., & Dzau, V. J. (2008). Paracrine mechanisms in adult stem cell signaling and therapy. *Circulation Research, 103*(11), 1204–1219.

Gopinath, S., Vanamala, S. K., Gondi, C. S., & Rao, J. S. (2010). Human umbilical cord blood derived stem cells repair doxorubicin-induced pathological cardiac hypertrophy in mice. *Biochemical and Biophysical Research Communications, 395*(3), 367–372.

Gregg, D., & Fisher, L. (1963). Blood supply to the heart. In *Handbook of physiology*. American Physiological Society.

Gyöngyösi, M., Blanco, J., Marian, T., et al. (2008). Serial noninvasive in vivo positron emission tomographic tracking of percutaneously intramyocardially injected autologous porcine mesenchymal stem cells modified for transgene reporter gene expression. *Circulation. Cardiovascular Imaging, 1*(2), 94–103.

Hamshere, S., Arnous, S., Choudhury, T., et al. (2015). Randomized trial of combination cytokine and adult autologous bone marrow progenitor cell administration in patients with non-ischaemic dilated cardiomyopathy: The REGENERATE-DCM clinical trial. *European Heart Journal, 36*(44), 3061–3069.

Hare, J. M., Traverse, J. H., Henry, T. D., et al. (2009). A randomized, double-blind, placebo-controlled, dose-escalation study of intravenous adult human mesenchymal stem cells (prochymal) after acute myocardial infarction. *Journal of the American College of Cardiology, 54*(24), 2277–2286.

Hare, J. M., Fishman, J. E., Gerstenblith, G., et al. (2012). Comparison of allogeneic vs autologous bone marrow–derived mesenchymal stem cells delivered by transendocardial injection in patients with ischemic cardiomyopathy: The POSEIDON randomized trial [published correction appears in JAMA. 2013 Aug 21;310(7):750. George, Richard [added]; Lardo, Albert [added]]. *Journal of the American Medical Association, 308*(22), 2369–2379.

Harrell, C. R., Gazdic, M., Fellabaum, C., et al. (2019). Therapeutic potential of amniotic fluid derived mesenchymal stem cells based on their differentiation capacity and immunomodulatory properties. *Current Stem Cell Research & Therapy, 14*, 327–336.

Hatzistergos, K. E., Quevedo, H., Oskouei, B. N., et al. (2010a). Bone marrow mesenchymal stem cells stimulate cardiac stem cell proliferation and differentiation. *Circulation Research, 107*(7), 913–922.

Hatzistergos, K. E., Quevedo, H., Oskouei, B. N., Hu, Q., Feigenbaum, G. S., Margitich, I. S., Mazhari, R., Boyle, A. J., Zambrano, J. P., Rodriguez, J. E., Dulce, R., Pattany, P. M., Valdes, D., Revilla, C., Heldman, A. W., McNiece, I., & Hare, J. M. (2010b). Bone marrow mesenchymal stem cells stimulate cardiac stem cell proliferation and differentiation. *Circulation Research, 107*, 913–922.

Heldman, A. W., DiFede, D. L., Fishman, J. E., et al. (2014). Transendocardial mesenchymal stem cells and mononuclear bone marrow cells for ischemic cardiomyopathy: The TAC-HFT randomized trial. *Journal of the American Medical Association, 311*(1), 62–73.

Heldring, N., Mäger, I., Wood, M. J., Le Blanc, K., & Andaloussi, S. E. (2015). Therapeutic potential of multipotent mesenchymal stromal cells and their extracellular vesicles. *Human Gene Therapy, 26*(8), 506–517.

Hong, S. J., Hou, D., Brinton, T. J., et al. (2014). Intracoronary and retrograde coronary venous myocardial delivery of adipose-derived stem cells in swine infarction lead to transient myocardial trapping with predominant pulmonary redistribution. *Catheterization and Cardiovascular Interventions, 83*(1), E17–E25.

Hoogduijn, M. J., Roemeling-van Rhijn, M., Engela, A. U., et al. (2013). Mesenchymal stem cells induce an inflammatory response after intravenous infusion. *Stem Cells and Development, 22*(21), 2825–2835.

Hou, D., Youssef, E. A., Brinton, T. J., et al. (2005a). Radiolabeled cell distribution after intramyocardial, intracoronary, and interstitial retrograde coronary venous delivery: Implications for current clinical trials. *Circulation, 112*(9 Suppl), I150–I156.

Hou, D., Youssef, E. A., Brinton, T. J., et al. (2005b). Radiolabeled cell distribution after intramyocardial, intracoronary, and interstitial retrograde coronary venous delivery: Implications for current clinical trials. *Circulation, 112*(9 Suppl), I150–I156.

https://www.who.int/health-topics/cardiovascular-diseases/#tab=tab_1

Huang, Z., Shen, Y., Sun, A., et al. (2013). Magnetic targeting enhances retrograde cell retention in a rat model of myocardial infarction. *Stem Cell Research & Therapy, 4*(6), 149.

Huang, P., Wang, L., Li, Q., et al. (2019). Combinatorial treatment of acute myocardial infarction using stem cells and their derived exosomes resulted in improved heart performance. *Stem Cell Research & Therapy, 10*(1), 300.

Ichihara, Y., Kaneko, M., Yamahara, K., et al. (2018). Self-assembling peptide hydrogel enables instant epicardial coating of the heart with mesenchymal stromal cells for the treatment of heart failure. *Biomaterials, 154*, 12–23.

Janowski, M., Lyczek, A., Engels, C., et al. (2013). Cell size and velocity of injection are major determinants of the safety of intracarotid stem cell transplantation. *Journal of Cerebral Blood Flow and Metabolism, 33*(6), 921–927.

Janssens, S., Dubois, C., Bogaert, J., et al. (2006). Autologous bone marrow-derived stem-cell transfer in patients with ST-segment elevation myocardial infarction: Double-blind, randomised controlled trial. *Lancet, 367*(9505), 113–121.

Jeong, J. O., Han, J. W., Kim, J. M., et al. (2011). Malignant tumor formation after transplantation of short-term cultured bone marrow mesenchymal stem cells in experimental myocardial infarction and diabetic neuropathy. *Circulation Research, 108*(11), 1340–1347.

Kanelidis, A. J., Premer, C., Lopez, J., Balkan, W., & Hare, J. M. (2017). Route of delivery modulates the efficacy of mesenchymal stem cell therapy for myocardial infarction: A meta-analysis of preclinical studies and clinical trials. *Circulation Research, 120*(7), 1139–1150.

Kang, S. K., Shin, I. S., Ko, M. S., Jo, J. Y., & Ra, J. C. (2012). Journey of mesenchymal stem cells for homing: Strategies to enhance efficacy and safety of stem cell therapy. *Stem Cells International, 2012*, 342968.

Karantalis, V., DiFede, D. L., Gerstenblith, G., et al. (2014). Autologous mesenchymal stem cells produce concordant improvements in regional function, tissue perfusion, and fibrotic burden when administered to patients undergoing coronary artery bypass grafting: The prospective randomized study of mesenchymal stem cell therapy in patients undergoing cardiac surgery (PROMETHEUS) trial. *Circulation Research, 114*(8), 1302–1310.

Kawada, H., Fujita, J., Kinjo, K., et al. (2004). Nonhematopoietic mesenchymal stem cells can be mobilized and differentiate into cardiomyocytes after myocardial infarction. *Blood, 104*(12), 3581–3587.

Kean, T. J., Lin, P., Caplan, A. I., & Dennis, J. E. (2013). MSCs: Delivery routes and engraftment, cell-targeting strategies, and immune modulation. *Stem Cells International, 2013*, 732742.

Kim, J. Y., Jeon, H. B., Yang, Y. S., Oh, W., & Chang, J. W. (2010). Application of human umbilical cord blood-derived mesenchymal stem cells in disease models. *The World Journal Stem Cells, 2*(2), 34–38.

Kocher, A. A., Schlechta, B., Gasparovicova, A., Wolner, E., Bonaros, N., & Laufer, G. (2007). Stem cells and cardiac regeneration. *Transplant International, 20*(9), 731–746.

Krause, K., Jaquet, K., Schneider, C., et al. (2009). Percutaneous intramyocardial stem cell injection in patients with acute myocardial in farction: First-Inman study. *Heart, 95*(14), 1145–1152.

Kyriakou, C., Rabin, N., Pizzey, A., et al. (2008). Factors that influence short-term homing of human bone marrow-derived mesenchymal stem cells in a xenogeneic animal model. *Haematologica, 93*, 1457–1465.

Lalu, M. M., Mazzarello, S., Zlepnig, J., et al. (2018). Safety and efficacy of adult stem cell therapy for acute myocardial infarction and ischemic heart failure (SafeCell heart): A systematic review and meta-analysis. *Stem Cells Translational Medicine, 7*(12), 857–866.

Le Blanc, K., & Mougiakakos, D. (2012). Multipotent mesenchymal stromal cells and the innate immune system. Nature reviews. *Immunology, 12*(5), 383–396.

Lee, C., Mitsialis, S. A., Aslam, M., et al. (2012). Exosomes mediate the cytoprotective action of mesenchymal stromal cells on hypoxia-induced pulmonary hypertension. *Circulation, 126*(22), 2601–2611.

Leistner, D. M., Fischer-Rasokat, U., Honold, J., et al. (2011). Transplantation of progenitor cells and regeneration enhancement in acute myocardial infarction (TOPCARE-AMI): Final 5-year results suggest long-term safety and efficacy. *Clinical Research in Cardiology, 100*(10), 925–934.

Levy, O., Kuai, R., Siren, E. M. J., et al. (2020). Shattering barriers toward clinically meaningful MSC therapies. *Science Advances, 6*(30), eaba6884.

Lim, M., Wang, W., Liang, L., et al. (2018). Intravenous injection of allogeneic umbilical cord-derived multipotent mesenchymal stromal cells reduces the infarct area and ameliorates cardiac function in a porcine model of acute myocardial infarction. *Stem Cell Research & Therapy, 9*(1), 129.

Liu, C. B., Huang, H., Sun, P., et al. (2016a). Human umbilical cord-derived mesenchymal stromal cells improve left ventricular function, perfusion, and remodeling in a porcine model of chronic myocardial ischemia. *Stem Cells Translational Medicine, 5*(8), 1004–1013.

Liu, S., Zhou, J., Zhang, X., et al. (2016b). Strategies to optimize adult stem cell therapy for tissue regeneration. *International Journal of Molecular Sciences, 17*(6), 982.

Liu, L., Jin, X., Hu, C. F., et al. (2017). Exosomes derived from mesenchymal stem cells rescue myocardial Ischaemia/reperfusion injury by inducing cardiomyocyte autophagy via AMPK and Akt pathways. *Cellular Physiology and Biochemistry, 43*(1), 52–68.

Llano, R., Epstein, S., Zhou, R., et al. (2009). Intracoronary delivery of mesenchymal stem cells at high flow rates after myocardial infarction improves distal coronary blood flow and decreases mortality in pigs. *Catheterization and Cardiovascular Interventions, 73*(2), 251–257.

Losordo, D. W., Schatz, R. A., White, C. J., Udelson, J. E., Veereshwarayya, V., Durgin, M., et al. (2007). Intramyocardial transplantation of autologous CD34+ stem cells for intractable angin a. *Circulation, 115*(25), 3165–3172.

Lu, L. L., Liu, Y. J., Yang, S. G., et al. (2006). Isolation and characterization of human umbilical cord mesenchymal stem cells with hematopoiesis-supportive function and other potentials. *Haematologica, 91*(8), 1017–1026.

Luger, D., Lipinski, M. J., Westman, P. C., et al. (2017). Intravenously delivered mesenchymal stem cells: Systemic anti-inflammatory effects improve left ventricular dysfunction in acute myocardial infarction and ischemic cardiomyopathy. *Circulation Research, 120*(10), 1598–1613.

Ma, T., Chen, Y., Chen, Y., et al. (2018). MicroRNA-132, delivered by mesenchymal stem cell-derived exosomes, promote angiogenesis in myocardial infarction. *Stem Cells International, 2018*, 3290372.

Markovic, B. S., Kanjevac, T., Harrell, C. R., et al. (2018). Molecular and cellular mechanisms involved in mesenchymal stem cell-based therapy of inflammatory bowel diseases. *Stem Cell Reviews, 14*(2), 153–165.

Mathiasen, A. B., Qayyum, A. A., Jørgensen, E., et al. (2015). Bone marrow-derived mesenchymal stromal cell treatment in patients with severe ischaemic heart failure: A randomized placebo-controlled trial (MSC-HF trial). *European Heart Journal, 36*(27), 1744–1753.

Mazhari, R., & Hare, J. M. (2007). Mechanisms of action of mesenchymal stem cells in cardiac repair: Potential influences on the cardiac stem cell niche. *Nature Clinical Practice. Cardiovascular Medicine, 4*(Suppl 1), S21–S26.

Menasché, P., Alfieri, O., Janssens, S., et al. (2008). The myoblast autologous grafting in ischemic cardiomyopathy (MAGIC) trial: First randomized placebo-controlled study of myoblast transplantation. *Circulation, 117*(9), 1189–1200.

Mewhort, H. E., Turnbull, J. D., Satriano, A., et al. (2016). Epicardial infarct repair with bioinductive extracellular matrix promotes vasculogenesis and myocardial recovery. *The Journal of Heart and Lung Transplantation, 35*(5), 661–670.

Meyer, G. P., Wollert, K. C., Lotz, J., et al. (2006). Intracoronary bone marrow cell transfer after myocardial infarction: Eighteen months' follow-up data from the randomized, controlled BOOST (BOne marrOw transfer to enhance ST-elevation infarct regeneration) trial. *Circulation, 113*(10), 1287–1294.

Moll, G., Ankrum, J. A., Kamhieh-Milz, J., et al. (2019). Intravascular mesenchymal stromal/stem cell therapy product diversification: Time for new clinical guidelines. *Trends in Molecular Medicine, 25*(2), 149–163.

Monsel, A., Zhu, Y. G., Gennai, S., Hao, Q., Liu, J., & Lee, J. W. (2014). Cell-based therapy for acute organ injury: Preclinical evidence and ongoing clinical trials using mesenchymal stem cells. *Anesthesiology, 121*(5), 1099–1121.

Mueller, S. M., & Glowacki, J. (2001). Age-related decline in the osteogenic potential of human bone marrow cells cultured in three-dimensional collagen sponges. *Journal of Cellular Biochemistry, 82*(4), 583–590.

Mushtaq, M., DiFede, D. L., Golpanian, S., et al. (2014). Rationale and design of the percutaneous stem cell injection delivery effects on Neomyogenesis in dilated cardiomyopathy (the POSEIDON-DCM study): A phase I/II, randomized pilot study of the comparative safety and efficacy of transendocardial injection of autologous mesenchymal stem cell vs. allogeneic mesenchymal stem cells in patients with non-ischemic dilated cardiomyopathy. *Journal of Cardiovascular Translational Research, 7*(9), 769–780.

Nagaya, N., Fujii, T., Iwase, T., et al. (2004). Intravenous administration of mesenchymal stem cells improves cardiac function in rats with acute myocardial infarction through angiogenesis and myogenesis. *American Journal of Physiology. Heart and Circulatory Physiology, 287*(6), H2670–H2676.

Narita, T., Shintani, Y., Ikebe, C., et al. (2013). The use of scaffold-free cell sheet technique to refine mesenchymal stromal cell-based therapy for heart failure. *Molecular Therapy, 21*(4), 860–867.

Nijboer, C. H., Kooijman, E., van Velthoven, C. T., et al. (2018). Intranasal stem cell treatment as a novel therapy for subarachnoid hemorrhage. *Stem Cells and Development, 27*(5), 313–325.

Ota, T., Patronik, N. A., Schwartzman, D., Riviere, C. N., & Zenati, M. A. (2008). Minimally invasive epi - cardial injections using a novel semiautonomous robotic device. *Circulation*, 118 115-S120.

Passier, R., van Laake, L. W., & Mummery, C. L. (2008). Stem-cell-based therapy and lessons from the heart. *Nature, 453*(7193), 322–329.

Patel, A. N., Mittal, S., Turan, G., et al. (2015). REVIVE trial: Retrograde delivery of autologous bone marrow in patients with heart failure. *Stem Cells Translational Medicine, 4*(9), 1021–1027.

Perea-Gil, I., Prat-Vidal, C., Gálvez-Montón, C., et al. (2016). A cell-enriched engineered myocardial graft limits infarct size and improves cardiac function: Pre-clinical study in the porcine myocardial infarction model. *JACC Basic Translational Science, 1*(5), 360–372.

Perin, E. C., & López, J. (2006). Methods of stem cell delivery in cardiac diseases. *Nature Clinical Practice. Cardiovascular Medicine, 1*, S110–S113.

Pohl, T., Giehrl, W., Reichart, B., et al. (2004). Retroinfusion-supported stenting in high-risk patients for percutaneous intervention and bypass surgery: Results of the prospective randomized myoprotect I study. *Catheterization and Cardiovascular Interventions, 62*(3), 323–330.

Price, M. J., Chou, C. C., Frantzen, M., et al. (2006). Intravenous mesenchymal stem cell therapy early after reperfused acute myocardial infarction improves left ventricular function and alters electrophysiologic properties. *International Journal of Cardiology, 111*(2), 231–239.

Prockop, D. J., & Olson, S. D. (2007). Clinical trials with adult stem/progenitor cells for tissue repair: let's not overlook some essential precautions. *Blood, 109*(8), 3147–3151.

Psaltis, P. J., Zannettino, A. C., Gronthos, S., & Worthley, S. G. (2010). Intramyocardial navigation and mapping for stem cell delivery. *Journal of Cardiovascular Translational Research, 3*(2), 135–146.

Qi, C. M., Ma, G. S., Liu, N. F., et al. (2008). Transplantation of magnetically labeled mesenchymal stem cells improves cardiac function in a swine myocardial infarction model. *Chinese Medical Journal, 121*(6), 544–550.

Raake, P., von Degenfeld, G., Hinkel, R., et al. (2004). Myocardial gene transfer by selective pressure-regulated retroinfusion of coronary veins: Comparison with surgical and percutaneous intramyocardial gene delivery. *Journal of the American College of Cardiology, 44*(5), 1124–1129.

Rice, M. J., Chou, C. C., Frantzen, M., et al. (2006). Intravenous mesenchymal stem cell therapy early after reperfused acute myocardial infarction improves left ventricular function and alters electrophysiologic properties. *International Journal of Cardiology, 111*(2), 231–239.

Roger, V. L., Go, A. S., Lloyd-Jones, D. M., et al. (2012). Executive summary: heart disease and stroke statistics--2012 update: a report from the American Heart Association [published correction appears in Circulation]. *Circulation, 125*(1), 188–197.

Ryan, J. M., Barry, F. P., Murphy, J. M., & Mahon, B. P. (2005). Mesenchymal stem cells avoid allogeneic rejection. *Journal of Inflammation (London), 2*, 8.

Saito, T., Kuang, J. Q., Bittira, B., Al-Khaldi, A., & Chiu, R. C. (2002). Xenotransplant cardiac chimera: Immune tolerance of adult stem cells. *The Annals of Thoracic Surgery, 74*(1), 19–24.

Salomon, C., Ryan, J., Sobrevia, L., et al. (2013). Exosomal signaling during hypoxia mediates microvascular endothelial cell migration and vasculogenesis. *PLoS One, 8*(7), e68451.

Samper, E., Diez-Juan, A., Montero, J. A., & Sepúlveda, P. (2013). Cardiac cell therapy: Boosting mesenchymal stem cells effects. *Stem Cell Reviews and Reports, 9*(3), 266–280.

Schuleri, K. H., Amado, L. C., Boyle, A. J., et al. (2008). Early improvement in cardiac tissue perfusion due to mesenchymal stem cells. *American Journal of Physiology. Heart and Circulatory Physiology, 294*(5), H2002–H2011.

Shabbir, A., Zisa, D., Suzuki, G., & Lee, T. (2009). Heart failure therapy mediated by the trophic activities of bone marrow mesenchymal stem cells: A noninvasive therapeutic regimen. *American Journal of Physiology. Heart and Circulatory Physiology, 296*(6), H1888–H1897.

Sherman, W., Martens, T. P., Viles-Gonzalez, J. F., & Siminiak, T. (2006). Catheter-based delivery of cells to the heart. *Nature Clinical Practice. Cardiovascular Medicine, 3*(Suppl 1), S57–S64.

Shi, Y., Wang, Y., Li, Q., et al. (2018). Immunoregulatory mechanisms of mesenchymal stem and stromal cells in inflammatory diseases. *Nature Reviews. Nephrology, 14*(8), 493–507.

Shin, E. Y., Wang, L., Zemskova, M., et al. (2018). Adenosine production by biomaterial-supported mesenchymal stromal cells reduces the innate inflammatory response in myocardial ischemia/reperfusion injury. *Journal of the American Heart Association, 7*(2), e006949.

Stenderup, K., Justesen, J., Clausen, C., & Kassem, M. (2003). Aging is associated with decreased maximal life span and accelerated senescence of bone marrow stromal cells. *Bone, 33*(6), 919–926.

Strauer, B. E. (1979). Myocardial oxygen consumption in chronic heart disease: Role of wall stress, hypertrophy and coronary reserve. *The American Journal of Cardiology, 44*, 730–740.

Strauer, B. E., Brehm, M., Zeus, T., et al. (2001). Intrakoronare, humane autologe Stammzelltransplantation zur Myokardregeneration nach Herzinfarkt [intracoronary, human autologous stem cell transplantation for myocardial regeneration following myocardial infarction]. *Deutsche Medizinische Wochenschrift, 126*(34–35), 932–938.

Suzuki, K., Murtuza, B., Beauchamp, J. R., et al. (2004). Role of interleukin-1beta in acute inflammation and graft death after cell transplantation to the heart. *Circulation, 110*(11 Suppl 1), II219–II224.

Tano, N., Narita, T., Kaneko, M., et al. (2014). Epicardial placement of mesenchymal stromal cell-sheets for the treatment of ischemic cardiomyopathy; in vivo proof-of-concept study. *Molecular Therapy, 22*(10), 1864–1871.

Teng, X., Chen, L., Chen, W., Yang, J., Yang, Z., & Shen, Z. (2015). Mesenchymal stem cell-derived exosomes improve the microenvironment of infarcted myocardium contributing to angiogenesis and anti-inflammation. *Cellular Physiology and Biochemistry, 37*(6), 2415–2424.

Tian, T., Chen, B., Xiao, Y., Yang, K., & Zhou, X. (2014). Intramyocardial autologous bone marrow cell transplantation for ischemic heart disease: A systematic review and meta-analysis of randomized controlled trials. *Atherosclerosis, 233*(2), 485–492.

Tigges, U., Komatsu, M., & Stallcup, W. B. (2013). Adventitial pericyte progenitor/mesenchymal stem cells participate in the restenotic response to arterial injury. *Journal of Vascular Research, 50*(2), 134–144.

Toma, C., Pittenger, M. F., Cahill, K. S., Byrne, B. J., & Kessler, P. D. (2002). Human mesenchymal stem cells differentiate to a cardiomyocyte phenotype in the adult murine heart. *Circulation, 105*(1), 93–98.

Tuma, J., Fernández-Viña, R., Carrasco, A., et al. (2011). Safety and feasibility of percutaneous retrograde coronary sinus delivery of autologous bone marrow mononuclear cell transplantation in patients with chronic refractory angina. *Journal of Translational Medicine, 9*, 183.

van Rhijn-Brouwer, F. C. C., Gremmels, H., Fledderus, J. O., & Verhaar, M. C. (2018). Mesenchymal stromal cell characteristics and regenerative potential in cardiovascular disease: Implications for cellular therapy. *Cell Transplantation, 27*(5), 765–785.

Viswanathan, C., Davidson, Y., Cooper, K., Tipnis, S., Pujari, G., & Kurian, V. M. (2010). Tansplantation of autologous bone marrow derived mesenchymal stem cells trans-epicardially in patients undergoing coronary bypass surgery. *Indian Heart Journal, 62*(1), 43–48.

Volarevic, V., Ljujic, B., Stojkovic, P., Lukic, A., Arsenijevic, N., & Stojkovic, M. (2011). Human stem cell research and regenerative medicine--present and future. *British Medical Bulletin, 99*, 155–168.

Volarevic, V., Gazdic, M., Simovic Markovic, B., et al. (2017). Mesenchymal stem cell-derived factors: Immuno-modulatory effects and therapeutic potential. *BioFactors, 43*(5), 633–644.

Volarevic, V., Markovic, B. S., Gazdic, M., et al. (2018). Ethical and safety issues of stem cell-based therapy. *International Journal of Medical Sciences, 15*(1), 36–45.

von Degenfeld, G., Raake, P., Kupatt, C., et al. (2003). Selective pressure-regulated retroinfusion of fibroblast growth factor-2 into the coronary vein enhances regional myocardial blood flow and function in pigs with chronic myocardial ischemia. *Journal of the American College of Cardiology, 42*(6), 1120–1128.

Vulliet, P. R., Greeley, M., Halloran, S. M., MacDonald, K. A., & Kittleson, M. D. (2004). Intra-coronary arterial injection of mesenchymal stromal cells and microinfarction in dogs. *Lancet, 363*(9411), 783–784.

Walczak, P., Zhang, J., Gilad, A. A., et al. (2008). Dual-modality monitoring of targeted intraarterial delivery of mesenchymal stem cells after transient ischemia. *Stroke, 39*(5), 1569–1574.

Wang, Y., Zhang, G., Hou, Y., et al. (2012). Transplantation of microencapsulated Schwann cells and mesenchymal stem cells augment angiogenesis and improve heart function. *Molecular and Cellular Biochemistry, 366*(1–2), 139–147.

Wang, Y., Chen, X., Cao, W., & Shi, Y. (2014). Plasticity of mesenchymal stem cells in immunomodulation: Pathological and therapeutic implications. *Nature Immunology, 15*(11), 1009–1016.

Wang, Q. L., Wang, H. J., Li, Z. H., Wang, Y. L., Wu, X. P., & Tan, Y. Z. (2017). Mesenchymal stem cell-loaded cardiac patch promotes epicardial activation and repair of the infarcted myocardium. *Journal of Cellular and Molecular Medicine, 21*(9), 1751–1766.

Watanabe, M., & Yavagal, D. R. (2016). Intra-arterial delivery of mesenchymal stem cells. *Brain Circulation, 2*(3), 114–117.

Wen, Z., Zheng, S., Zhou, C., Yuan, W., Wang, J., & Wang, T. (2012). Bone marrow mesenchymal stem cells for post-myocardial infarction cardiac repair: microRNAs as novel regulators. *Journal of Cellular and Molecular Medicine, 16*(4), 657–671.

White, I. A., Sanina, C., Balkan, W., & Hare, J. M. (2016). Mesenchymal stem cells in cardiology. *Methods in Molecular Biology, 1416*, 55–87.

Wollert, K. C., Meyer, G. P., Müller-Ehmsen, J., et al. (2017). Intracoronary autologous bone marrow cell transfer after myocardial infarction: The BOOST-2-randomised placebo-controlled clinical trial. *European Heart Journal, 38*(39), 2936–2943.

Wu, J., Li, J., Zhang, N., & Zhang, C. (2011). Stem cell-based therapies in ischemic heart diseases: A focus on aspects of microcirculation and inflammation. *Basic Research in Cardiology, 106*(3), 317–324.

Xie, X. H., Wang, X. L., He, Y. X., et al. (2012). Promotion of bone repair by implantation of cryopreserved bone marrow-derived mononuclear cells in a rabbit model of steroid-associated osteonecrosis. *Arthritis and Rheumatism, 64*(5), 1562–1571.

Yavagal, D. R., Lin, B., Raval, A. P., et al. (2014). Efficacy and dose-dependent safety of intra-arterial delivery of mesenchymal stem cells in a rodent stroke model. *PLoS One, 9*(5), e93735.

Ye, X., & Zhang, C. (2017). Effects of hyperlipidemia and cardiovascular diseases on proliferation, differentiation and homing of mesenchymal stem cells. *Current Stem Cell Research & Therapy, 12*(5), 377–387.

Yokoi, H., Kinoshita, T., & Zhang, S. (2005). Dynamic reassembly of peptide RADA16 nanofiber scaffold. *Proceedings of the National Academy of Sciences of the United States of America, 102*(24), 8414–8419.

Yokoyama, S., Fukuda, N., Li, Y., et al. (2006). A strategy of retrograde injection of bone marrow mononuclear cells into the myocardium for the treatment of ischemic heart disease. *Journal of Molecular and Cellular Cardiology, 40*(1), 24–34.

Zhang, S. (2003). Fabrication of novel biomaterials through molecular self-assembly. *Nature Biotechnology, 21*(10), 1171–1178.

Zisa, D., Shabbir, A., Suzuki, G., & Lee, T. (2009). Vascular endothelial growth factor (VEGF) as a key therapeutic trophic factor in bone marrow mesenchymal stem cell-mediated cardiac repair. *Biochemical and Biophysical Research Communications, 390*(3), 834–838.

Zuba-Surma, E. K., Adamiak, M., & Dawn, B. (2015). Chapter 5 – Stem cell extracellular vesicles: A novel cell-based therapy for cardiovascular diseases. In *Mesenchymal stem cell derived exosomes* (pp. 93–117). Academic.

Zuk, P. A., Zhu, M., Ashjian, P., et al. (2002). Human adipose tissue is a source of multipotent stem cells. *Molecular Biology of the Cell, 13*(12), 4279–4295.

Adv Exp Med Biol - Innovations in Cancer Research and Regenerative Medicine (2023) 4: 101–116
https://doi.org/10.1007/5584_2023_769

Published online: 30 March 2023

Intravenous Infusion of Exosomes Derived from Human Adipose Tissue-Derived Stem Cells Promotes Angiogenesis and Muscle Regeneration: An Observational Study in a Murine Acute Limb Ischemia Model

Hue Thi Doan, Phuc Van Pham, and Ngoc Bich Vu

Abstract

Introduction Recent studies have demonstrated that adipose tissue-derived stem cell (ADSC) transplantation could promote neoangiogenesis in various ischemic diseases. However, as whole cells, ADSCs have some disadvantages, such as shipping and storage issues, high costs, and controversies related to the fates of grafted cells in the recipients. Therefore, this study aimed to investigate the effects of intravenously infused exosomes purified from human ADSCs on ischemic disease in a murine hindlimb ischemia model.

Methods ADSCs were cultured in exosome-free medium for 48 h before the conditioned medium was collected for exosome isolation by ultracentrifugation. The murine ischemic hindlimb models were created by cutting and burning the hindlimb arteries. Exosomes were intravenously infused into murine models (ADSC-Exo group), with phosphate-buffered saline (PBS) used as a placebo (PBS group). Treatment efficacy was determined using a murine mobility assay (frequency of pedaling in water per 10 s), peripheral blood oxygen saturation (SpO_2 index), and the recovery of vascular circulation by trypan blue staining. The formation of blood vessels was shown by X-ray. Expression levels of genes related to angiogenesis and muscle tissue repair were quantified by quantitative reverse-transcription polymerase chain reaction. Finally, H&E staining was used to determine the histological structure of muscle in the treatment and placebo groups.

Supplementary Information: The online version contains supplementary material available at https://doi.org/10.1007/5584_2023_769.

H. T. Doan
Faculty of Biological Sciences, Nong Lam University, Ho Chi Minh City, Vietnam

P. Van Pham
Vietnam National University, Ho Chi Minh City, Vietnam

Stem Cell Institute, University of Science, Ho Chi Minh City, Vietnam

N. B. Vu (✉)
Vietnam National University, Ho Chi Minh City, Vietnam

Stem Cell Institute, University of Science, Ho Chi Minh City, Vietnam

Laboratory of Stem Cell Research and Application, University of Science, Ho Chi Minh City, Vietnam
e-mail: ngocvu@sci.edu.vn; vbngoc@hcmus.edu.vn

Results The rates of acute limb ischemia in the PBS and ADSC-Exo injection groups were 66% (9/16 mice) and 43% (6/14 mice), respectively. The mobility of the limbs 28 days after surgery was significantly different between the ADSC-Exo treatment group (41 ± 1 times/10 s) and the PBS group (24 ± 1 times/10 s; $n = 3$; $p < 0.05$). Peripheral blood oxygen saturation 21 days after treatment was 83.83% ± 2.02% in the PBS group and 83% ± 1.73% in the ADSC-Exo treatment group, and the difference was not statistically significant ($n = 3$, $p > 0.05$). On day 7 after treatment, the time required to stain the toes after trypan blue injection was 20.67 ± 12.5 s and 85 ± 7.09 s in the ADSC-Exo and PBS groups, respectively ($n = 3$, $p < 0.05$). On day 3 after the operation, the expression of genes promoting angiogenesis and muscle remodeling, such as *Flk1*, *Vwf*, *Ang1*, *Tgfb1*, *Myod*, and *Myf5*, was increased 4–8 times in the ADSC-Exo group compared with the PBS group. No mice in either group died during the experimental period.

Conclusions These results revealed that intravenous infusion of human ADSC-derived exosomes is a safe and effective method to treat ischemic disease, especially hindlimb ischemia, by promoting angiogenesis and muscle regeneration.

Keywords

Acute hindlimb ischemia · Adipose derived stem cells · Angiogenesis · Exosomes · Muscle regeneration

Abbreviations

ADSCs	Human adipose stem cells
ASCD-exo	Human adipose stem cell-derived exosomes
MSC	Mesenchymal stem cells
PBS	Phosphate buffer saline
VEGF	Vascular endothelial growth factor

1 Introduction

Ischemia is a common disease caused by a restriction in blood supply to any tissues, causing a shortage of oxygen and nutrients required for cell metabolism. Ischemia can occur in various tissues and cause extreme conditions, such as cardiac ischemia, brain ischemia, and limb ischemia. Currently, approaches used to treat ischemia include anticoagulant therapy, thrombolysis, embolectomy, surgical revascularization, and partial amputation (Fluck et al., 2020; Hage et al., 2018; Kempe et al., 2014; Morrison, 2006; Yeager et al., 1992). These treatments provide some benefits for patients with certain types of ischemia. With the rapid development of stem cell therapy, stem cell transplantation has also been used to treat some ischemic diseases (Bang, 2016; Bartunek et al., 2020; Kawabori et al., 2020; Lasala et al., 2010; Misra et al., 2012).

In a meta-analysis, Attar et al. (2021) analyzed the effectiveness of mesenchymal stem cell (MSC) transplantation for acute myocardial infarction in 468 patients (Attar et al., 2021). The results showed that transplantation of MSCs after acute myocardial infarction induced a significant increase in LVEF (Attar et al., 2021). In another meta-analysis, Boncoraglio et al. (2019) analyzed seven randomized controlled trials that used stem cell transplantation to treat ischemic stroke and found that overall, stem cell transplantation was associated with better clinical outcomes (Boncoraglio et al., 2019). Similarly, Xie et al. (2018) confirmed that stem cell therapy is effective in critical limb ischemia (Xie et al., 2018). In almost all clinical trials, MSCs have been used. Indeed, MSCs have been shown to promote wound healing in ischemic diseases via anti-inflammatory as well as immunomodulatory mechanisms (Gao et al., 2019). Some of the main mechanisms by which MSCs function in ischemic disease treatment were comprehensively reviewed by Van Nguyen et al. (2021). MSCs produce factors that modulate the immune response at ischemic sites, including interleukin 6 (IL-6), IL-1 alpha, IL-1 beta, IL-10, and IL-12. Some angiogenic factors produced by MSCs include hepatocyte growth factor (HGF), vascular endothelial growth factor (VEGF), transforming growth factor beta 1 (TGF-β1), and fibroblast growth factor 21, which trigger neoangiogenesis. These factors can also rescue some injured cells at ischemic sites via anti-apoptotic mechanisms (Van Nguyen et al., 2021).

However, stem cell transplantation is associated with high costs, and stem cells are difficult to ship and store. Therefore, stem cell transplantation is not commonly used in the treatment of ischemic diseases. To address the costs, in addition to some novel strategies to reduce cell processing, shipping, and storage costs, exosomes derived from MSCs have been used as cell-free biological agents that can reduce costs due to their easy storage and usage. In recent years, exosomes have been used to treat cerebral ischemia (Li et al., 2021; Liu et al., 2021), myocardial ischemia (Ju et al., 2018a; Zhao et al., 2019), and hindlimb ischemia (Ju et al., 2018b) in preclinical models. Some mechanisms by which exosomes elicit therapeutic effects in ischemic diseases have been discovered, including modulating microglial M1/M2 phenotypes (Liu et al., 2021), suppressing the inflammatory response (Li et al., 2021), modifying the polarization status of macrophages (Zhao et al., 2019), and delivering miR-210 to ischemic sites (Cheng et al., 2020). However, in publications currently in the literature, exosomes were directly injected into ischemic sites. Although exosomes can be directly delivered to ischemic sites using this method, in some cases of complex ischemic diseases that affect wide regions of the body or require surgery-guided injection, local injection is difficult to impossible. Therefore, this study aimed to investigate the effects of MSC-derived exosomes in ischemic disease using a murine hindlimb ischemia model and a single dose of intravenous infusion of adipose tissue-derived stem cell (ADSC)-derived exosomes.

2 Methods

2.1 Adipose Tissue-Derived Stem Cell Expansion and Conditioned Medium Collection

ADSCs were obtained from a stem cell bank (SCICellbank, HCMC, VN) and thawed and expanded in MSCCult I medium (Regenmedlab, Ho Chi Minh City, VN) in a T175 flask. When the cell confluence reached 70%, the MSCCult I medium was replaced by MSCCult MV (Regenmedlab, HCMC, VN), which is free of exosomes, for 48 h. The conditioned medium was then collected for exosome isolation.

2.2 Isolation of Exosomes Derived from Adipose Tissue-Derived Stem Cells

The conditioned medium was sequentially centrifuged at 2000 × g for 15 min, 10,000 × g for 30 min, and 100,000 × g for 70 min at 4 °C. The supernatant was then discarded. To remove residual soluble factors, pelleted ADSC-derived exosomes (ADSC-Exos) were washed with phosphate-buffered saline (PBS) once by centrifugation at 100,000 × g for 60 min at 4 °C. The purified ADSC-Exos were suspended in 150 μl of PBS in a 1.5 mL Eppendorf tube and stored at −86 °C or 4 °C for 1 week.

2.3 ADSC-Exo Characterization

ADSC-Exos were characterized based on the expression of CD9, CD63, and CD81 on the surface of the exosomes using flow cytometry.

In addition, the size of exosomes was determined using a nanosize meter and stem capture.

To determine the surface expression of CD9, CD63, and CD81, exosomes were incubated with 8 nm latex beads overnight. The latex beads were then centrifuged at 1500 × g for 5 min to remove the unbound exosomes. The latex beads were stained with anti-CD9-PE, anti-CD63-FITC, and anti-CD81-APC for 20 min at room temperature. The beads were centrifuged to remove unbound antibodies and then used to analyze the expression of CD9, CD63, and CD81 by flow cytometry. The data were analyzed and visualized using FlowJo software.

2.4 Exosome Quantity

A protein standard curve was prepared using 0, 0.125, 0.25, 0.5, 1.5, and 2 μg/mL bovine serum albumin, and the ADSC-Exo quantity was determined using this standard curve. Briefly, 10 μL of ADSC-Exos was loaded in technical triplicate in a 96-well plate, and 200 μL of Bradford dye was added to each well. The results were read with a spectrophotometer (DTX 880, Beckman Coulter) at 595 nm for 10 min.

2.5 Acute Limb Ischemia Mouse Model

The Animal Experimental Ethical Committee of the Stem Cell Institute, University of Science (VNUHCM), Ho Chi Minh City, Vietnam, approved all animal experiments in this study. An acute limb ischemia mouse model was established according to protocols published by Ngoc et al. using male mice over 6 months of age (Dao et al., 2016). Briefly, 6-month-old male mice were anesthetized with zoletil and xylazine at 5 mg/kg. The location of vascular ligation was the area above and below the femoral triangle. The central artery and vein were excised, followed by excision of the lateral vessels. The vessel was occluded using double knots. After dissection, a silk suture was passed underneath the proximal end of the femoral artery. The morphological characteristics, movement, and histological changes of mice that underwent vascular ligation were monitored continuously for 28 days.

2.6 ADSC-Exo Intravenous Infusion in Acute Limb Ischemic Mice

Acute hindlimb ischemic mice were randomized into two groups: the ADSC-Exo group (14 mice) and the placebo group (14 mice). In the ADSC-Exo group, 100 μg of exosomes in 150 μl PBS was directly infused into the tail vein. The mice in the placebo group were infused with 150 μl PBS in the same manner. Both exosomes and PBS were infused as soon as the models were created. All mice were housed and cared for under the same conditions.

2.7 Evaluation of Recovery of Injured Limbs

2.7.1 Hindlimb Morphology

Hindlimb morphology was evaluated based on the color changes and degree of ischemic damage following the guidelines by Goto et al. (2006):

Grade 0: Normal limb without swelling, purple color, or necrosis.
Grade I: Purple-black color, necrosis extending from the nail to the toe.
Grade II: Necrosis from the toe extending to the end of the foot.
Grade III: Necrosis to the knee joint (knee loss).
Grade IV: Necrosis to the hip joint.

Mice were considered ischemic if they exhibited grade I–IV injuries. Mice classified into grades 0, I, and II were used to determine the peripheral capillary oxygen saturation (SpO_2) index using a trypan blue assay to evaluate the blood circulation and motility of the mouse limbs in water.

For the trypan blue assay, 0.15 ml of 0.4% trypan blue was directly injected into the tail vein. The time required for trypan blue to circulate to the toes and stain the limbs as observed by the naked eye was recorded using a timer. The motility of the mouse limbs was determined by counting the pedaling frequency of the limbs while swimming in water for 10 s.

2.7.2 Hematoxylin-Eosin Staining

The ischemic muscle tissues were collected on days 3, 7, and 28 for hematoxylin-eosin (H&E) staining to reveal the histological structure. Samples were soaked in 4% paraformaldehyde mixed in PBS, pH 7.2–7.4, then kept at 4 °C to preserve the tissue before staining. Tissue section slides were stained with H&E and assessed by microscopy (Carl Zeiss, Oberkochen, Germany).

2.7.3 X-Ray Imaging

Mice were deeply anesthetized. Next, the chest cavity was opened to reveal the internal organs. An Easy-Load II Pump Head (Masterflex) was used to deliver heparin (40 IU/ml) into the left ventricle at a rate of 2–3 ml/min over 2.5 min. Next, the portal vein of the liver was cut to allow blood to drain out. Once the internal organs turned white, PBS was pumped at the same speed for 2.5 min. Next, 1 g/ml $BaSO_4$ was pumped, and once it was observed that $BaSO_4$ had entered the circuit (approximately 0.5 min), the pump was stopped. Images were taken with an InVivo Extreme I (Bruker).

2.8 Quantitative Reverse-Transcription Polymerase Chain Reaction

For RNA extraction, mouse thigh muscle tissue at the knot site was separated from the mouse body on day 3 and the easy-BLUE Total RNA Extraction Kit (iNtRON Biotechnology) was used to extract RNA according to the manufacturer's guidelines. The concentration of RNA was measured using a BioPhotometer Plus (Eppendorf). The following primers were used: *Ang1* (angiopoietin 1), *Flt1* (VEGF receptor 1 [VEGFR1]), *Flk1* (VEGFR2), *Cdh5* (VE-cadherin), *Hgf* (HGF), *Mmp2* (matrix metallopeptidase 2 or type 4 collagenase), *Myod*, *Myf5*, *Tgfb1* (TGF-β1), *Pecam1* (CD31), *Vwf* (Von Willebrand factor), and *Gapdh* (glyceraldehyde 3-phosphate dehydrogenase), which was used as a housekeeping gene (all from Viet Nam Applied Biology and Technical Joint Stock Company, Ho Chi Minh City, Vietnam). The primer sequences used in this study are provided in the Supplementary File. RT-PCR was set up according to the manufacturer's guidelines using the Luna Universal Probe One-Step RT-qPCR mix (New England Biolabs). Data were analyzed with GraphPad Prism software, and reactions were performed in triplicate. Data were normalized to *Gapdh* mRNA expression for each condition and were quantified relative to the corresponding gene expression from the PBS injection sample, which was standardized to 1.

2.9 Statistical Analysis

Statistical analyses were performed using Student's t-test. All data are presented as mean ± SD, and $p < 0.05$ was considered statistically significant. Data were analyzed with GraphPad Prism 8.0 software.

3 Results

3.1 Exosome Characteristics

The obtained exosomes expressed CD63 (96.5%), CD81 (98.8%), and CD9 (87.1%) (Fig. 1a). Exosome purity was observed using a nanosize meter. The results showed that 30–150 nm particles comprised 90% of all particles in the solution. Under a transmission electron microscope, exosomes with particular shapes, such as cups, were easily observed (Fig. 1b).

3.2 Safety of Intravenous Infusion of Exosomes

The safety of intravenous infusion of exosomes was evaluated 48 h after infusion and when the experiments were finished. At both 48 h after infusion and 1 month after infusion (the end of the experiment), 100% of the mice in both groups were alive.

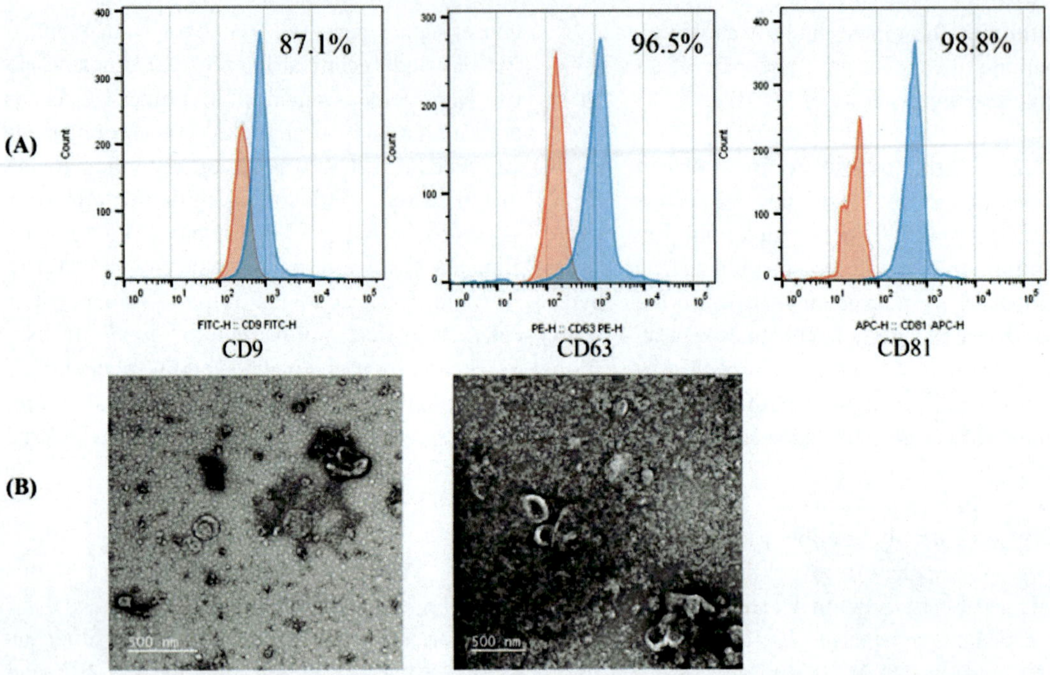

Fig. 1 Exosomes from human ADSCs. (**a**) Exosomes expressed CD9, CD63, and CD81. (**b**) The exosomes were observed under TEM with cup shapes

3.3 Efficacy of Intravenous Infusion of Exosomes to Treat Hindlimb Ischemic Disease in Mice

3.3.1 Limb Morphology After Treatment

On the first day of treatment, in the PBS group, 36% (5/14 mice) exhibited grade 1 ischemia and 64% (9/14 mice) had grade 0 ischemia, while in the ADSC-Exo group, 100% had grade 0 ischemia (Fig. 2).

On the second day, in the PBS group, 57% (8/14) of the mice had grade 0, 21% (3/14) had grade 1, 14% (2/14) had grade 3, and 7% (1/14) had grade 4 ischemia (Fig. 2). In the ADSC-Exo group, only 7% (1/14) of the mice had grade 2 ischemia, and the remaining 93% of mice had grade 0 ischemia. On the next day, 25% (4/16) of the mice in the PBS group had grade 4 ischemia, while in the ADSC-Exo group, no mice had grade 4 ischemia.

On day 28, the rate of complete recovery of ischemic limbs (grade 0) or minor injury to the limb (grade 1) in the treatment group was 71% (10/14 mice), which was significantly higher than in the PBS group (57%; 8/14). Notably, 29% (4/14) of mice in the PBS group had grade 4 ischemia, but 0% of mice in the ADSC-Exo group had grade 4 ischemia ($p < 0.01$; Fig. 2).

3.3.2 Saturation of Peripheral Oxygen

The SpO_2 index is one of the first physiological indicators to be evaluated to determine the extent of damage in an ischemic mouse model. On the first day after surgery, the SpO_2 index in both groups combined dropped from 96.71% ± 5.88% before surgery to 68,56.12% ± 6.97% after surgery to induce acute limb ischemia. On the third day after treatment, the SpO_2 index of mice in the ADSC-Exo group climbed to 84.63% ± 3.26%, while the SpO_2 index was maintained at 70.12% ± 9.47% in the PBS group ($p < 0.05$), and the values for both treatment groups were lower than in normal mice ($p < 0.05$; Fig. 3). On day 7 post-treatment, the SpO_2 index of the PBS group was 74.36% ± 6.17%, which was significantly lower than that of the ADSC-Exo group (82.83% ± 2.93%; $p < 0.05$). On day 14, the SpO_2 index in the ADSC-Exo group (88.67% ± 0.58%) remained higher than that in the

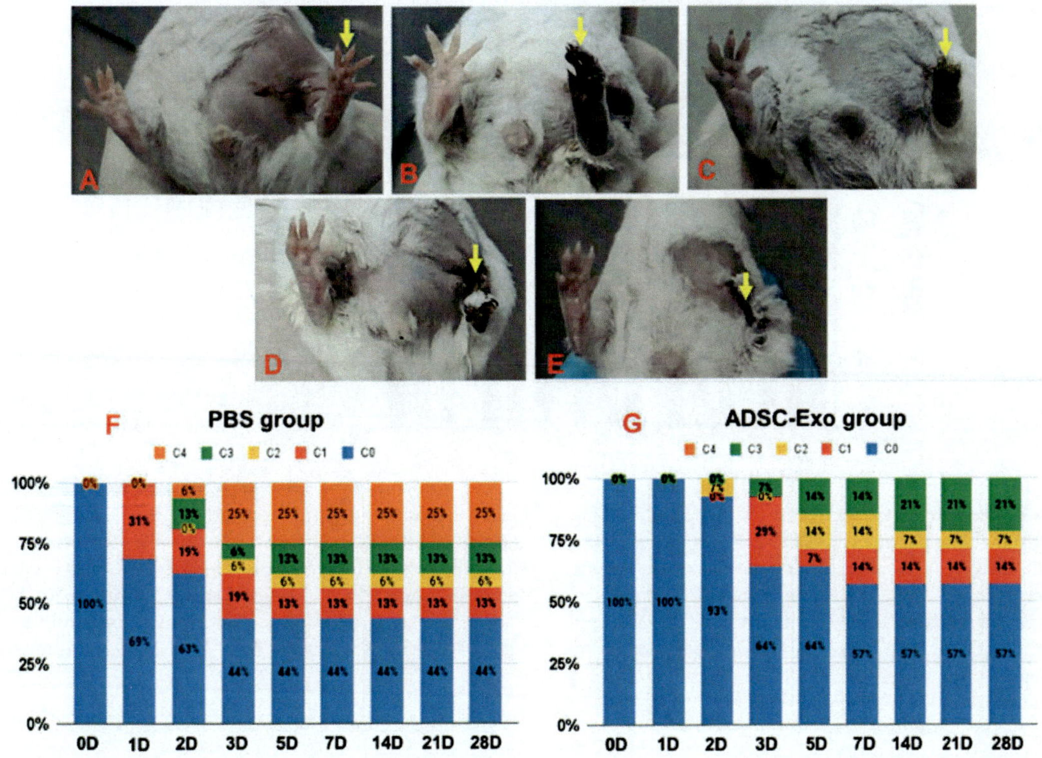

Fig. 2 Ischemic limb morphology and ischemic limb recovery in the ADSC-Exo group and PBS group. (**a**–**e**) Some kinds of ischemic limbs were classified according to Goto et al. The recovery of ischemic limbs was based on their morphology from day 0 to day 28 in the PBS group (**f**) and ADSC-Exo group (**g**). C1: Grade 1; C2: Grade 2; C3: Grade 3; C4: Grade 4; D: day

PBS group (81.67% ± 4.36%). However, on day 21, these values were not significantly different between the groups, as the SpO_2 index was approximately 85% in both groups. All data were recorded in mice that recovered and had a limb morphology of grade 0, 1, or 2.

3.3.3 Pedal Frequency

Three days after surgery, the limb mobility of mice in both experimental groups was severely affected due to ischemia. The limb pedaling frequency of mice was 16 ± 1 times/10 s in the ADSC-Exo group and 14 ± 4 times/10 s in the PBS group (Fig. 4), which was significantly lower than that in the normal group (42 ± 5 times/10 s; $p < 0.001$). On day 7, the mobility of the limbs of mice in the ADSC-Exo group was increased significantly (29 ± 1 times/10 s) compared with that in the PBS group (19 ± 1 times/10 s; $p < 0.001$).

On day 28, there was no significant difference in limb mobility between the ADSC-Exo and normal groups (41 ± 1 vs. 42 ± 5 times/10 s, respectively; $p > 0.05$). However, the mobility of the limbs of mice in the PBS group was 24 ± 6 times/10 s, which was significantly lower than that in both ADSC-Exo-treated and normal mice ($p < 0.05$; Fig. 4).

3.3.4 Vascular Circulation

Three days after surgery, the time required to observe trypan blue staining of toes in mice in the ADSC-Exo group was 60 ± 20.22 s (Fig. 5c, g), which was significantly faster than in the PBS group (181 ± 38.94 s), but still very slow compared with the normal group (12.33 ± 8.74 s; $p < 0.05$; Fig. 5b, g). On day 7, the time required to stain toes was reduced in both the ADSC-Exo and PBS groups (20.67 ± 12.5 s and 85 ± 7.09 s, respectively;

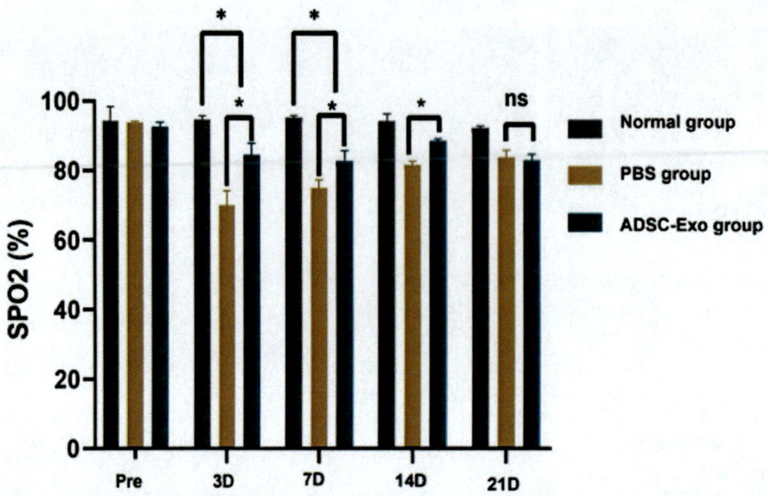

Fig. 3 The SpO_2 index was measured in the limbs of mice in groups (normal mice, ADSC-Exo group, and PBS group) before and after treatment. SpO_2 in the three groups (normal mice, ADSC-Exo group, and PBS group) was similar for all mice before creating the model and treatment. SpO_2 in the treatment group with exosomes increased from day 3 to day 14 and was significantly higher than that in the PBS group ($p < 0.05$). However, on day 21, the SpO_2 indexes were similar between the ADSC-Exo and PBS groups ($p > 0.05$). All data were recorded in only grade 0, 1, and 2 mice. * = difference at 5% significance level, *ns* no statistically significant difference, *Pre* before surgery, *D* date; $N = 3$ in each group; *p*-value <0.05

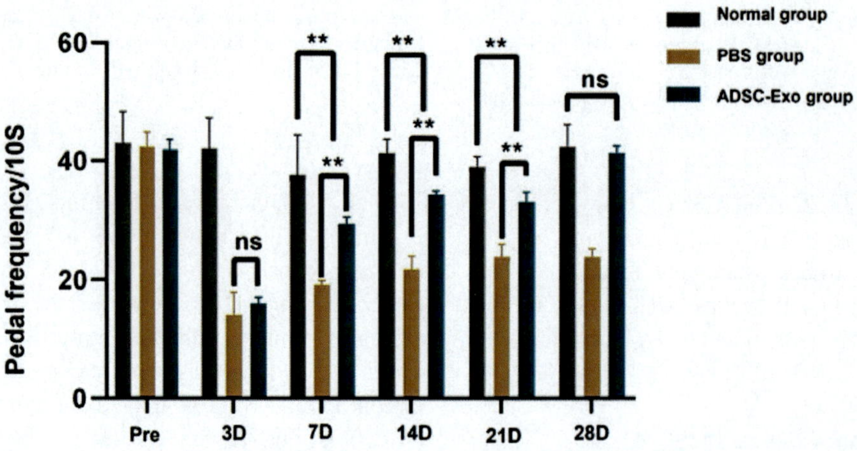

Fig. 4 Recovery of ischemic limbs via pedal frequency recorded in treatment and placebo compared to normal mice. The pedal frequency was recorded in grade 0, 1, and 2 mice for all groups. On day 28, in mice with grades 0, 1, and 2, the pedal frequency of mice in the treatment group was similar to that of normal mice ($p > 0.05$) but significantly higher than that in the PBS group ($p < 0.05$). ** difference at 1% significance level, *ns* no statistically significant difference, *Pre* before surgery, *D* date, *T* test, $N = 3$ in each group, *p*_value <0.001

$p < 0.05$; Fig. 5e, g). On this day, the time required to stain toes in the ADSC-Exo group was similar to that in the normal mice (20.67 ± 12.5 s and 12.33 ± 8.74 s, respectively; $p > 0.05$).

3.3.5 Hematoxylin-Eosin Staining

The normal mouse muscle histological structure is closely arranged, with the nucleus close to the cell membrane and easily observed muscle

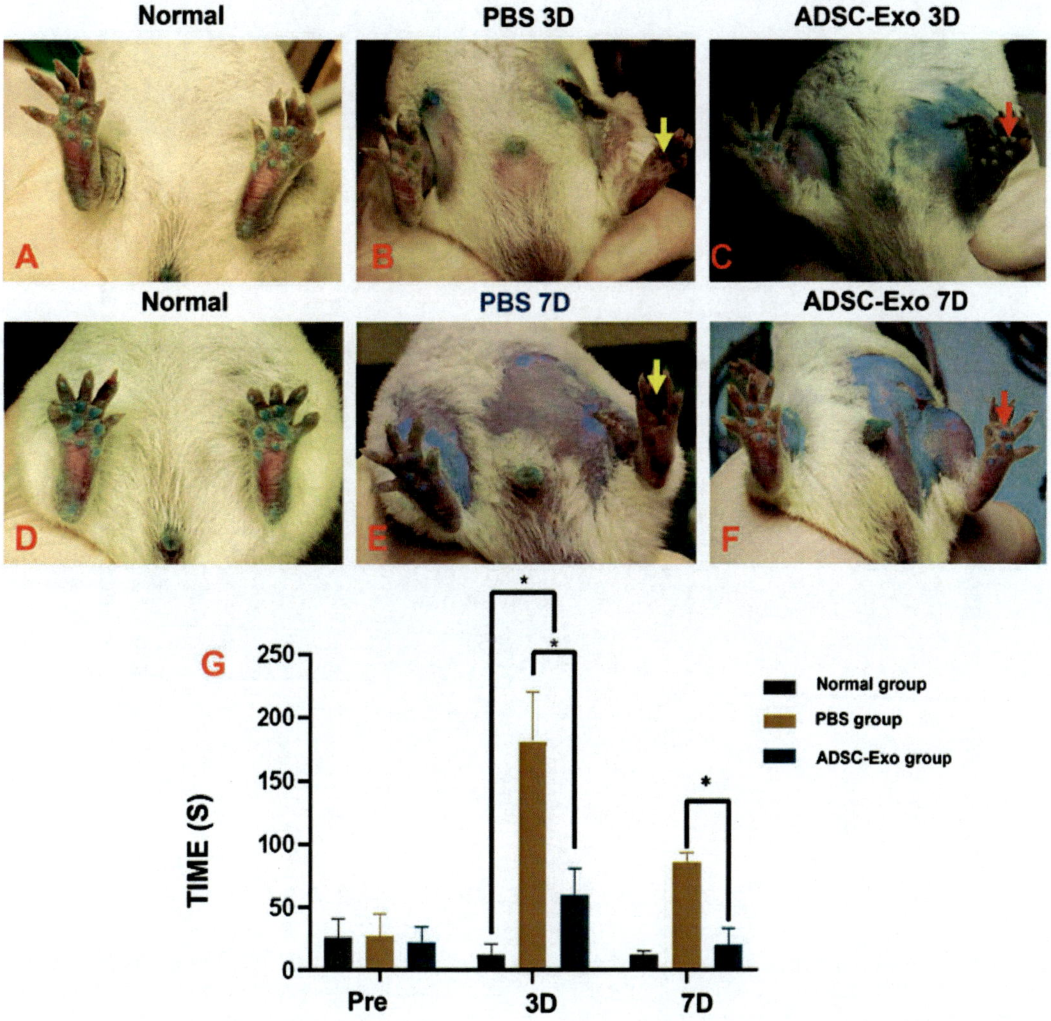

Fig. 5 Trypan blue staining for toes and limb after trypan blue injection into tail vein (**a**–**d**, and **f**) and time for staining of trypan blue in some groups (**g**). The trypan blue staining times on day 7 in mice in the ADSC-Exo group were not significantly different from those in normal mice but were much lower than those in the PBS group. * difference at 5% significance level, *ns* no statistically significant difference, *Pre* before surgery, *D*: date, $N = 3$ in each lot, p-value <0.05. Yellow arrow: ischemic leg in PBS group, the red arrow indicates ischemic leg in ADSC-Exo group. Time for the toes to become blue because of the trypan blue dye in seconds (s)

bundles (Fig. 6a, b). The cytoplasm is uniformly pink and the nucleus is violet. These are signs of a normal, healthy muscle bundle. Meanwhile, the mice that underwent vascular ligation had loose muscle tissue structures and loose muscle bundles 3 days after surgery, as recorded in both the ADSC-Exo and PBS groups (Fig. 6c, d). Moreover, some inflammatory cells were detected at ischemic sites in both groups. This may be how the mouse body responds to acute ischemia. This was also noted in previous studies by (Sachdev et al., 2014).

On days 7 and 28 (Fig. 6e–h), the tissue structure had gradually returned to its normal form in mice in both groups. However, in the PBS group, the tissue structure was composed of many wide interstitial spaces and discrete muscle bundles and was less tight than in the ADSC-Exo group.

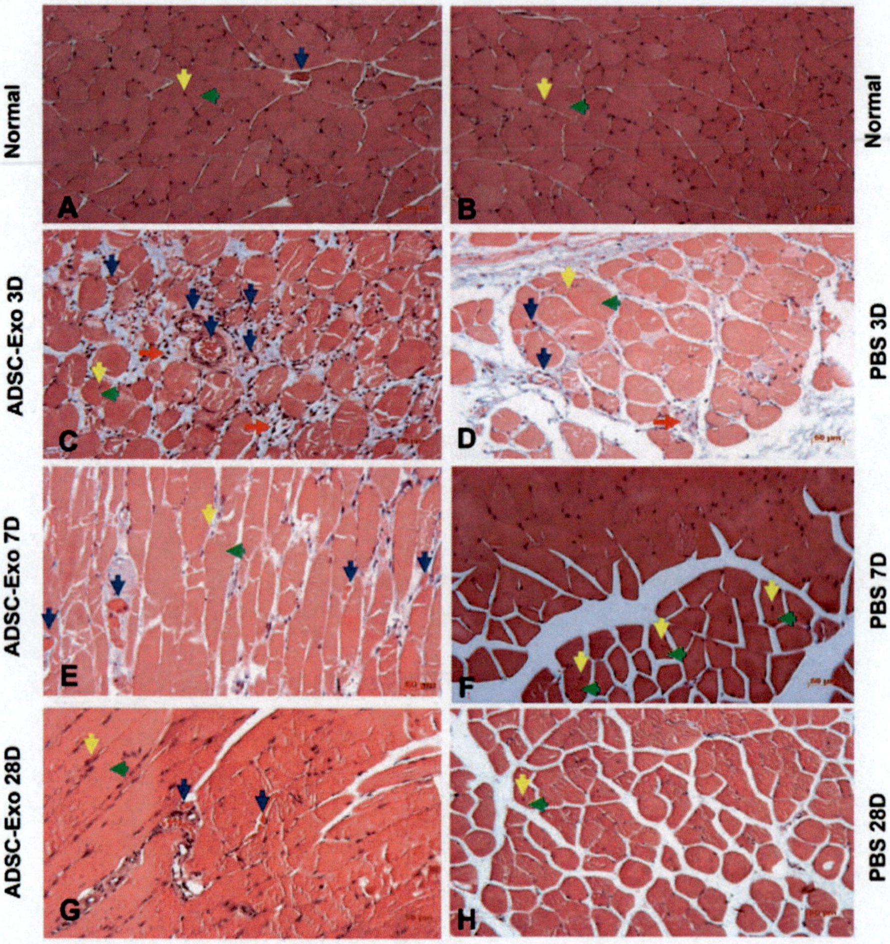

Fig. 6 H&E staining of muscle tissue from normal mice, ADSC-Exos, and PBS-treated mice. On day 28, the muscle of mice treated with ADSC-Exos recovered with histological morphology similar to that in normal mice but different from that in the PBS group. Yellow arrows point to the cell nucleus, green arrows to the cytoplasm, blue arrows to blood vessels, and red arrows to inflamed cells. Images were taken at 20X

3.3.6 X-Ray

On day 3, angiogenesis was observed in both experimental groups (Fig. 7a, b). By imaging, mice in the PBS group showed more vascular growth than the ADSC-Exo group at the onset of ischemia. However, on day 7, the ADSC-Exo group (Fig. 7c) exhibited significantly more blood vessels than the PBS group (Fig. 7d). On day 28, the mice in both experimental groups (Fig. 7e, f) had relatively few visible blood vessels. The blood vessels that developed in the two experimental groups (Fig. 7) were mainly preexisting arteries. Some of the blood vessels originated from the section of the femoral artery that had been cut, while others developed from the coccyx artery to the level of the limb. In most mice, these two types of arteries developed simultaneously. This was also noted by Arima et al. (Arima et al., 2018). In addition, the study noted that 28 days after surgery, blood vessels were forming in both experimental groups. The PBS group (Fig. 7b, d) exhibited formation of blood vessels, but not enough to save the limb from necrosis.

Twenty-eight days after ADSC-Exo injection, at the site of vascular ablation (Fig. 7a), the

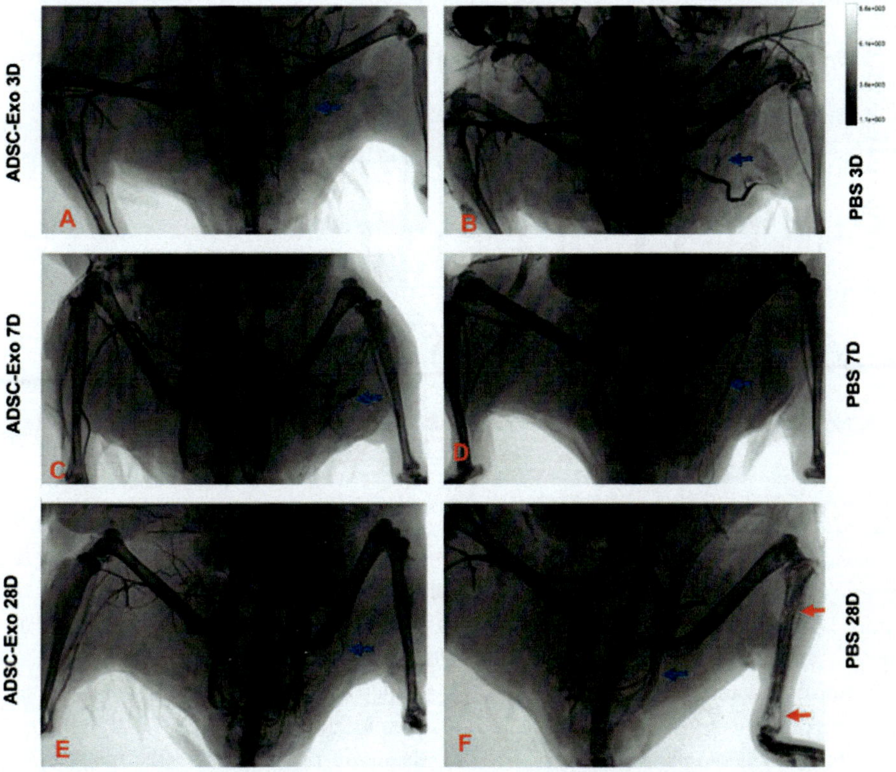

Fig. 7 X-ray capture with $BaSO_4$ injection in mice of both treatment and placebo groups. The results showed that new blood vessels formed in mice with exosome intravenous infusion, compared to the placebo group

descending femoral artery branched out into small branches supplying blood to the extremities. The results were similar to those observed 7 days after exosome injection (Fig. 7c). Due to the newly formed blood vessels supplying nutrients and oxygen to the mouse extremities, injury recovery was stronger in the ADSC-Exo group (Fig. 7c). In PBS-injected mice, blood vessels were discontinuous, with few small branches (Fig. 7b). As a result, these mice had necrosis of the feet (Fig. 7d). In summary, ADSC-Exos stimulated angiogenesis from existing blood vessels that then branched out into small branches that supplied blood to each area of the mouse limb.

3.3.7 Gene Expression

Gene expression was assessed in thigh muscle tissue from 3 ADSC-Exo–injected and 3 PBS-injected mice on day 3. The results showed that expression of the angiogenesis-promoting gene *Ang1* was 4- to eight-fold higher in the ADSC-Exo–injected animals than in the PBS-injected animals ($p < 0.05$; Fig. 8). In the remaining ADSC-Exo-injected animals (Fig. 8), the expression levels of other angiogenic genes, such as *Flk1* and *Vwf*, were dominant and were 6 to 8 times higher than those of the PBS-injected animals ($p < 0.05$). Muscle-related genes such as *Myod* and *Myf5* also showed 5- to 8-fold increased gene expression in ADSC-Exo-injected mice compared with PBS-injected mice ($p < 0.05$).

4 Discussion

This study aimed to evaluate the safety and efficacy of intravenous infusion of exosomes derived from human ADSCs to treat ischemic diseases in a murine model of acute hindlimb ischemia. The results indicated that exosome infusion significantly promotes the regeneration of ischemic

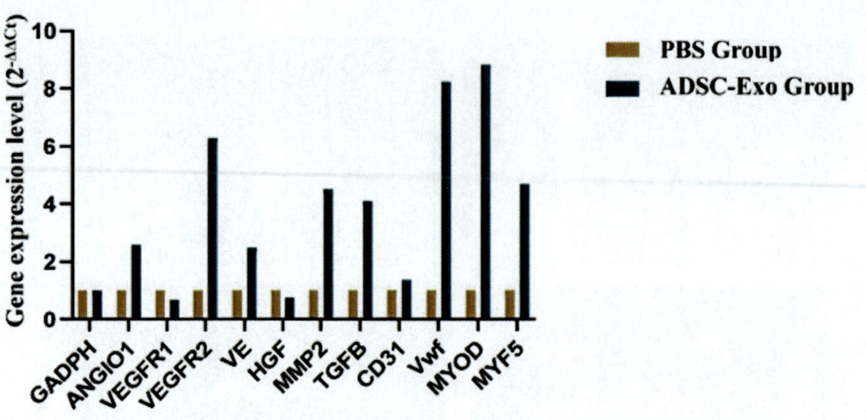

Fig. 8 Expression levels of angiogenesis- and myogenesis-related genes in cells derived from ischemic sites in mice of groups

sites by triggering the expression of factors related to angiogenesis as well as muscle regeneration. These results suggest that exosomes from ADSCs could be delivered via intravenous infusion to support ischemic tissue regeneration.

To generate exosomes for subsequent experiments, exosomes were isolated and purified from the conditioned medium of ADSC culture. The exosomes were characterized by the expression of CD9, CD63, and CD81 and were primarily cup-shaped with a diameter of 30–150 nm, as described in previous publications (An et al., 2019; Song et al., 2022; Zhou et al., 2022). Exosomes were quantified to ensure that the appropriate dose was used for intravenous treatments.

Exosomes were intravenously infused into the tail vein of acute hindlimb ischemic mouse models, with PBS used as a placebo. The safety of infusion was monitored by counting the number of dead mice over 28 days. The results confirmed that exosome infusion was safe, as no mice died during the 28 days of the experiment. The treatment efficacy was carefully evaluated using several assays and methods, from physical examinations (limb morphology, mobility of limbs in water) to preclinical tests (SpO_2 index, trypan blue assay, H&E staining, gene expression, X-ray imaging). All results from these assays suggested that exosome infusion can promote tissue regeneration via both new blood vessel formation and muscle regeneration.

Overall, the results demonstrated that exosome infusion could elicit therapeutic effects on acute hindlimb ischemia. The first effect of treatment assessed was its ability to rescue the hindlimb from necrosis (grade 3 and grade 4 ischemia) and increase the rate of complete recovery of the ischemic hindlimb (grade 0, 1, and 2 ischemia). Indeed, on day 28 after treatment, 57% of the ADSC-Exo-treated mice had grade 0 ischemia, with 0% of the mice having grade 4 ischemia, while only 44% of the mice in the placebo group had grade 0 ischemia and 25% had grade 4 ischemia.

The second effect of exosome treatment we assessed was its triggering of angiogenesis, which occurred faster in exosome-treated mice than in mice that underwent self-recovery (in the placebo group). The SpO_2 index supported this observation. On day 3, the SpO_2 recorded in the mice treated with exosomes climbed to 84.63% $\pm$ 3.26%, while this value was 70.30% $\pm$ 4.13% in placebo-treated mice. Likewise, on day 14, this value was higher in ADSC-Exo-treated mice than in placebo mice. However, on day 21, the self-recovering mice exhibited the same SpO_2 values as the ADSC-Exo-treated mice. This means that in some mice with high self-recovery, angiogenesis was promoted to rescue the ischemic limb; however, this process was slower than in mice infused with exosomes. After 21 days, essential new blood vessels were formed in both the treatment and placebo groups and the SpO_2 increased toward normal values. These results were supported by a trypan blue staining assay, which was used to demonstrate blood circulation. After injecting trypan blue into the tail vein, the circulation carries this dye to all blood vessels and stains the toes and limbs with a blue color that can be observed by the naked eye. If only a short amount of time is required for dyes injected into the tail vein to circulate to limbs and stain the toes, this indicates smooth circulation. On days 3 and 7, the time required to stain the toes with trypan blue was significantly reduced in the ADSC-Exo-injected mice compared to the placebo-treated mice. These results confirmed that the blood circulation of exosome-treated mice was better than that of placebo-treated mice. Improved angiogenesis in ADSC-Exo-treated mice was further supported by X-ray imaging with $BaSO_4$ to visualize the blood vessels. The X-ray images showed that exosomes stimulated angiogenesis from existing blood vessels, which branched out into small branches that supplied blood to each area of the mouse limb. Among the placebo-treated mice, angiogenesis was recorded in some self-recovering mice; however, the density of blood vessels was lower than in the treatment group. Based on these observations, it is clear that exosome infusion can strongly trigger angiogenesis.

The last effect of exosome infusion we measured was its ability to trigger muscle regeneration. One of the extreme effects of ischemia is muscle injury that reduces limb activity. As a result of angiogenesis, muscle regeneration was observed in both treated mice and some self-recovered mice in the placebo group. However, it is clear from the results that exosomes strongly stimulated muscle regeneration in the treatment group, enabling rapid recovery of limb activity. In recovered mice in the treatment group, the pedaling frequency was similar to that in normal mice on day 28, while in recovered mice in the PBS group, although the SpO_2 and blood circulation were similar to normal mice, the pedaling frequency was significantly lower than in the treatment group and the normal mouse group. These results revealed that the angiogenesis observed in placebo-treated mice could create new blood vessels to rescue the limbs, but may not have been sufficient to regenerate injured muscles.

H&E staining of muscle provided further evidence supporting the role of exosomes in muscle regeneration. On day 28, the muscle structure of ADSC-Exo-treated mice was almost regenerated, with a structure similar to that of normal mice, while in the PBS group, the muscle contained many wide interstitial spaces, discrete muscle bundles, and less tightness. This may be why the pedaling frequency was so low in the PBS-treated mice.

Finally, RT-PCR demonstrated that exosome infusion boosts the expression of some genes related to angiogenesis and muscle regeneration. The expression of *Ang1*, *Flk1*, and *Vwf*, which significantly contribute to angiogenesis, was much higher in ADSC-Exo-treated than placebo-treated mice. Similarly, the expression of *Myod* and *Myf5*, which control muscle regeneration, was much higher in the treatment group than in the placebo group.

Hence, intravenous infusion can significantly contribute to ischemia disease treatment by triggering angiogenesis and muscle regeneration. According to Persson et al., angiogenesis is a complex and sequential process (Persson & Buschmann, 2011). This process is divided into three steps: vasculogenesis, angiogenesis, and arteriogenesis. Vasculogenesis is a process influenced by genetic programs in which vascular structures are formed from endothelial stem cells (angioblasts) that proliferate into de novo endothelial cells. Angiogenesis is the formation of blood vessels from preexisting vascular structures and pericytes. Unlike angiogenic processes, which are established due to changes in tissue oxygenation, arteriogenesis is mainly affected by changes in blood flow. These processes are repair mechanisms for damaged tissue (Camussi et al., 2010). In the present study, exosomes mainly stimulated angiogenesis. Notably, it has previously been reported that exosomes can trigger angiogenesis (Han et al., 2019; Hu et al., 2019; Kato et al., 2021).

Kato et al. (2021) proposed that ADSCs can induce angiogenesis in mice by delivering exosomal microRNAs (Kato et al., 2021). Some microRNAs that can contribute to angiogenesis were identified in ADSC-derived exosomes, including microRNA-125a-3p (Pi et al., 2022), microRNA-181b/TRPM7 (Yang et al., 2018), and the EGR-1/lncRNA-SENCR/DKC1/VEGF-A axis (Sun et al., 2022). According to Zhu et al., recovery of ischemia and mitigation of myocardial infarction occur through the microRNA-31/FIH1/HIF-1α pathway (Zhu et al., 2022). When miR-31 inhibits FIH1 (factor inhibiting hypoxia-inducible factor-1), it releases HIF-1α (hypoxia-inducible factor-1α) from inactivation, thereby promoting angiogenesis (Zhu et al., 2022). In addition, some miRNAs found in angiogenic exosomes have been studied, such as miR-210, miR-23a-3p, miR-424, let-7f, miR-30b, miR-30c, miR-126, miR- 21, miR-132, miR-130a-3p, miR-214, miR-378, miR-126, and miR-133 (Moghiman et al., 2021). In addition to using exosomes from conventional ADSCs, exosomes can also be obtained from cells that are stimulated by hypoxia, nitric oxide (Du et al., 2017), or lipopolysaccharide (Wu et al., 2021). These factors also influence the composition and efficacy of exosome treatment. This is a new direction of development to improve the therapeutic effect of exosomes. Proteomic analysis revealed that MSC-derived exosomes contain growth factors, cytokines, signaling molecules, and adhesion molecules, such as VEGF, TGFB1, and IL-8, which have been shown to contribute to angiogenic activity (Lai et al., 2012).

In addition to their therapeutic effects on angiogenesis, exosomes have been shown to promote muscle regeneration (Byun et al., 2021; Luo et al., 2021; Nakamura et al., 2015). Nakamura et al. (2015) showed that exosomes from MSCs can promote muscle regeneration via miRNAs such as miR-494 (Nakamura et al., 2015). Byun et al. (2021) showed that treatment with ADSCs could preserve the shapes and sizes of muscle bundles (Byun et al., 2021). In a recent publication, Luo et al. (2021) showed that exosomes from MSCs can promote muscle healing following contusion through macrophage polarization (Luo et al., 2021).

The study design limited the generalizability of the results, and only one dose of ADSC-derived exosomes was used in this study. Indeed, a gradient of exosome concentrations could be used in future studies to explore the concentration-dependent effects of exosomes. Furthermore, in this study, a comparison of the effects of intravenous infusion and local injection of exosomes was not carried out.

5 Conclusion

This study explored the therapeutic effects of ADSC-derived exosomes in the treatment of ischemic disease by intravenous infusion. The results confirmed that intravenous infusion of exosomes could have therapeutic effects in acute hindlimb ischemia models. Exosomes not only triggered angiogenesis, but also boosted skeletal muscle regeneration. More importantly, the study showed that exosome infusion helped rescue limb necrosis. The effects of exosomes were achieved by stimulating the upregulation of angiogenic and myogenic genes. The results of this study suggest that exosome infusion is a viable new approach to treating some ischemic diseases.

Acknowledgments We thank SCI Biobank for providing mesenchymal stem cells used in this study.

Author's Contributions All the authors equally contributed in this work. All the authors read and approved the final version of the manuscript for submission.

Availability of Data and Materials Data and materials used and/or analyzed during the current study are available from the corresponding author on reasonable request.

Ethics Approval and Consent to Participate All animal experiments were approved by the *Institutional Animal* Care and Use *Committee*, Stem Cell Institute (HCMC, VN).

Consent for Publication Not applicable.

Competing Interests The authors declare that they have no competing interests.

Funding This research is funded by Vietnam National University, Ho Chi Minh City (VNU-HCM) under grant number 562-2020-18-03.

References

An, Y., et al. (2019). Exosomes from Adipose-Derived Stem Cells (ADSCs) overexpressing miR-21 promote vascularization of endothelial cells. *Scientific Reports, 9*(1), 12861.

Arima, Y., et al. (2018). Evaluation of collateral source characteristics with 3-dimensional analysis using micro-X-ray computed tomography. *Journal of the American Heart Association, 7*(6), e007800.

Attar, A., et al. (2021). Mesenchymal stem cell transplantation after acute myocardial infarction: A meta-analysis of clinical trials. *Stem Cell Research & Therapy, 12*(1), 600.

Bang, O. Y. (2016). Clinical trials of adult stem cell therapy in patients with ischemic stroke. *Journal of Clinical Neurology, 12*(1), 14–20.

Bartunek, J., et al. (2020). Cardiopoietic stem cell therapy in ischaemic heart failure: Long-term clinical outcomes. *ESC Heart Failure, 7*(6), 3345–3354.

Boncoraglio, G. B., et al. (2019). Stem cell transplantation for ischemic stroke. *Cochrane Database of Systematic Reviews, 5*(5), Cd007231.

Byun, S. E., et al. (2021). Skeletal muscle regeneration by the exosomes of adipose tissue-derived mesenchymal stem cells. *Current Issues in Molecular Biology, 43*(3), 1473–1488.

Camussi, G., et al. (2010). Exosomes/microvesicles as a mechanism of cell-to-cell communication. *Kidney International, 78*(9), 838–848.

Cheng, H., et al. (2020). Hypoxia-challenged MSC-derived exosomes deliver miR-210 to attenuate post-infarction cardiac apoptosis. *Stem Cell Research & Therapy, 11*(1), 224.

Dao, T. T.-T., et al. (2016). Human adipose-derived mesenchymal stem cell could participate in angiogenesis in a mouse model of acute hindlimb ischemia. *Biomedical Research and Therapy, 3*(8), 1–10.

Du, W., et al. (2017). Enhanced proangiogenic potential of mesenchymal stem cell-derived exosomes stimulated by a nitric oxide releasing polymer. *Biomaterials, 133*, 70–81.

Fluck, F., et al. (2020). Current treatment options in acute limb ischemia. In *RöFo-Fortschritte auf dem Gebiet der Röntgenstrahlen und der bildgebenden Verfahren*. © Georg Thieme Verlag KG.

Gao, W. H., et al. (2019). Effectiveness of umbilical cord mesenchymal stem cells in patients with critical limb ischemia. *Medicina Clínica (Barcelona), 153*(9), 341–346.

Goto, T., et al. (2006). Search for appropriate experimental methods to create stable hind-limb ischemia in mouse. *The Tokai Journal of Experimental and Clinical Medicine, 31*(3), 128–132.

Hage, A. N., et al. (2018). Acute limb ischemia therapies: When and how to treat endovascularly. In *Seminars in interventional radiology*. Thieme Medical Publishers.

Han, Y., et al. (2019). Exosomes from hypoxia-treated human adipose-derived mesenchymal stem cells enhance angiogenesis through VEGF/VEGF-R. *The International Journal of Biochemistry & Cell Biology, 109*, 59–68.

Hu, X., et al. (2019). Effect of adipose-derived stem cell derived exosomes on angiogenesis after skin flap transplantation in rats. *Zhongguo Xiu Fu Chong Jian Wai Ke Za Zhi, 33*(12), 1560–1565.

Ju, C., et al. (2018a). Transplantation of cardiac mesenchymal stem cell-derived exosomes promotes repair in ischemic myocardium. *Journal of Cardiovascular Translational Research, 11*(5), 420–428.

Ju, C., et al. (2018b). Transplantation of cardiac mesenchymal stem cell-derived exosomes for angiogenesis. *Journal of Cardiovascular Translational Research, 11*(5), 429–437.

Kato, T., et al. (2021). Treatment with adipose-derived regenerative cells enhances ischemia-induced angiogenesis via exosomal microRNA delivery in mice. *Nagoya Journal of Medical Science, 83*(3), 465–476.

Kawabori, M., et al. (2020). Clinical trials of stem cell therapy for cerebral ischemic stroke. *International Journal of Molecular Sciences, 21*(19), 7380.

Kempe, K., et al. (2014). Results of surgical management of acute thromboembolic lower extremity ischemia. *Journal of Vascular Surgery, 60*(3), 702–707.

Lai, R. C., et al. (2012). Proteolytic potential of the MSC exosome proteome: Implications for an exosome-mediated delivery of therapeutic proteasome. *International Journal of Proteomics, 2012*, 971907.

Lasala, G. P., et al. (2010). Combination stem cell therapy for the treatment of severe limb ischemia: Safety and efficacy analysis. *Angiology, 61*(6), 551–556.

Li, S., et al. (2021). Dental pulp stem cell-derived exosomes alleviate cerebral ischaemia-reperfusion injury through suppressing inflammatory response. *Cell Proliferation, 54*(8), e13093.

Liu, X., et al. (2021). Bone marrow mesenchymal stem cell-derived exosomes attenuate cerebral ischemia-reperfusion injury-induced neuroinflammation and pyroptosis by modulating microglia M1/M2 phenotypes. *Experimental Neurology, 341*, 113700.

Luo, Z., et al. (2021). Bone marrow stromal cell-derived exosomes promote muscle healing following contusion through macrophage polarization. *Stem Cells and Development, 30*(3), 135–148.

Misra, V., et al. (2012). Stem cell therapy in ischemic stroke: Role of IV and intra-arterial therapy. *Neurology, 79*(13 Supplement 1), S207–S212.

Moghiman, T., et al. (2021). Therapeutic angiogenesis with exosomal microRNAs: An effectual approach for the treatment of myocardial ischemia. *Heart Failure Reviews, 26*(1), 205–213.

Morrison, H. L. (2006). Catheter-directed thrombolysis for acute limb ischemia. In *Seminars in interventional radiology*. Copyright© 2006 by Thieme Medical Publishers, Inc.

Nakamura, Y., et al. (2015). Mesenchymal-stem-cell-derived exosomes accelerate skeletal muscle regeneration. *FEBS Letters, 589*(11), 1257–1265.

Persson, A. B., & Buschmann, I. R. (2011). Vascular growth in health and disease. *Frontiers in Molecular Neuroscience, 4*, 14.

Pi, L., et al. (2022). Exosomal microRNA-125a-3p from human adipose-derived mesenchymal stem cells promotes angiogenesis of wound healing through inhibiting PTEN. *Molecular and Cellular Biochemistry, 477*(1), 115–127.

Sachdev, U., et al. (2014). MyD88 and TRIF mediate divergent inflammatory and regenerative responses to skeletal muscle ischemia. *Physiological Reports, 2*(5), e12006.

Song, K., et al. (2022). Exosomes from tendon derived stem cells promote tendon repair through miR-144-3p-regulated tenocyte proliferation and migration. *Stem Cell Research & Therapy, 13*(1), 80.

Sun, Y., Ju, Y., & Fang, B. (2022). Exosomes from human adipose-derived mesenchymal stromal/stem cells accelerate angiogenesis in wound healing: Implication of the EGR-1/lncRNA-SENCR/DKC1/VEGF-A axis. *Human Cell, 35*(5), 1375–1390.

Van Nguyen, T. T., Vu, N. B., & Van Pham, P. (2021). Mesenchymal stem cell transplantation for ischemic diseases: Mechanisms and challenges. *Tissue Engineering and Regenerative Medicine, 18*(4), 587–611.

Wu, S.-C., et al. (2021). Increased angiogenesis by exosomes secreted by adipose-derived stem cells upon lipopolysaccharide stimulation. *International Journal of Molecular Sciences, 22*(16), 8877.

Xie, B., et al. (2018). Autologous stem cell therapy in critical limb ischemia: A meta-analysis of randomized controlled trials. *Stem Cells International, 2018*, 7528464.

Yang, Y., et al. (2018). Exosomes secreted by adipose-derived stem cells contribute to angiogenesis of brain microvascular endothelial cells following oxygen-glucose deprivation in vitro through MicroRNA-181b/TRPM7 Axis. *Journal of Molecular Neuroscience, 65*(1), 74–83.

Yeager, R. A., et al. (1992). Surgical management of severe acute lower extremity ischemia. *Journal of Vascular Surgery, 15*(2), 385–393.

Zhao, J., et al. (2019). Mesenchymal stromal cell-derived exosomes attenuate myocardial ischaemia-reperfusion injury through miR-182-regulated macrophage polarization. *Cardiovascular Research, 115*(7), 1205–1216.

Zhou, Y., et al. (2022). Human adipose-derived mesenchymal stem cells-derived exosomes encapsulated in pluronic F127 hydrogel promote wound healing and regeneration. *Stem Cell Research & Therapy, 13*(1), 407.

Zhu, D., et al. (2022). Exosomes from adipose-derived stem cells alleviate myocardial infarction via microRNA-31/FIH1/HIF-1α pathway. *Journal of Molecular and Cellular Cardiology, 162*, 10–19.

Part II

Advances in Tissue Engineering

Adv Exp Med Biol - Innovations in Cancer Research and Regenerative Medicine (2023) 4: 119–135
https://doi.org/10.1007/5584_2022_706

Published online: 8 April 2022

Bone Using Stem Cells for Maxillofacial Bone Disorders: A Systematic Review and Meta-analysis

Ebrahim Eini, Azadeh Ghaemi, and Fakher Rahim

Abstract

Due to economic, cultural, environmental, and social factors, the prevalence of maxillofacial bone disorders varies in different parts of the world. The present meta-analysis was conducted to assess the efficacy and safety of different type of stem cells–based scaffolds and their construction methods in maxillofacial bone disorders. We searched major indexing databases, including PubMed/Medline, ISI Web of Science, Scopus, Embase, and Cochrane Central without any language, study region, or type restrictions. A systematic search of articles published up to July 2021 was done. Of the 428 studies found through initial searches, 36 met the inclusion criteria. After applying the exclusion criteria, the main properties of 32 articles on 643 animals and 4 experimental studies on 52 patients (age range from 43 to 74 years) included in this meta-analysis. Our pooled analysis showed that stem cells–based scaffolds significantly improved the bone regeneration and formation in maxillofacial bone disorders (Prevalence: 0.54; 95% CI: 0.43, 0.64, $P < 00001$, $I^2 = 90\ 2$). According to the results of these studies, in most studies, bone marrow–derived mesenchymal stem cells (BMSCs) have been used to regenerate bone, and these cells are still the gold standard in bone tissue engineering, a growth factor that is one of the three sides of the tissue engineering triangle. Bone morphogenetic proteins (BMP) especially BMP2 and platelet-rich plasma (PRP) are the most widely used growth factor and scaffold respectively. Platelet-rich plasma (PRP) is used as a scaffold and since it contains proteins, it also used as a growth factor and can be a stimulant of ossification. It seems that the future perspective of bone tissue engineering is to use the prototyping rapid method to build a composite and patient-specific scaffold from CT and MRI images, along with genetically modified stem cells.

Keywords

Maxillofacial bone disorders · Meta-analysis · Scaffolds · Stem cells-based

E. Eini
MSD, Department of Orthodontics, School of Dentistry, Ahvaz Jundishapur University of Medical Sciences, Ahvaz, Iran

A. Ghaemi
MSD, Private Practitioner, Ahvaz, Iran

F. Rahim (✉)
Metabolomics and Genomics Research Center, Endocrinology and Metabolism Molecular–Cellular Sciences Institute, Tehran University of Medical Sciences, Tehran, Iran

Health Research Institute, Thalassemia and Hemoglobinopathies Research Centre, Ahvaz Jundishapur University of Medical Sciences, Ahvaz, Iran
e-mail: Bioinfo2003@gmail.com

Abbreviations

ADSCs	Adipose-derived mesenchymal stem cells
AKP	Alkaline phosphatase
BMSCs	Bone marrow–derived mesenchymal stem cells
CAS	Carbonate apatite scaffold
CDHA	Ca-deficient hydroxyapatite
CS	Collagen sponge
EMD	Enamel matrix derivatives
FHA	Fluoro-hydroxyapatite
GS	Gelatin sponge
HA/TCP	Hydroxyapatite/tricalcium phosphate
hDPSC	Human dental pulp stem cell
HIF-1α	Hypoxia-inducible factor-1α
MOOSE	Meta-analyses Of Observational Studies in Epidemiology
n-HA/PA	Nano-hydroxyapatite/polyamide
NON	Nonorganic natural
OMF	Oro-maxillofacial
ON	Organic natural
PC	Porous *Ceramic*
PGDF	Platelet-derived growth factor
PHAs	*Polyhydroxyalkanoates*
PLLA	Poly L-lactic acid
PP	Porous polymer
PRISMA	Preferred Reporting Items for Systematic reviews and Meta-Analyses
PRP	Platelet-rich plasma
SYRCLE	Systematic Review Center for Laboratory animal Experimentation
VEGFA	Vascular endothelial growth factor A

1 Background

1.1 Overview

Due to the different and multiple functions of bone tissue in the body, any change in its structure following an injury, lesion or disease affects the balance of the body and a person's life. Following a lesion, bone has an innate ability to heal on its own. In fractures and small bone injuries, the healing process is performed by the body so that following a bone injury, stem cells and stromal cells in the tissue, along with macrophages and other cells in the bone tissue, including osteoblasts and osteoclasts, which work in unison to promote a cascade of restorative functions to repair bone damage (Raggatt et al., 2014). However, these capabilities have limitations so that in large fractures and defects such as accidents, congenital cleft palate, misplaced bone fusion, periodontal disorders, and bone loss due to trauma, the body healing process could be not effective (Csaki et al., 2008). The risks of this graft are infections and immune reactions that may lead to graft rejection. Allografts are transplants that are usually taken from humans with different genotypes after their death. Therefore, in order to use this graft, or to avoid immune reactions and infection transmission, the target tissue must be completely sterilized, which also destroys many tissue components during sterilization. Limitations of this type of transplant include donor deficiency and subsequent infections (Verrier et al., 2016). It also carries the risk of transmitting diseases such as AIDS and hepatitis from the donor to the recipient (Oryan et al., 2014).

1.2 Introduction

Due to economic, cultural, environmental, and social factors, the prevalence of maxillofacial bone disorders varies in different parts of the world (Lee, 2012). These disorders may be isolated or associated with other injuries including the head, spine, chest, abdomen, pelvis, and limbs. In addition, such injuries are so severe that they are often the result of undesirable complex treatments (Adeyemo et al., 2005). Maxillofacial bone injuries results *from* trauma, periodontal disease, and *congenital abnormalities*, and *cause* alveolar *bone* loss (Titsinides et al., 2019). Maxillofacial bone injuries do not heal *spontaneously* and require a replacement bone graft (Wang & Yeung, 2017). Autogenous bone graft is still considered as the gold standard for repairing these bone lesions (Misch, 2010). But disadvantages such as painful surgery, risk of infection, bleeding, nerve damage and loss of function have prompted researchers to look for ways to replace it (de Grado et al., 2018).

Tissue engineering is a multidisciplinary field of principles and applications of engineering and life sciences methods, in order to fundamentally understand the relationship between structure and function in natural and diseased tissues (Almouemen et al., 2019). This discipline is a combination of cells, engineering, materials, and suitable physical and chemical factors that aim to maintain the stable state of the tissue, improve the function of the target tissue, or replace the biological function of the tissue.

Stem cells have high application and potential for cell therapy and tissue engineering due to their self-healing and self-renewal properties and their ability to differentiate into different cells, including osteoblasts under the influence of host tissue or culture medium (Han et al., 2019). For these reasons, the stem cells are being used in the field of reproductive medicine and tissue engineering (Mahla, 2016). Bone tissue engineering necessarily requires three components, including bone progenitor cells, bone growth factor, and scaffolding for adhesion and cell function (Amini et al., 2012).

Scaffolds are needed to transfer cells to the lesion site, which plays a key role against progenitor cells in tissue engineering (Chan & Leong, 2008). Scaffolds should initially provide a suitable extracellular matrix for cell growth and differentiation, and should also be able to restore proper tissue function (O'Brien, 2011).

The present meta-analysis was conducted to assess the efficacy and safety of different type of stem cells–based scaffolds and their construction methods in maxillofacial bone disorders.

1.3 Organic/Nonorganic Natural Scaffold

Organic natural scaffold are polymers made from materials found in nature such as plants, insects, or animals. Their advantages include biocompatibility, mechanical properties similar to natural tissue, and low inflammatory reactions, while disadvantages of these polymers include low mechanical strength (Song et al., 2018). These scaffold also support cell binding, proliferation, and differentiation (Chen & Liu, 2016). Organic natural scaffolds are classified into three groups, which are as follows:

1. Polymers that are derived from proteins, such as collagen, gelatin, and silk.
2. Polymers of polysaccharide origin, obtained from various sources such as plant, animal, and microbial. These polymers are nontoxic, react with living cells, have good biocompatibility, and are usually less expensive than other biological materials. These include chitosan, alginate, and hyaluronic.
3. The third group of polymers is comprised of *polyhydroxyalkanoates* (PHAs), which are synthesized by bacteria. These polymers have attracted a lot of attention because they can be produced from a variety of renewable sources. These polymers have high biodegradability, biocompatibility, and elasticity, and generally produce a good cell adhesion (Sharma et al., 2014).

1.4 Synthetic Scaffolds

These scaffolds are polymers made under controlled conditions; therefore, their degradability and mechanical properties can be controlled. These materials have low biological properties and flexibility (Dhandayuthapani et al., 2011).

1.5 Bio-ceramic Scaffolds

Metals and alloys are widely used in dentistry and orthopedics to repair broken bones as implants. Titanium alloys, cobalt base alloys, and L316 stainless steel are the most common alloys used in implants. Metal implants are like a double-edged blade. On the one hand, they help regenerate tissue, and after being placed in the body environment, they are surrounded by fibrous tissue, which leads to undesirable properties such as reduced implant mechanical properties and reaction. Immune ions are woven and implanted in the joint. Also, the use of metal implants to repair bone damage leads to the release of harmful ions, toxic ions, and their accumulation in various organs of the body, which increases the potential risk of cancer (Vasconcelos et al., 2016). For example,

L316 stainless steel, when planted in a physiological environment with high concentrations of chlorine, corrodes the environment and releases toxic elements such as nickel, chromium, and cobalt into the body, which are highly carcinogenic (Brown et al., 2015; Raza et al., 2016). To this end, researchers have stepped up their efforts to obtain new and biodegradable materials to prevent these problems. Bio-ceramics are a class of ceramic materials that are used to repair and strengthen parts of defective or damaged tissue and organs in the human body (Salgado et al., 2004). Bio-ceramics are divided into two groups based on their origin, natural (including hydroxyapatite, coral) and synthetic (including synthetic hydroxyapatite, calcium triphosphate and bioactive glass) (LeGeros et al., 2008).

2 Methods

This systematic review and meta-analysis was conducted according to the Meta-analyses Of Observational Studies in Epidemiology (MOOSE) (Stroup et al., 2000) and Preferred Reporting Items for Systematic reviews and Meta-Analyses (PRISMA) (Liberati et al., 2009) guidelines.

2.1 Search Strategy

We searched major indexing databases, including PubMed/Medline, ISI Web of Science (WOS), Scopus, Embase, and Cochrane Central, using ((((((bone formation[Title/Abstract]) OR (bone defect[Title/Abstract])) OR ((((Bone Regenerations[Title/Abstract]) OR (Regeneration, Bone [Title/Abstract])) OR (Regenerations, Bone [Title/Abstract])) OR (Osteoconduction[Title/Abstract]))) OR (bone regeneration[MeSH Terms])) OR (bone tissue engineering[Title/Abstract])) OR (((((((Broken Bones[Title/Abstract]) OR (Bone, Broken[Title/Abstract])) OR (Broken Bone[Title/Abstract])) OR (Bone Fractures[Title/Abstract])) OR (Bone Fracture [Title/Abstract])) OR (Fracture, Bone[Title/Abstract])))) OR ("Fractures, Bone"[Mesh]), without any language, study region, or type restrictions. A systematic search of articles published up to July 2021 was done.

2.2 Inclusion Criteria

Studies considering animal models or humans reporting bore formation after using simultaneous any type of stem, cells such as bone marrow–derived mesenchymal stem cells (BMSCs) or adipose-derived mesenchymal stem cells (ADSCs), and so on with scaffold in maxillofacial bone disorders, were included.

All types of animal studies were selected in this review irrespective of type, sex, and age.

2.3 Study Selections

After removing duplicated studies, two authors (EE and FR) independently screened titles and abstracts of potential papers considering predefined inclusion and exclusion criteria. Any disagreements were resolved by either reevaluation of the source article or consulting third author (ME).

2.4 Data Extraction

Information, including author's name, publication year, country, age, sample size, study design, and final findings are extracted from the included studies.

2.5 Methodological Quality Assessment

Two authors (FR and EE) independently conducted the methodological quality assessment with special consideration to potential sources of risk of bias. The methodological quality of the included studies was assessed based on the 10-item based Systematic Review Center for Laboratory animal Experimentation (SYRCLE) Animal Experiment Bias Risk Assessment Tool (Hooijmans et al., 2014). Any disagreement was resolved either by double-checking the reference paper or discussion with a third author.

2.6 Ethical Consideration

Ethical committee approval and informed consent were not essential due to working on previously published studies.

2.7 Statistical Analysis

We retrieved rate ratio (RR) with 95% confidence interval (CI) from the eligible studies calculated with the random-effect or fixed-effect models depending on the level of heterogeneity, to evaluate the association of bone formation rate with the stem cells–based scaffolds in maxillofacial bone disorders (Lathyris & Haidich, 2021). Afterward, we measured heterogeneity across studies using Cochran's Q statistics and I^2 test. When I^2 values (more than 50%) showed a high heterogeneity sensitivity and subgroup analyses were performed to discover the source of the heterogeneity.

3 Results

3.1 Study Characteristics

Of the 428 studies found through initial searches, 36 met the inclusion criteria (Fig. 1). After applying the exclusion criteria, the main properties of 32 articles on 643 animals and 4 experimental studies on 52 patients (age range from 43 to 74 years) are included in this meta-analysis (Table 1).

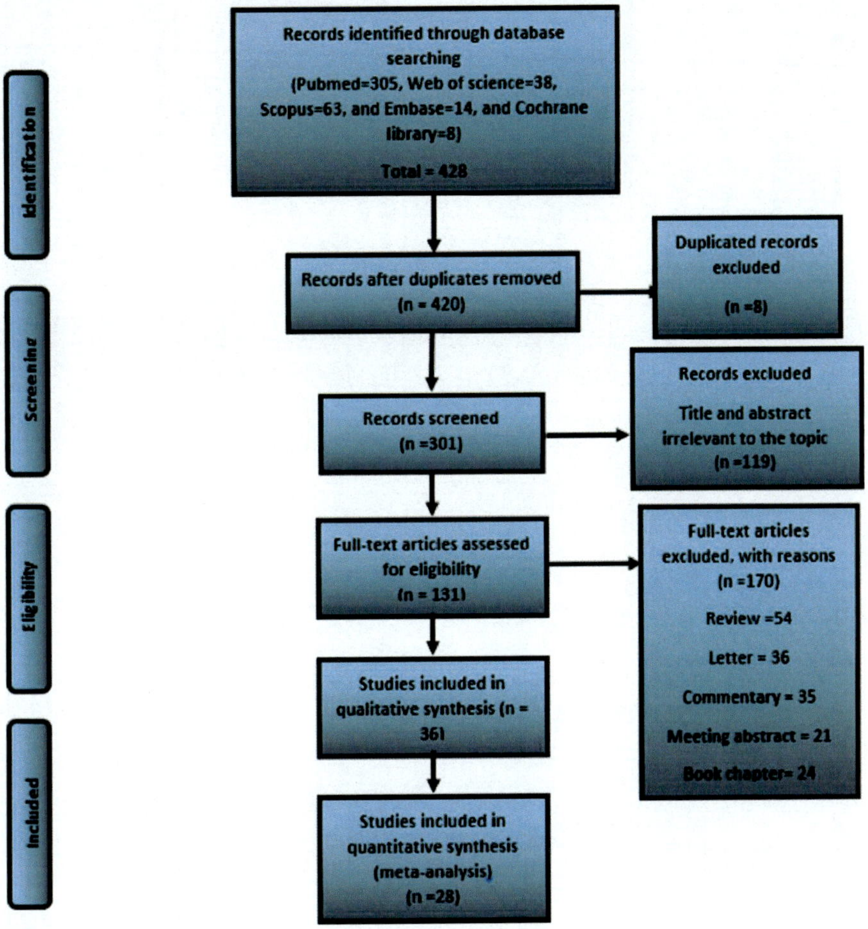

Fig. 1 Flowchart of study selection process

Table 1 Basic characteristics of the included studies

Author, year/ country	Cell type	Scaffold	Scaffold type	Growth factor	Animals/human	Location of the lesion	Duration	Outcome (bone formation)
Preclinical studies								
Lee et al. (2021)/ Korea	ADSCs	βTCP	PC	–	10, 36-months-old healthy beagles dogs	Mandible	8 weeks	Improved ossification
Zhang et al. (2020)/USA	BMSCs	βTCP	PC	VEGFA	17 Sprague Dawley male rats (280–330 g)	Mandible	8 weeks	17.05%
Prahasanti et al. (2020)/Indonesia	BMSCs	CAS	ON	BMP2 BMP7	14 healthy, male, 12- to 16-week-old Wistar rats (200–250 g)	Mandible	8 weeks	Enhance gene expression during bone remodeling
Lopez et al. (2018)/USA	BMSCs	βTCP	PC	–	5 skeletally mature New Zealand white rabbits (3.5 kg)	Mandible	8 weeks	54.3
Moser et al. 2017/ Germany	BMSCs	βTCP	Mixed	BMP2	24 adult male Wistar rats	Mandible	26 weeks	10.4%
Tee et al. 2016/ USA	BMSCs	βTCP	PC	VEGFA BMP2	12, 4-month-old domestic pigs	Mandible	12 weeks	Promising for the reconstruction of defects
Lee et al. (2015)/ Korea	ADSCs	PLGA	PC	BMP2	28, Lewis rat pups	Mandible	12 weeks	50%
Yun et al. (2014)/ Korea	BMSCs	βTCP	PC	BMP2	6, 5–6-months-old beagle dogs (8.0–10.0 kg)	Mandible	8 weeks	16.01%
Alfotawei et al. (2014)/UK	BMSCs	βTCP	PC	BMP2	8 male New Zealand white rabbits (3 to 4 kg)	Mandible	12 weeks	51.6%
Kohgo et al. (2011)/Japan	BMSCs	PRP	ON	PGDF	7 hybrid dogs (mean age: 2 year)	Mandible	8 weeks	55.63%
Duan et al. (2011)/USA	hDPSC	Silk	NON	EMD	24, 8-week-old male nude mice (25–30 g)	Mandible	2 weeks	58.53%
Zou et al. (2011)/China	BMSCs	βTCP	PC	BMP2	5, 6-week-old male Fisher 344 rats (50 ± 5 g)	Mandible	8 weeks	31.83%
Tsumanuma et al. (2011)/Japan	BMSCs	βTCP, CS	Mixed	–	4 healthy male beagle dogs (10 kg)	Periodontal	8 weeks	70.1%
Li et al. (2010)/China	BMSCs	HA/PA	Mixed	BMP7	54 healthy New Zealand white rabbits (about 3.0 kg)	Mandible	16 weeks	85.54%
Yuan et al. (2010)/China	BMSCs	Coral	NON	BMP2	24 healthy adult mongrel dogs, aged 16 months old (19.7 kg)	Mandible	32 weeks	30%
Mankani et al. (2006)/USA	BMSCs	HA/TCP	ON	BMP2	Three-month-old immunocompromised Bg-Nu-Xid female mice	Mandible	8 weeks	78%

Kim et al. (2009)/ Korea	BMSCs	βTCP	PC	–	4, male beagle dogs (12–15 kg)	Mandible	8 weeks	40.17%
Zheng et al. (2009)/China	hDPSC	βTCP	PC	–	16 inbred female minipigs (4–6 months, 20–30 kg)	Mandible	24 weeks	83.1%
Pieri et al. (2009)/ Italy	BMSCs	FHA	PC	–	8 adult minipigs (18 months, 45 ± 4.5 kg)	Mandible	12 weeks	45.28%
Guo et al. (2009)/China	BMSCs	CDHA	PC	–	12 healthy New Zealand white rabbits (3.0 kg)	Mandible	12 weeks	81%
Jafarian et al. (2008)/Iran	BMSCs	βTCP	PC	–	4 adult mongrel dogs (20–30 kg)	Mandible	6 weeks	65.78%
Wang et al. (2007)/China	BMSCs	HA/PA	Mixed	BMP2	76 New Zealand white male rabbits (3.0 kg)	Mandible	12 weeks	80%
Mylonas et al. (2007)/USA	BMSCs	βTCP	PC	–	Six male and two female 1-year-old beagle dogs	Mandible	7 weeks	58.25%
Ren et al. 2007/China	BMSCs	PLLA, PLA-PEG and PLGA	PP	–	12 mature white New Zealand rabbits (2.5 kg)	Mandible	12 weeks	81.9%
Yuan et al. (2007)/China	BMSCs	βTCP	PC	AKP	16, 18-month adult mongrel healthy dogs (20.9 kg)	Mandible	32 weeks	well repair the defects
He et al. (2007)/China	BMSCs	βTCP	PC	–	6, 1–2-year-old Beagle dogs (10 to 16 kg)	Mandible	12 weeks	Mean height of the reconstructed bone of 18.54 mm
Wu et al. (2006)/China	BMSCs	βTCP	PC	–	6, 10-month-old male beagle canines (11–13 kg)	Mandible	12 weeks	Possess the ability to form new bone tissue
Seto et al. (2006)/ Japan	BMSCs	βTCP	Mixed	BMP2	5 healthy adult Japanese monkeys (7–11 kg)	Mandible	24 weeks	A reliable method for the reconstruction of segmental bone defects
Ito et al. (2005)/ Japan	BMSCs	PRP	ON	BMP2	5 hybrid dogs (mean age: 2 years)	Mandible	12 weeks	17%
Marei et al. (2005)/Egypt	BMSCs	PLA-PEG	PP	–	36 New Zealand White male rabbits (6–8 weeks; 1–1.2 kg)	Mandible	6 weeks	74.94%
Al Salihi (2004)/ Malaysia	BMSCs	Coral	NON	BMP2	5 weeks old, Sprague Dawley 4 male rats	Mandible	12 weeks	57%
Yamada et al. (2004)/Japan	BMSCs	PRP	ON	–	4 adult hybrid dogs (mean age: 2 years)	Mandible	8 weeks	61.4%

(continued)

Table 1 (continued)

Author, year/ country	Cell type	Scaffold	Scaffold type	Growth factor	Animals/human	Location of the lesion	Duration	Outcome (bone formation)
Clinical studies								
Yamada et al. (2013)/Japan	BMSCs	PRP	ON	PGDF	23 patients (mean age: 60.4 years, and a range of 43–74 years)	Maxilla	24 weeks	
d'Aquino et al. (2009)/Italy	hDPSC	CS	ON	–	17 patients	OMF	48 weeks	70%
Shayesteh et al. (2008)/Iran	BMSCs	βTCP	PC	–	7 patients (4 males and 3 females)	Maxilla	12 weeks	41.34%
Smiler et al. (2007)/USA	BMSCs	Alloplastic	ON	BMP2	5 patients	Maxilla	28 weeks	45%

ADSCs Adipose-derived stem cells, *BMSCs* Bone mesenchymal stem cells, *GS* gelatin sponge, *HIF-1α* hypoxia-inducible factor-1α, *PRP* platelet-rich plasma, *PGDF* platelet-derived growth factor, *ON* Organic natural, *NON* Nonorganic natural, *OMF* oro-maxillofacial, *CS* collagen sponge, *hDPSC* human dental pulp stem cell, *EMD* enamel matrix derivatives, *PP* Porous polymer, *PLLA* poly (L-lactic acid), *PC* Porous Ceramic, *FHA* fluorohydroxyapatite, *CDHA* Ca-deficient hydroxyapatite, *n-HA/PA* nano-hydroxyapatite/polyamide, *VEGFA* vascular endothelial growth factor A, *CAS* carbonate apatite scaffold, *AKP* alkaline phosphatase, *HA/TCP* hydroxyapatite/tricalcium phosphate

3.2 Meta-analysis

Our pooled analysis showed that stem cells–based scaffolds significantly improved the bone regeneration and formation in maxillofacial bone disorders (Prevalence: 0.54; 95% CI: 0.43, 0.64, $P < 00001$, $I^2 = 90$, Fig. 2). Most of the used stem cell types were BMSCs with a frequency of 85.7% (24/28) followed by ADSCs (Table 2). Both types of stem cells were associated with significant bone regeneration and formation (Fig. 3).

3.3 Methodological Quality

All 36 studies described the same baselines of each group. All of the studies described animals that were randomly selected in the evaluation result. No incomplete data was reported, and the included studies did selectively report results. However, the following aspects were not clear: sufficient allocation concealment, randomly placing the animals during the experiment, blinded researchers, the result evaluator adopting the blind method, and no other bias (Fig. 4).

3.4 Publication Bias

The funnel plots give us a visual observation of the publication bias of two comparisons, including SMD of immunologic factors (Fig. 5a) and SMD of clinical scores (Fig. 5b). Obviously, two symmetrical figures showed that the publication bias isn't large.

4 Discussion

In this meta-analysis, scaffolding used in the reconstruction of jaw and facial bone lesions, it was found that there is still a long way to go to achieve the ideal treatment, although scaffolding

Study or Subgroup	Prevalence	SE	Weight	Prevalence IV, Random, 95% CI
Alfotawei 2014	0.52	0.1	3.7%	0.52 [0.32, 0.72]
Al-Salihi 2004	0.57	0.13	3.4%	0.57 [0.32, 0.82]
d'Aquino 2009	0.7	0.11	3.6%	0.70 [0.48, 0.92]
Duan 2011	0.59	0.1	3.7%	0.59 [0.39, 0.79]
Guo 2009	0.81	0.11	3.6%	0.81 [0.59, 1.03]
Ito 2005	0.17	0.08	3.9%	0.17 [0.01, 0.33]
Jafarian 2008	0.66	0.12	3.5%	0.66 [0.42, 0.90]
Kim 2009	0.4	0.12	3.5%	0.40 [0.16, 0.64]
Kohgo 2011	0.56	0.12	3.5%	0.56 [0.32, 0.80]
Lee 2015	0.5	0.09	3.8%	0.50 [0.32, 0.68]
Li 2010	0.86	0.05	4.1%	0.86 [0.76, 0.96]
Lopez 2018	0.54	0.11	3.6%	0.54 [0.32, 0.76]
Mankani 2006	0.75	0.07	4.0%	0.75 [0.61, 0.89]
Marei 2005	0.75	0.07	4.0%	0.75 [0.61, 0.89]
Moser 2017	0.1	0.06	4.0%	0.10 [-0.02, 0.22]
Mylonas 2007	0.58	0.1	3.7%	0.58 [0.38, 0.78]
Pieri 2009	0.45	0.11	3.6%	0.45 [0.23, 0.67]
Ren 2007	0.82	0.1	3.7%	0.82 [0.62, 1.02]
Shayesteh 2008	0.41	0.13	3.4%	0.41 [0.16, 0.66]
Smiler 2007	0.45	0.13	3.4%	0.45 [0.20, 0.70]
Tsumanuma 2011	0.7	0.11	3.6%	0.70 [0.48, 0.92]
Wang 2007	0.8	0.05	4.1%	0.80 [0.70, 0.90]
Yuan 2010	0.3	0.09	3.8%	0.30 [0.12, 0.48]
Yun 2014	0.16	0.07	4.0%	0.16 [0.02, 0.30]
Zhang 2020	0.17	0.09	3.8%	0.17 [-0.01, 0.35]
Zheng 2009	0.83	0.09	3.8%	0.83 [0.65, 1.01]
Zou 2011	0.32	0.1	3.7%	0.32 [0.12, 0.52]
Total (95% CI)			**100.0%**	**0.54 [0.43, 0.64]**

Heterogeneity: $Tau^2 = 0.07$; $Chi^2 = 252.26$, $df = 26$ ($P < 0.00001$); $I^2 = 90\%$
Test for overall effect: $Z = 10.14$ ($P < 0.00001$)

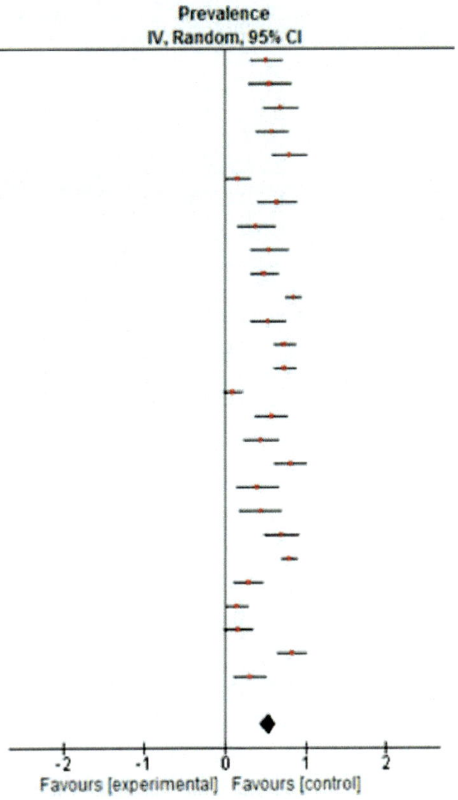

Fig. 2 Forrest plot of rate ratio of bone formation due to using stem cells–based scaffolds in maxillofacial bone disorders

Table 2 Analysis of subgroups according to the type of scaffold, stem cells, study targets, and growth factors

Subgroups	Number	Frequency (%)
Type of scaffolds		
βTCP	21	58.3
PRP	4	11.1
Coral	2	5.6
HA/PA	2	5.6
CAS	1	2.8
Silk	1	2.8
CDHA	1	2.8
Alloplastic	1	2.8
FHA	1	2.8
Mixed	2	5.6
Stem cell types		
BMSCs	32	85.7
ADSCs	4	14.3
Study target	13	
Animals	7	36.1
Dogs	6	19.4
Rabbit	3	16.7
Rat	2	8.3
Pig	1	5.6
Mice	4	2.8
Monkey		11.1
Human		
Growth factors		
BMP2	13	36.1
BMP7	2	5.6
VEGFA	2	5.6
PGDF	1	2.8
EMD	1	2.8
AKP	1	2.8
PGDF	1	2.8
Missing	15	41.7
Total	36	100

ADSCs Adipose-derived stem cells, *BMSCs* Bone mesenchymal stem cells, *GS* gelatin sponge, *HIF-1α* hypoxia-inducible factor-1α, *PRP* platelet-rich plasma, *PGDF* platelet-derived growth factor, *ON* Organic natural, *NON* Nonorganic natural, *OMF* oro-maxillofacial, *CS* collagen sponge, *hDPSC* human dental pulp stem cell, *EMD* enamel matrix derivatives, *PP* Porous polymer, *PLLA* poly (L-lactic acid), *PC* Porous Ceramic, *FHA* fluorohydroxyapatite, *CDHA* Ca-deficient hydroxyapatite, *n-HA/PA* nano-hydroxyapatite/polyamide, *VEGFA* vascular endothelial growth factor A, *CAS* carbonate apatite scaffold, *AKP* alkaline phosphatase, *HA/TCP* hydroxyapatite/tricalcium phosphate

is a key factor in the success of tissue engineering. According to the results of these studies, in most studies, BMSCs have been used to regenerate bone, and these cells are still the gold standard in bone tissue engineering, a growth factor that is one of the three sides of the tissue engineering triangle. BMP, especially BMP2, is the most widely used, or inherent growth factor of scaffolds, and PRP, which is also used as a scaffold, contains proteins that can It is a stimulant of ossification and is used as a growth factor. Of natural organic scaffolds, PRP and sponge collagen are used more than other natural organic scaffolds and collagen seems to have more satisfactory results.

Our analysis showed that stem cells–based scaffolds significantly improved the bone regeneration. In line with our findings Iaquinta et al.

Study or Subgroup	Prevalence	SE	Weight	Prevalence IV, Random, 95% CI
1.2.1 BMSCs				
Alfotawei 2014	0.52	0.1	3.6%	0.52 [0.32, 0.72]
Al-Salihi 2004	0.57	0.13	3.3%	0.57 [0.32, 0.82]
Guo 2009	0.81	0.11	3.5%	0.81 [0.59, 1.03]
Ito 2005	0.17	0.08	3.7%	0.17 [0.01, 0.33]
Jafarian 2008	0.66	0.12	3.4%	0.66 [0.42, 0.90]
Kim 2009	0.4	0.12	3.4%	0.40 [0.16, 0.64]
Kohgo 2011	0.56	0.12	3.4%	0.56 [0.32, 0.80]
Lee 2021	0.54	0.11	3.5%	0.54 [0.32, 0.76]
Li 2010	0.86	0.05	4.0%	0.86 [0.76, 0.96]
Lopez 2018	0.54	0.11	3.5%	0.54 [0.32, 0.76]
Mankani 2006	0.75	0.07	3.8%	0.75 [0.61, 0.89]
Marei 2005	0.75	0.07	3.8%	0.75 [0.61, 0.89]
Moser 2017	0.1	0.06	3.9%	0.10 [-0.02, 0.22]
Mylonas 2007	0.58	0.1	3.6%	0.58 [0.38, 0.78]
Pieri 2009	0.45	0.11	3.5%	0.45 [0.23, 0.67]
Ren 2007	0.82	0.1	3.6%	0.82 [0.62, 1.02]
Shayesteh 2008	0.41	0.13	3.3%	0.41 [0.16, 0.66]
Smiler 2007	0.45	0.13	3.3%	0.45 [0.20, 0.70]
Tsumanuma 2011	0.7	0.11	3.5%	0.70 [0.48, 0.92]
Wang 2007	0.8	0.05	4.0%	0.80 [0.70, 0.90]
Yuan 2010	0.3	0.09	3.7%	0.30 [0.12, 0.48]
Yun 2014	0.16	0.07	3.8%	0.16 [0.02, 0.30]
Zhang 2020	0.17	0.09	3.7%	0.17 [-0.01, 0.35]
Zou 2011	0.32	0.1	3.6%	0.32 [0.12, 0.52]
Subtotal (95% CI)			**85.7%**	**0.52 [0.40, 0.63]**
Heterogeneity: Tau² = 0.07; Chi² = 240.23, df = 23 (P < 0.00001); I² = 90%				
Test for overall effect: Z = 8.93 (P < 0.00001)				
1.2.2 ADSCs				
d'Aquino 2009	0.7	0.11	3.5%	0.70 [0.48, 0.92]
Duan 2011	0.59	0.1	3.6%	0.59 [0.39, 0.79]
Lee 2015	0.5	0.09	3.7%	0.50 [0.32, 0.68]
Zheng 2009	0.83	0.09	3.7%	0.83 [0.65, 1.01]
Subtotal (95% CI)			**14.3%**	**0.65 [0.51, 0.80]**
Heterogeneity: Tau² = 0.01; Chi² = 7.34, df = 3 (P = 0.06); I² = 59%				
Test for overall effect: Z = 8.62 (P < 0.00001)				
Total (95% CI)			**100.0%**	**0.54 [0.44, 0.64]**
Heterogeneity: Tau² = 0.06; Chi² = 252.28, df = 27 (P < 0.00001); I² = 89%				
Test for overall effect: Z = 10.46 (P < 0.00001)				
Test for subgroup differences: Chi² = 2.10, df = 1 (P = 0.15), I² = 52.4%				

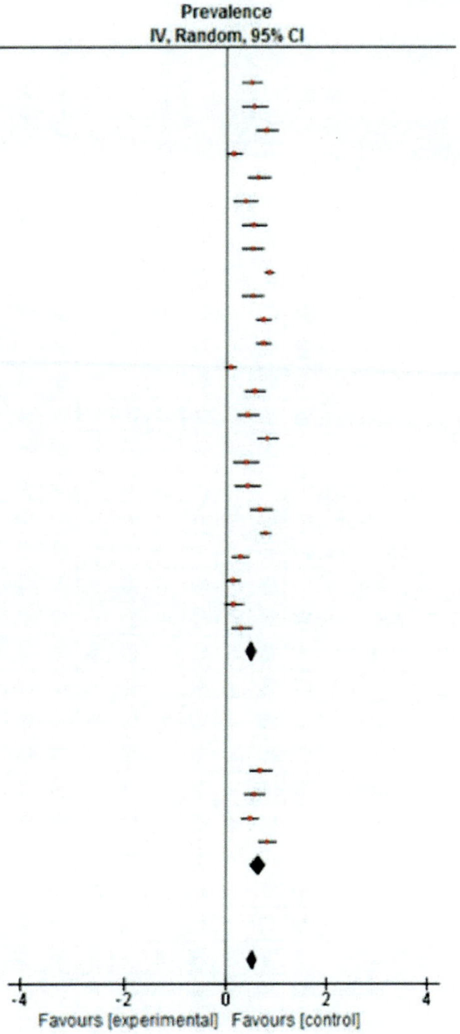

Fig. 3 Forrest plot of rate ratio of bone formation among different stem cell subgroups using stem cells–based scaffolds in maxillofacial bone disorders

described bone structure and the osteogenic differentiation of stem cells in the context of bone repair and regeneration, and described the role of MSCs for bone regeneration in the tissue engineering field and their recent advances in clinical implications (Iaquinta et al., 2019). Recently, Xia et al. addressed the recent advances in using stem cells in applications of bone scaffolds, and discussed stem cell recruitment based on scaffold features as promising techniques for the treatment of bone defects (Xia et al., 2021). Moreover, Dong et al. showed that MSC-based tissue-engineered scaffolds significantly enhanced osteogenesis in animal model of bone defect compared with MSC alone (Dong et al., 2020).

One of the main components in tissue engineering is cell. The two main sources of cells are the target tissue cells and the stem cells. Limited availability and low proliferation of cells in a tissue restricts their use for tissue engineering. Due to the characteristics of stem cells, including self-renewal and differentiation, these cells are one of the main sources used in the field of tissue engineering (Goetzke et al., 2018). Among stem cells, mesenchymal stem cells are the focus of attention due to their unique properties. MSCs,

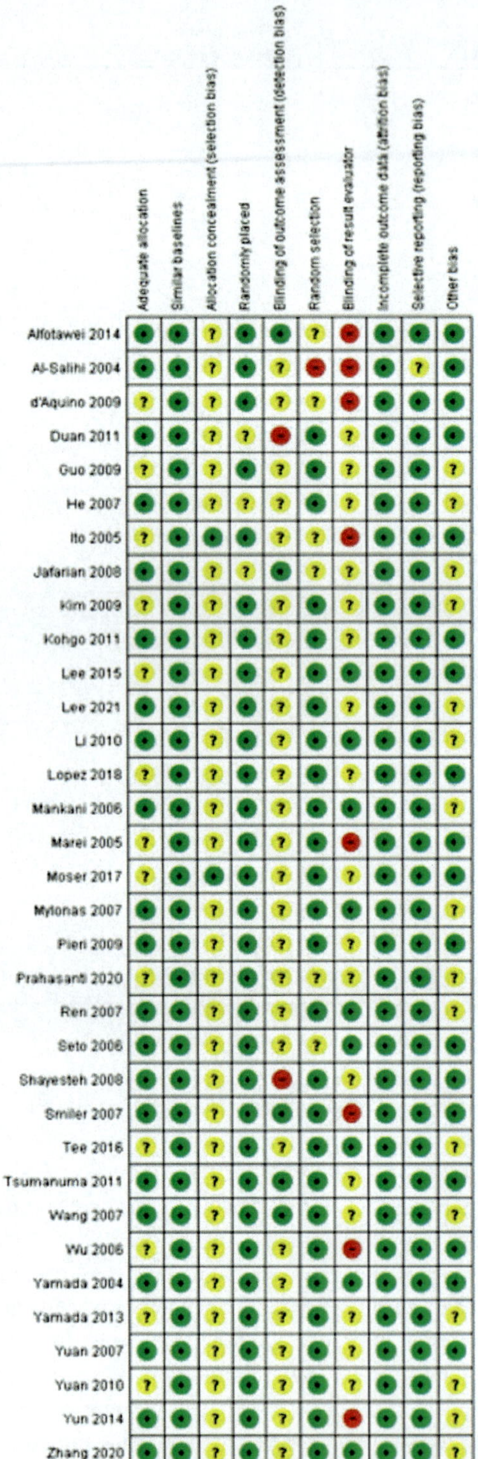

Fig. 4 The risk of bias assessment graph of the included studies

with their ability to differentiate into several mesoderm classes, are obtained from different tissues such as bone marrow, adipose tissue, or synovium (da Silva Meirelles et al., 2006). Findings from the present meta-analysis showed that most of the used stem cell types used for bone regeneration were BMSCs, which have been shown to retain their ability to differentiate and proliferate over a long period of time (Gugjoo et al., 2019).

Coral and silver are the most widely used natural organic scaffolds, and the results obtained from the use of corals have been better than silver (Neto & Ferreira, 2018). The result of this meta-analysis pointed out that among the polymer industrial scaffolding, poly(lactic-co-glycolic acid) (PLGA) has the most application and good results have been obtained from its application. Our findings were in line with Zhao et al. study, which showed that PLGA impose suitable biocompatibility, degradability, mechanical properties, and competences to improve bone regeneration (Zhao et al., 2021). Moreover, tricalcium phosphates/hydroxyapatites (TCP/HA) also seem to be more successful than other industrial ceramic scaffolds (Shuai et al., 2013). Also, in line with former studies, the present meta-analysis highlights this fact that Better results can be obtained from combined scaffolding (PCL/hydroxyapatite) compared with the pure PCL scaffold (Donnaloja et al., 2020). The use of composite scaffolds in the reconstruction of the jaw, face, and skull bones, although generally lower than other scaffolds, has increased in recent years. Scaffolding preparations from the extracellular matrix of tissues to some extent affect the surface morphology and mechanical properties of the scaffold. Various studies have shown that the efficiency and effectiveness of a particular detoxification method depends to a large extent on the characteristics of the tissue in question. Therefore, according to these results, it is difficult to balance the complete elimination of cells and maintain tissue biomechanics (Gilpin & Yang, 2017; Chen & Liu, 2016).

In line with previous studies (Huang et al., 2020; Chocholata et al., 2019, Prochor & Gryko, 2020), the best results have been obtained from coral scaffolds and secondarily from composite scaffolds with more than 80% ossification.

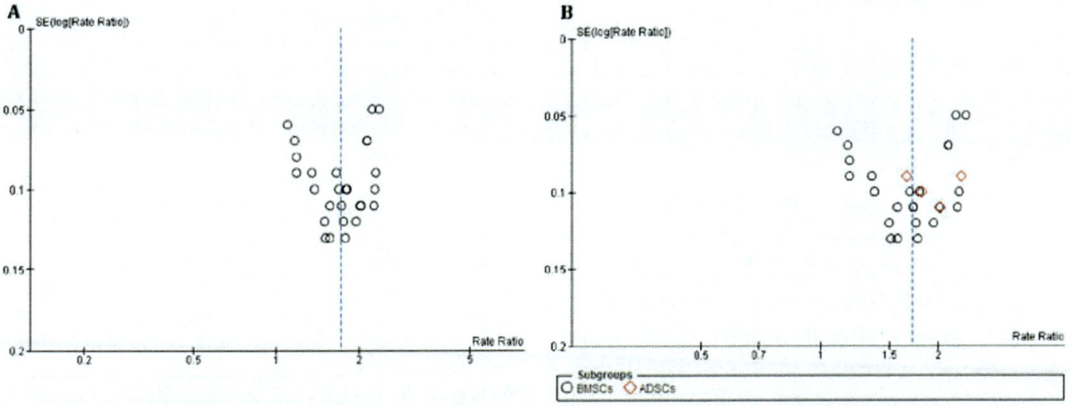

Fig. 5 Funnel plot of clinical scores across included studies to assess the source of heterogeneity among included studies, (**a**) overall; (**b**) subgroups

To achieve the desired results in bone tissue engineering, there are factors that studies should pay attention to. These factors include the following: (1) Clinical studies: So far, limited clinical studies have been performed on humans and the number of samples in each study has been very small. (2) Animal studies: Most studies have been done on the reconstruction of small lesions in small animals. To achieve results that can be generalized to humans, larger animals with similar human-like structures and immune systems must be examined. (3) Control groups: Proper selection of control groups can help to better interpret the results and show the effect of each variable well. (4) Time: Most studies have been done in a short time. Following the sample for longer periods of time can improve the results as well as show the possible side effects of the material. (5) Tissue engineering trio: In a small number of studies, all three components of tissue engineering have been placed next to each other. It is necessary to pay attention to what factors are needed to simulate natural restoration and that these factors should be released gradually into the environment. (6) Evaluation of results: There are different ways to present the results of a study, but it should be noted that the study of factors such as angiogenesis and immunological reactions, which are important factors in bone regeneration, must be done. Also, providing qualitative ossification results can help to compare similar studies.

5 Clinical Implication

The science of dentistry has made significant progress in recent years. In particular, the fields of implantology (dental implants) and periodontology (gingival surgery) have undergone many changes and have brought many innovations to the dental community. Some of these innovations have emerged in the field of microsurgery and microscopy with new and effective techniques. From the consequences of these innovations, it can be mentioned that we will be able to perform implant operations in the patient's mouth with multiple accuracy. In this process, in addition to using magnifying glasses and modern surgical microscopes, microsurgery tools subtle operations on the eyes, brain, and nerves are also used. In modern implantology, removing a tooth from its cavity plays a key role and is of great importance. Classical extraction methods are usually associated with damage to the bones and gums around the dental cavity, which can cause more severe resorption of the jaw in this area, disrupting the placement of the implant, and even preventing it from settling. Therefore, in these conditions, first the damage to the bone and gums should be reconstructed over time and then the implant should be placed. We are pleased to announce that we can use new tools and techniques to minimize the potential for these injuries and tissue damage.

6 Conclusion

Tissue engineering, on the other hand, although it can be theoretically applied in dentistry; However, its clinical application has not been widespread due to its cost-effectiveness and considerations for cell manipulation. But, on the other hand, its prospects attract dentists and researchers. After more than 20 years of tissue engineering, the ideal scaffold has not been designed. Most studies have been done on the reconstruction of small lesions and features such as angiogenesis and bone physiology have been less considered. It seems that the future perspective of bone tissue engineering is to use the prototyping rapid method to build a composite and patient-specific scaffold from CT and MRI images, along with genetically modified stem cells.

Competing Interests The authors declare that they have no competing interests.

References

Adeyemo, W. L., Ladeinde, A. L., Ogunlewe, M. O., & James, O. (2005). Trends and characteristics of oral and maxillofacial injuries in Nigeria: A review of the literature. *Head & Face Medicine, 1*(1), 7. https://doi.org/10.1186/1746-160X-1-7

Alfotawei, R., Naudi, K. B., Lappin, D., Barbenel, J., Di Silvio, L., Hunter, K., McMahon, J., & Ayoub, A. (2014). The use of TriCalcium Phosphate (TCP) and stem cells for the regeneration of osteoperiosteal critical-size mandibular bony defects, an in vitro and preclinical study. *Journal of Cranio-Maxillo-Facial Surgery, 42*(6), 863–869. https://doi.org/10.1016/j.jcms.2013.12.006

Almouemen, N., Kelly, H. M., & O'Leary, C. (2019). Tissue engineering: Understanding the role of biomaterials and biophysical forces on cell functionality through computational and structural biotechnology analytical methods. *Computational and Structural Biotechnology Journal, 17*, 591–598. https://doi.org/10.1016/j.csbj.2019.04.008

Al-Salihi, K. A. (2004). Tissue-engineered bone via seeding bone marrow stem cell derived osteoblasts into coral: a rat model. *The Medical Journal of Malaysia, 59*(Suppl B), 200–201.

Amini, A. R., Laurencin, C. T., & Nukavarapu, S. P. (2012). Bone tissue engineering: Recent advances and challenges. *Critical Reviews in Biomedical Engineering, 40*(5), 363–408. https://doi.org/10.1615/critrevbiomedeng.v40.i5.10

Brown, R. P., Fowler, B. A., Fustinoni, S., & Nordberg, M. (2015). Chapter 5 – Toxicity of metals released from implanted medical devices. In G. F. Nordberg, B. A. Fowler, & M. Nordberg (Eds.), *Handbook on the toxicology of metals* (4th ed., pp. 113–122). Academic Press.

Chan, B. P., & Leong, K. W. (2008). Scaffolding in tissue engineering: General approaches and tissue-specific considerations. *European Spine Journal, 17*(Suppl 4), 467–479. https://doi.org/10.1007/s00586-008-0745-3

Chen, F. M., & Liu, X. (2016). Advancing biomaterials of human origin for tissue engineering. *Progress in Polymer Science, 53*, 86–168. https://doi.org/10.1016/j.progpolymsci.2015.02.004

Chocholata, P., Kulda, V., & Babuska, V. (2019). Fabrication of scaffolds for bone-tissue regeneration. *Materials (Basel), 12*(4). https://doi.org/10.3390/ma12040568

Csaki, C., Schneider, P. R., & Shakibaei, M. (2008). Mesenchymal stem cells as a potential pool for cartilage tissue engineering. *Annals of Anatomy, 190*(5), 395–412. https://doi.org/10.1016/j.aanat.2008.07.007

da Silva Meirelles, L., Chagastelles, P. C., & Nardi, N. B. (2006). Mesenchymal stem cells reside in virtually all post-natal organs and tissues. *Journal of Cell Science, 119*(Pt 11), 2204–2213. https://doi.org/10.1242/jcs.02932

d'Aquino, R., De Rosa, A., Lanza, V., Tirino, V., Laino, L., Graziano, A., Desiderio, V., Laino, G., & Papaccio, G. (2009). Human mandible bone defect repair by the grafting of dental pulp stem/progenitor cells and collagen sponge biocomplexes. *European Cells & Materials, 18*, 75–83. https://doi.org/10.22203/ecm.v018a07

de Grado, F., Gabriel, L. K., Idoux-Gillet, Y., Wagner, Q., Musset, A.-M., Benkirane-Jessel, N., Bornert, F., & Offner, D. (2018). Bone substitutes: A review of their characteristics, clinical use, and perspectives for large bone defects management. *Journal of Tissue Engineering, 9*, 2041731418776819–2041731418776819. https://doi.org/10.1177/2041731418776819

Dhandayuthapani, B., Yoshida, Y., Maekawa, T., & Sakthi Kumar, D. (2011). Polymeric scaffolds in tissue engineering application: A review. *International Journal of Polymer Science, 2011*, 290602. https://doi.org/10.1155/2011/290602

Dong, R., Bai, Y., Dai, J., Deng, M., Zhao, C., Tian, Z., Zeng, F., Liang, W., Liu, L., & Dong, S. (2020). Engineered scaffolds based on mesenchymal stem cells/preosteoclasts extracellular matrix promote bone regeneration. *Journal of Tissue Engineering, 11*, 2041731420926918. https://doi.org/10.1177/2041731420926918

Donnaloja, F., Jacchetti, E., Soncini, M., & Raimondi, M. T. (2020). Natural and synthetic polymers for bone scaffolds optimization. *Polymers, 12*(4). https://doi.org/10.3390/polym12040905

Duan, X., Tu, Q., Zhang, J., Ye, J., Sommer, C., Mostoslavsky, G., Kaplan, D., Yang, P., & Chen, J. (2011). Application of induced pluripotent stem (iPS) cells in periodontal tissue regeneration. *Journal of Cellular Physiology, 226*(1), 150–157. https://doi.org/10.1002/jcp.22316

Gilpin, A., & Yang, Y. (2017). Decellularization strategies for regenerative medicine: From processing techniques to applications. *BioMed Research International, 2017*, 9831534–9831534. https://doi.org/10.1155/2017/9831534

Goetzke, R., Sechi, A., De Laporte, L., Neuss, S., & Wagner, W. (2018). Why the impact of mechanical stimuli on stem cells remains a challenge. *Cellular and Molecular Life Sciences, 75*(18), 3297–3312. https://doi.org/10.1007/s00018-018-2830-z

Gugjoo, M. B., Amarpal, Makhdoomi, D. M., & Sharma, G. T. (2019). Equine mesenchymal stem cells: properties, sources, characterization, and potential therapeutic applications. *Journal of Equine Veterinary Science, 72*, 16–27. https://doi.org/10.1016/j.jevs.2018.10.007

Guo, H., Su, J., Wei, J., Kong, H., & Liu, C. (2009). Biocompatibility and osteogenicity of degradable Ca-deficient hydroxyapatite scaffolds from calcium phosphate cement for bone tissue engineering. *Acta Biomaterialia, 5*(1), 268–278. https://doi.org/10.1016/j.actbio.2008.07.018

Han, Y., Li, X., Zhang, Y., Han, Y., Chang, F., & Ding, J. (2019). Mesenchymal stem cells for regenerative medicine. *Cell, 8*(8), 886. https://doi.org/10.3390/cells8080886

He, Y., Zhang, Z.-y., Zhu, H.-g., Qiu, W., Jiang, X., & Guo, W. (2007). Experimental study on reconstruction of segmental mandible defects using tissue engineered bone combined bone marrow stromal cells with three-dimensional tricalcium phosphate. *Journal of Craniofacial Surgery, 18*(4).

Hooijmans, C. R., Rovers, M. M., de Vries, R. B., Leenaars, M., Ritskes-Hoitinga, M., & Langendam, M. W. (2014). SYRCLE's risk of bias tool for animal studies. *BMC Medical Research Methodology, 14*, 43. https://doi.org/10.1186/1471-2288-14-43

Huang, Q., Liu, Y., Ouyang, Z., & Feng, Q. (2020). Comparing the regeneration potential between PLLA/Aragonite and PLLA/Vaterite pearl composite scaffolds in rabbit radius segmental bone defects. *Bioactive Materials, 5*(4), 980–989. https://doi.org/10.1016/j.bioactmat.2020.06.018

Iaquinta, M. R., Mazzoni, E., Bononi, I., Rotondo, J. C., Mazziotta, C., Montesi, M., Sprio, S., Tampieri, A., Tognon, M., & Martini, F. (2019). Adult stem cells for bone regeneration and repair. *Frontiers in Cell and Developmental Biology, 7*(268). https://doi.org/10.3389/fcell.2019.00268

Ito, K., Yamada, Y., Nagasaka, T., Baba, S., & Ueda, M. (2005). Osteogenic potential of injectable tissue-engineered bone: A comparison among autogenous bone, bone substitute (Bio-oss), platelet-rich plasma, and tissue-engineered bone with respect to their mechanical properties and histological findings. *Journal of Biomedical Materials Research. Part A, 73*(1), 63–72. https://doi.org/10.1002/jbm.a.30248

Jafarian, M., Eslaminejad, M. B., Khojasteh, A., Mashhadi Abbas, F., Dehghan, M. M., Hassanizadeh, R., & Houshmand, B. (2008). Marrow-derived mesenchymal stem cells-directed bone regeneration in the dog mandible: A comparison between biphasic calcium phosphate and natural bone mineral. *Oral Surgery, Oral Medicine, Oral Pathology, Oral Radiology, and Endodontics, 105*(5), e14–e24. https://doi.org/10.1016/j.tripleo.2008.01.010

Kim, S. H., Kim, K. H., Seo, B. M., Koo, K. T., Kim, T. I., Seol, Y. J., Ku, Y., Rhyu, I. C., Chung, C. P., & Lee, Y. M. (2009). Alveolar bone regeneration by transplantation of periodontal ligament stem cells and bone marrow stem cells in a canine peri-implant defect model: A pilot study. *Journal of Periodontology, 80*(11), 1815–1823. https://doi.org/10.1902/jop.2009.090249

Kohgo, T., Yamada, Y., Ito, K., Yajima, A., Yoshimi, R., Okabe, K., Baba, S., & Ueda, M. (2011). Bone regeneration with self-assembling peptide nanofiber scaffolds in tissue engineering for osseointegration of dental implants. *The International Journal of Periodontics & Restorative Dentistry, 31*(4), e9–e16.

Lathyris, D., & Haidich, A. B. (2021). Meta-analysis of diagnostic test accuracy: Just another meta-analysis? *Intensive & Critical Care Nursing*, 103028. https://doi.org/10.1016/j.iccn.2021.103028

Lee, K. (2012). Global trends in maxillofacial fractures. *Craniomaxillofacial Trauma & Reconstruction, 5*(4), 213–222. https://doi.org/10.1055/s-0032-1322535

Lee, M. K., DeConde, A. S., Lee, M., Walthers, C. M., Sepahdari, A. R., Elashoff, D., Grogan, T., Bezouglaia, O., Tetradis, S., St John, M., & Aghaloo, T. (2015). Biomimetic scaffolds facilitate healing of critical-sized segmental mandibular defects. *American Journal of Otolaryngology, 36*(1), 1–6. https://doi.org/10.1016/j.amjoto.2014.06.007

Lee, J. S., Park, T. H., Ryu, J. Y., Kim, D. K., Eun Jung, O., Kim, H. M., Shim, J.-H., Yun, W.-S., Huh, J. B., Moon, S. H., Kang, S. S., & Chung, H. Y. (2021). Osteogenesis of 3D-printed PCL/TCP/bdECM scaffold using adipose-derived stem cells aggregates; an experimental study in the Canine Mandible. *International Journal of Molecular Sciences, 22*(11), 5409. https://doi.org/10.3390/ijms22115409

LeGeros, R., Daculsi, G, & Legeros, J. (2008). Bioactive bioceramics. pp. 153–181.

Li, J., Li, Y., Ma, S., Gao, Y., Zuo, Y., & Hu, J. (2010). Enhancement of bone formation by BMP-7 transduced MSCs on biomimetic nano-hydroxyapatite/polyamide composite scaffolds in repair of mandibular defects. *Journal of Biomedical Materials Research. Part A, 95*(4), 973–981. https://doi.org/10.1002/jbm.a.32926

Liberati, A., Altman, D. G., Tetzlaff, J., Mulrow, C., Gøtzsche, P. C., Ioannidis, J. P., Clarke, M., Devereaux, P. J., Kleijnen, J., & Moher, D. (2009). The PRISMA statement for reporting systematic reviews and meta-analyses of studies that evaluate health care interventions: Explanation and elaboration.

Journal of Clinical Epidemiology, 62(10), e1–e34. https://doi.org/10.1016/j.jclinepi.2009.06.006

Lopez, C. D., Diaz-Siso, J. R., Witek, L., Bekisz, J. M., Cronstein, B. N., Torroni, A., Flores, R. L., Rodriguez, E. D., & Coelho, P. G. (2018). Three dimensionally printed bioactive ceramic scaffold osseoconduction across critical-sized mandibular defects. *The Journal of Surgical Research, 223*, 115–122. https://doi.org/10.1016/j.jss.2017.10.027

Mahla, R. S. (2016). Stem cells applications in regenerative medicine and disease therapeutics. *International Journal of Cell Biology, 2016*, 6940283–6940283. https://doi.org/10.1155/2016/6940283

Mankani, M. H., Kuznetsov, S. A., Wolfe, R. M., Marshall, G. W., & Robey, P. G. (2006). In vivo bone formation by human bone marrow stromal cells: reconstruction of the mouse calvarium and mandible. *Stem Cells, 24*(9), 2140–2149. https://doi.org/10.1634/stemcells.2005-0567

Marei, M. K., Nouh, S. R., Saad, M. M., & Ismail, N. S. (2005). Preservation and regeneration of alveolar bone by tissue-engineered implants. *Tissue Engineering, 11*(5–6), 751–767. https://doi.org/10.1089/ten.2005.11.751

Misch, C. (2010). Autogenous bone: Is it still the gold standard? *Implant Dentistry, 19*, 361. https://doi.org/10.1097/ID.0b013e3181f8115b

Moser, N., Lohse, N., Golstein, J., Kauffmann, P., Sven, B., Epple, M., & Schliephake, H. (2017). Do we need retarded delivery of bone growth factors in facial bone repair? An experimental study in rats. *European Cells & Materials, 34*, 162–179. https://doi.org/10.22203/eCM.v034a11

Mylonas, D., Vidal, M. D., De Kok, I. J., Moriarity, J. D., & Cooper, L. F. (2007). Investigation of a thermoplastic polymeric carrier for bone tissue engineering using allogeneic mesenchymal stem cells in granular scaffolds. *Journal of Prosthodontics, 16*(6), 421–430. https://doi.org/10.1111/j.1532-849X.2007.00218.x

Neto, A. S., & Ferreira, J. M. F. (2018). Synthetic and Marine-derived Porous scaffolds for bone tissue engineering. *Materials (Basel), 11*(9). https://doi.org/10.3390/ma11091702

O'Brien, F. J. (2011). Biomaterials & scaffolds for tissue engineering. *Materials Today, 14*(3), 88–95. https://doi.org/10.1016/S1369-7021(11)70058-X

Oryan, A., Alidadi, S., Moshiri, A., & Maffulli, N. (2014). Bone regenerative medicine: Classic options, novel strategies, and future directions. *Journal of Orthopaedic Surgery and Research, 9*(1), 18–18. https://doi.org/10.1186/1749-799X-9-18

Pieri, F., Lucarelli, E., Corinaldesi, G., Fini, M., Aldini, N. N., Giardino, R., Donati, D., & Marchetti, C. (2009). Effect of mesenchymal stem cells and platelet-rich plasma on the healing of standardized bone defects in the alveolar ridge: A comparative histomorphometric study in minipigs. *Journal of Oral and Maxillofacial Surgery, 67*(2), 265–272. https://doi.org/10.1016/j.joms.2008.06.036

Prahasanti, C., Nugraha, A. P., Saskianti, T., Suardita, K., Riawan, W., & Ernawati, D. S. (2020). Exfoliated human deciduous tooth stem cells incorporating carbonate apatite scaffold enhance BMP-2, BMP-7 and attenuate MMP-8 expression during initial alveolar bone remodeling in wistar rats (rattus norvegicus). *Clinical, Cosmetic and Investigational Dentistry, 12*, 79–85. https://doi.org/10.2147/CCIDE.S245678

Prochor, P., & Gryko, A. (2020). Numerical analysis of the influence of porosity and pore geometry on functionality of scaffolds designated for orthopedic regenerative medicine. *Materials (Basel), 14*(1), 109. https://doi.org/10.3390/ma14010109

Raggatt, L. J., Wullschleger, M. E., Alexander, K. A., Wu, A. C., Millard, S. M., Kaur, S., Maugham, M. L., Gregory, L. S., Steck, R., & Pettit, A. R. (2014). Fracture healing via periosteal callus formation requires macrophages for both initiation and progression of early endochondral ossification. *The American Journal of Pathology, 184*(12), 3192–3204. https://doi.org/10.1016/j.ajpath.2014.08.017

Raza, M. R., Ahmad, F., Muhamad, N., Sulong, A. B., Omar, M. A., Akhtar, M. N., & Aslam, M. (2016). Effects of solid loading and cooling rate on the mechanical properties and corrosion behavior of powder injection molded 316 L stainless steel. *Powder Technology, 289*, 135–142. https://doi.org/10.1016/j.powtec.2015.11.063

Ren, J., Ren, T., Zhao, P., Huang, Y., & Pan, K. (2007). Repair of mandibular defects using MSCs-seeded biodegradable polyester porous scaffolds. *Journal of Biomaterials Science. Polymer Edition, 18*(5), 505–517. https://doi.org/10.1163/156856207780852578

Salgado, A. J., Coutinho, O. P., & Reis, R. L. (2004). Bone tissue engineering: State of the art and future trends. *Macromolecular Bioscience, 4*(8), 743–765. https://doi.org/10.1002/mabi.200400026

Seto, I., Marukawa, E., & Asahina, I. (2006). Mandibular reconstruction using a combination graft of rhBMP-2 with bone marrow cells expanded in vitro. *Plastic and Reconstructive Surgery, 117*(3), 902–908. https://doi.org/10.1097/01.prs.0000200069.81973.49

Sharma, S., Srivastava, D., Grover, S., & Sharma, V. (2014). Biomaterials in tooth tissue engineering: A review. *Journal of Clinical and Diagnostic Research, 8*(1), 309–315. https://doi.org/10.7860/jcdr/2014/7609.3937

Shayesteh, Y. S., Khojasteh, A., Soleimani, M., Alikhasi, M., Khoshzaban, A., & Ahmadbeigi, N. (2008). Sinus augmentation using human mesenchymal stem cells loaded into a beta-tricalcium phosphate/hydroxyapatite scaffold. *Oral Surgery, Oral Medicine, Oral Pathology, Oral Radiology, and Endodontics, 106*(2), 203–209. https://doi.org/10.1016/j.tripleo.2007.12.001

Shuai, C., Li, P., Liu, J., & Peng, S. (2013). Optimization of TCP/HAP ratio for better properties of calcium phosphate scaffold via selective laser sintering. *Materials Characterization, 77*, 23–31. https://doi.org/10.1016/j.matchar.2012.12.009

Smiler, D., Soltan, M., & Lee, J. W. (2007). A histomorphogenic analysis of bone grafts augmented with adult stem cells. *Implant Dentistry, 16*(1), 42–53. https://doi.org/10.1097/ID.0b013e3180335934

Song, R., Murphy, M., Li, C., Ting, K., Soo, C., & Zheng, Z. (2018). Current development of biodegradable polymeric materials for biomedical applications. *Drug Design, Development and Therapy, 12*, 3117–3145. https://doi.org/10.2147/DDDT.S165440

Stroup, D. F., Berlin, J. A., Morton, S. C., Olkin, I., Williamson, G. D., Rennie, D., Moher, D., Becker, B. J., Sipe, T. A., & Thacker, S. B. (2000). Meta-analysis of observational studies in epidemiology: a proposal for reporting. Meta-analysis Of Observational Studies in Epidemiology (MOOSE) group. *JAMA, 283*(15), 2008–2012. https://doi.org/10.1001/jama.283.15.2008

Tee, B. C., Desai, K. G., Kennedy, K. S., Sonnichsen, B., Kim, D. G., Fields, H. W., Mallery, S. R., Schwendeman, S. P., & Sun, Z. (2016). Reconstructing jaw defects with MSCs and PLGA-encapsulated growth factors. *American Journal of Translational Research, 8*(6), 2693–2704.

Titsinides, S., Agrogiannis, G., & Karatzas, T. (2019). Bone grafting materials in dentoalveolar reconstruction: A comprehensive review. *The Japanese Dental Science Review, 55*(1), 26–32. https://doi.org/10.1016/j.jdsr.2018.09.003

Tsumanuma, Y., Iwata, T., Washio, K., Yoshida, T., Yamada, A., Takagi, R., Ohno, T., Lin, K., Yamato, M., Ishikawa, I., Okano, T., & Izumi, Y. (2011). Comparison of different tissue-derived stem cell sheets for periodontal regeneration in a canine 1-wall defect model. *Biomaterials, 32*(25), 5819–5825. https://doi.org/10.1016/j.biomaterials.2011.04.071

Vasconcelos, D. M., Santos, S. G., Lamghari, M., & Barbosa, M. A. (2016). The two faces of metal ions: From implants rejection to tissue repair/regeneration. *Biomaterials, 84*, 262–275. https://doi.org/10.1016/j.biomaterials.2016.01.046

Verrier, S., Alini, M., Alsberg, E., Buchman, S. R., Kelly, D., Laschke, M. W., Menger, M. D., Murphy, W. L., Stegemann, J. P., Schütz, M., Miclau, T., Stoddart, M. J., & Evans, C. (2016). Tissue engineering and regenerative approaches to improving the healing of large bone defects. *European Cells & Materials, 32*(87–110), 10.22203/ecm.v032a06.

Wang, W., & Yeung, K. W. K. (2017). Bone grafts and biomaterials substitutes for bone defect repair: A review. *Bioactive Materials, 2*(4), 224–247. https://doi.org/10.1016/j.bioactmat.2017.05.007

Wang, H., Li, Y., Zuo, Y., Li, J., Ma, S., & Cheng, L. (2007). Biocompatibility and osteogenesis of biomimetic nano-hydroxyapatite/polyamide composite scaffolds for bone tissue engineering. *Biomaterials, 28*(22), 3338–3348. https://doi.org/10.1016/j.biomaterials.2007.04.014

Wu, W., Chen, X., Mao, T., Chen, F., & Feng, X. (2006). Bone marrow-derived osteoblasts seeded into porous beta-tricalcium phosphate to repair segmental defect in canine's mandibula. *Ulusal Travma ve Acil Cerrahi Dergisi, 12*(4), 268–276.

Xia, B., Deng, Y., Lv, Y., & Chen, G. (2021). Stem cell recruitment based on scaffold features for bone tissue engineering. *Biomaterials Science, 9*(4), 1189–1203. https://doi.org/10.1039/D0BM01591A

Yamada, Y., Ueda, M., Naiki, T., Takahashi, M., Hata, K., & Nagasaka, T. (2004). Autogenous injectable bone for regeneration with mesenchymal stem cells and platelet-rich plasma: Tissue-engineered bone regeneration. *Tissue Engineering, 10*(5–6), 955–964. https://doi.org/10.1089/1076327041348284

Yamada, Y., Nakamura, S., Ueda, M., & Ito, K. (2013). Osteotome technique with injectable tissue-engineered bone and simultaneous implant placement by cell therapy. *Clinical Oral Implants Research, 24*(4), 468–474. https://doi.org/10.1111/j.1600-0501.2011.02353.x

Yuan, J., Cui, L., Zhang, W. J., Liu, W., & Cao, Y. (2007). Repair of canine mandibular bone defects with bone marrow stromal cells and porous β-tricalcium phosphate. *Biomaterials, 28*(6), 1005–1013. https://doi.org/10.1016/j.biomaterials.2006.10.015

Yuan, J., Zhang, W. J., Liu, G., Wei, M., Qi, Z. L., Liu, W., Cui, L., & Cao, Y. L. (2010). Repair of canine mandibular bone defects with bone marrow stromal cells and coral. *Tissue Engineering. Part A, 16*(4), 1385–1394. https://doi.org/10.1089/ten.TEA.2009.0472

Yun, P.-Y., Kim, Y.-K., Jeong, K.-I., Park, J.-C., & Choi, Y.-J. (2014). Influence of bone morphogenetic protein and proportion of hydroxyapatite on new bone formation in biphasic calcium phosphate graft: Two pilot studies in animal bony defect model. *Journal of Cranio-Maxillofacial Surgery, 42*(8), 1909–1917. https://doi.org/10.1016/j.jcms.2014.07.011

Zhang, W., Shi, W., Shaohua, W., Kuss, M., Jiang, X., Untrauer, J. B., Reid, S. P., & Duan, B. (2020). 3D printed composite scaffolds with dual small molecule delivery for mandibular bone regeneration. *Biofabrication, 12*(3), 035020. https://doi.org/10.1088/1758-5090/ab906e

Zhao, D., Zhu, T., Li, J., Cui, L., Zhang, Z., Zhuang, X., & Ding, J. (2021). Poly(lactic-co-glycolic acid)-based composite bone-substitute materials. *Bioactive Materials, 6*(2), 346–360. https://doi.org/10.1016/j.bioactmat.2020.08.016

Zheng, Y., Liu, Y., Zhang, C. M., Zhang, H. Y., Li, W. H., Shi, S., Le, A. D., & Wang, S. L. (2009). Stem cells from deciduous tooth repair mandibular defect in swine. *Journal of Dental Research, 88*(3), 249–254. https://doi.org/10.1177/0022034509333804

Zou, D., Zhang, Z., He, J., Zhu, S., Wang, S., Zhang, W., Zhou, J., Xu, Y., Huang, Y., Wang, Y., Han, W., Zhou, Y., Wang, S., You, S., Jiang, X., & Huang, Y. (2011). Repairing critical-sized calvarial defects with BMSCs modified by a constitutively active form of hypoxia-inducible factor-1α and a phosphate cement scaffold. *Biomaterials, 32*(36), 9707–9718. https://doi.org/10.1016/j.biomaterials.2011.09.005

Adv Exp Med Biol - Innovations in Cancer Research and Regenerative Medicine (2023) 4: 137–163
https://doi.org/10.1007/5584_2022_707

Published online: 8 April 2022

Tissue Engineering for Tracheal Replacement: Strategies and Challenges

Asmak Abdul Samat, Zuratul Ain Abdul Hamid, Mariatti Jaafar @ Mustapha, and Badrul Hisham Yahaya

Abstract

The critical feature in trachea replacement is to provide a hollow cylindrical framework that is laterally stable and longitudinally flexible, facilitating cartilage and epithelial tissue formation. Despite advanced techniques and sources of materials used, most inherent challenges are related to the complexity of its anatomy. Limited blood supply leads to insufficient regenerative capacity for cartilage and epithelium. Natural and synthetic scaffolds, different types of cells, and growth factors are part of tissue engineering approaches with varying outcomes. Pre-vascularization remains one of the crucial factors to expedite the regenerative process in tracheal reconstruction. This review discusses the challenges and strategies used in tracheal tissue engineering, focusing on scaffold implantation in clinical and preclinical studies conducted in recent decades.

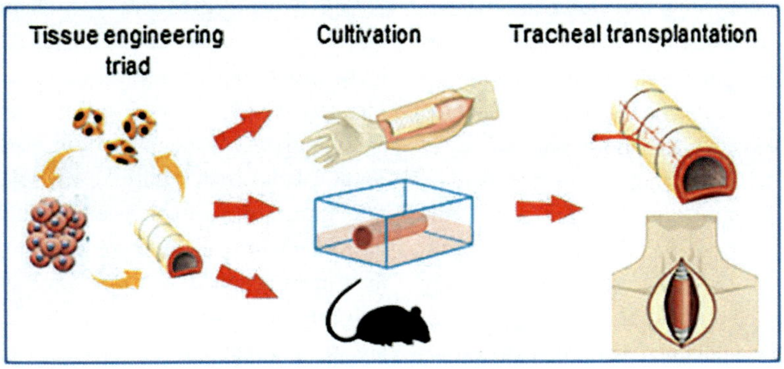

Mariatti Jaafar @ Mustapha
School of Materials and Mineral Resources Engineering, Universiti Sains Malaysia, Penang, Malaysia

A. A. Samat
Lung Stem Cell and Gene Therapy Group, Regenerative Medicine Cluster, Advanced Medical and Dental Institute (IPPT), Universiti Sains Malaysia, Penang, Malaysia

Fundamental Dental and Medical Sciences, Kulliyyah of Dentistry, International Islamic University Malaysia, Kuantan, Pahang, Malaysia

Z. A. A. Hamid
School of Materials and Mineral Resources Engineering, Universiti Sains Malaysia, Penang, Malaysia

B. H. Yahaya (✉)
Lung Stem Cell and Gene Therapy Group, Regenerative Medicine Cluster, Advanced Medical and Dental Institute (IPPT), Universiti Sains Malaysia, Penang, Malaysia
e-mail: badrul@usm.my

Keywords

Cartilage · Epithelium · Tissue engineering · Trachea replacement · Vascularization

1 Introduction

Tissue and organ transplant in patients with diseased, injured, or terminally failed organs may save lives and improve their quality of life. However, global shortages of tissues and organs remain an obstacle for organ transplantation, and it has been a significant worldwide public health problem for many years. According to the Organ Procurement and Transplantation Network, United States Department of Health and Human Services, the number of patients on the national transplant waiting list until July 2019 has increased to more than one hundred thousand. Out of the number, two-thirds are above 50, while almost 2,000 are below 18, and only one-third of the total numbers received organ transplants (US Health Resources and Services Administration, 2019). Advances in tissue engineering provide promising alternative approaches to assemble functional constructs that repair, preserve, or enhance defective tissues or organs. Cells expanded in the three-dimensional (3D) scaffolds serve as an adhesive substrate for the implanted cells and provide a mechanical framework leading to the formation of new organs (Viola et al., 2003). The approach was initially founded to deal with the critically unmet demand of the markedly growing number of organ transplant patients' waiting lists. However, there is only a small percentage of the donated organs (Giwa et al., 2017; Jones & Bes, 2012). Promising findings have been reported in reconstructing and substituting multiple tissues and organs, including the skin, heart, kidney, and liver (Knoepfler, 2015; Bailey et al., 2014; Tang et al., 2016).

The indication for tracheal replacement or reconstruction surgery is extensive tracheal injuries caused by multiple pathologies originating from primary or secondary neoplasms, infections, congenital malformation, or trauma results. It is conducted in cases of failed reconnection of the trachea through the direct anastomosis. Although the incidence is considered uncommon, its occurrence is life-threatening (George et al., 2005; Mueller, 2018). In general, extensive tracheal lesions of greater than half of the tracheal length in adults or one-third of small children indicate the need for a tracheal reconstruction or replacement (Grillo, 2002; Kucera et al., 2007). In this condition, the mechanical stress induces dehiscence at the wound site, contributing to severe and fatal postoperative complications (Grillo, 2002). Repairing long-segment tracheal defects is a big obstacle in healthcare; thus, seeking an optimal alternative for tracheal transplantation is in demand.

The available sources of tracheal replacements range from autologous tissue transplants, allografts, xenografts, or prostheses (Lee et al., 2018). While the approaches in certain instances have shown progress, there are still substantial pitfalls. Attempts to replace partial structures of the airways are restricted by persistent infection, collapsed graft, vascular erosion, or the need for lifelong immunosuppression (Chiang et al., 2016). Advanced progress in tissue engineering offers more excellent prospects for tracheal reconstruction. A wide range of techniques and materials has been reported, including decellularized tissues and biosynthetic materials. Some of the issues may be alleviated; however, the clinical success of tissue-engineered tracheal grafts remains speculative and yet to reach its theoretical capacity. Major problems involving multiple hospitalizations and further procedures are still notable (Chiang et al., 2016; Elliott et al., 2012; GonfiOtti et al., 2014; Hamilton et al., 2015). After several decades, clinical reconstruction of the trachea remains elusive, and the reasons are multifactorial. Cartilage stability, vascularization, and epithelization are the inherent challenges in reconstructing the trachea (Delaere & Van Raemdonck, 2016).

This chapter discusses the strategies used in tracheal tissue engineering and the challenges of reconstructing and replacing the trachea, focusing on scaffold implantation in clinical and preclinical studies conducted in recent decades.

2 Current Strategies in Tracheal Replacement

Several tissue sources and tracheal repair techniques are used to restore the long-segmental tracheal defect. The tissues can be transplanted from one site to another inside the same patient, or one individual to another (allograft), or xenogeneic (cross-species transplantation). Xenogeneic transplantation, on the other hand, will not be explored in this review. For an extended period, prosthetic scaffolds have been utilized as tracheal substitutes with various results. Each method has its own set of benefits and drawbacks. The types of tracheal transplant, tissues, mechanisms, advantages, and disadvantages of each are summarized in Fig. 1.

Donor tissues can be vascularized or non-vascularized (free flap), fresh or frozen, with complex marker histocompatibility in autograft and allograft procedures (Haykal et al., 2014). The advantage of the autogenous free flap is the alleviation of immunosuppressive therapy. The tissues used are bronchial flaps and extrathoracic muscle flaps such as the latissimus dorsi, serratus anterior, and pectoralis major (Chen, 2010; Ris et al., 2008). Dissolution and stenosis can result from non-vascularized tissues, while improved benefits have been reported as tissues are vascularized (Haykal et al., 2014).

Tissue engineering is replacing tissues and organs by isolating and cultivating cells outside of the body. The tissues are eventually seeded

Graft type	Autologous	Allogenic	Prosthetic
	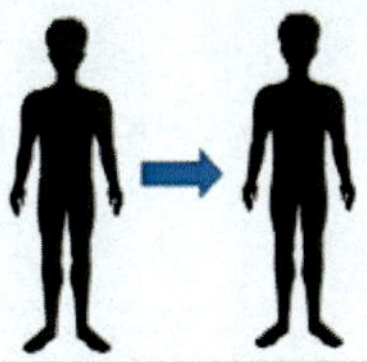	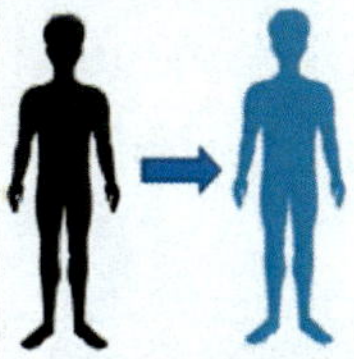	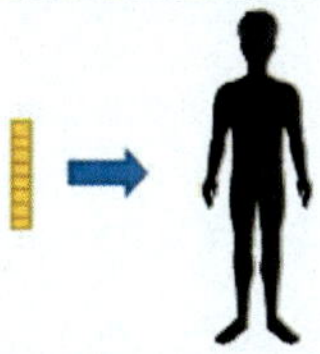
Type of tissue	Aortic, omentum, pulmonary	Trachea, aortic, omentum	Natural, synthetic (silicone, metal, polymers, hydrogel etc)
Mechanism	Tubularized autologous tissue with or without vascularization, supported by a stent and/or additional support with autologous tissues	Donor trachea undergo several weeks of heterotopic revascularization prior to implantation as a tracheal conduit	Natural decellularized donor tissues or synthetic scaffold seeded with autologous or allogenic cells. In theory, the cells will proliferate and differentiate and to repopulate the scaffold
Advantages	No need for immunosuppression	Structural and mechanical properties or native tissue Fresh or can be cryopreserved	No need for immunosuppression Possibility of growth potential
Disadvantages	Technically difficult Donor site morbidity	Requires donor Several weeks of heterotopic revascularization Requires period of immunosuppression	Difficult to revascularize *Ex vivo* cell seeding and/or requires cell line expansion

Fig. 1 Three graft types used in tracheal replacement are autologous, allogenic, and tissue-engineered, summarized with the mechanisms, advantages, and disadvantages

into a functionalized scaffold before implantation (Haykal et al., 2014). Three main components for a successful tissue engineering construct, often referred to as 'tissue engineering triad', are the matrix or materials used as templates or scaffold, types of cells seeded, and suitable signals such as biophysical cues and chemical mediators. This triad synchronizes to ultimately regenerate a functional tissue or promote regeneration of the original tissue (O'Brien, 2011; Mhanna & Hasan, 2017). In general, a scaffold with previously harvested host cells is implanted to allow the cells to differentiate into mature cells, repopulate, and gradually replace the scaffold by depositing a new extracellular matrix (ECM), which occurs either in a bioreactor or after implantation. Cell–scaffold interaction has a significant influence on tissue engineering (Murphy et al., 2013). The interaction promotes cell reaction through cell proliferation and differentiation, while the biomaterial microarchitecture can stimulate chemoattraction, adhesion, and migration, leading to matrix deposition (Zeltinger et al., 2001; Keogh et al., 2010; Ravindran et al., 2010). Furthermore, the inclusion of growth factors as scaffold surface modifiers integrates biomaterial with pro-adhesive sequences and enhances the biomaterial's surface structure (Klimek & Ginalska, 2020). Figure 2 shows the schematic diagram of the tissue engineering triad.

This section reviews some tissue-engineered tracheal transplantation approaches reported in clinical and preclinical research in recent decades.

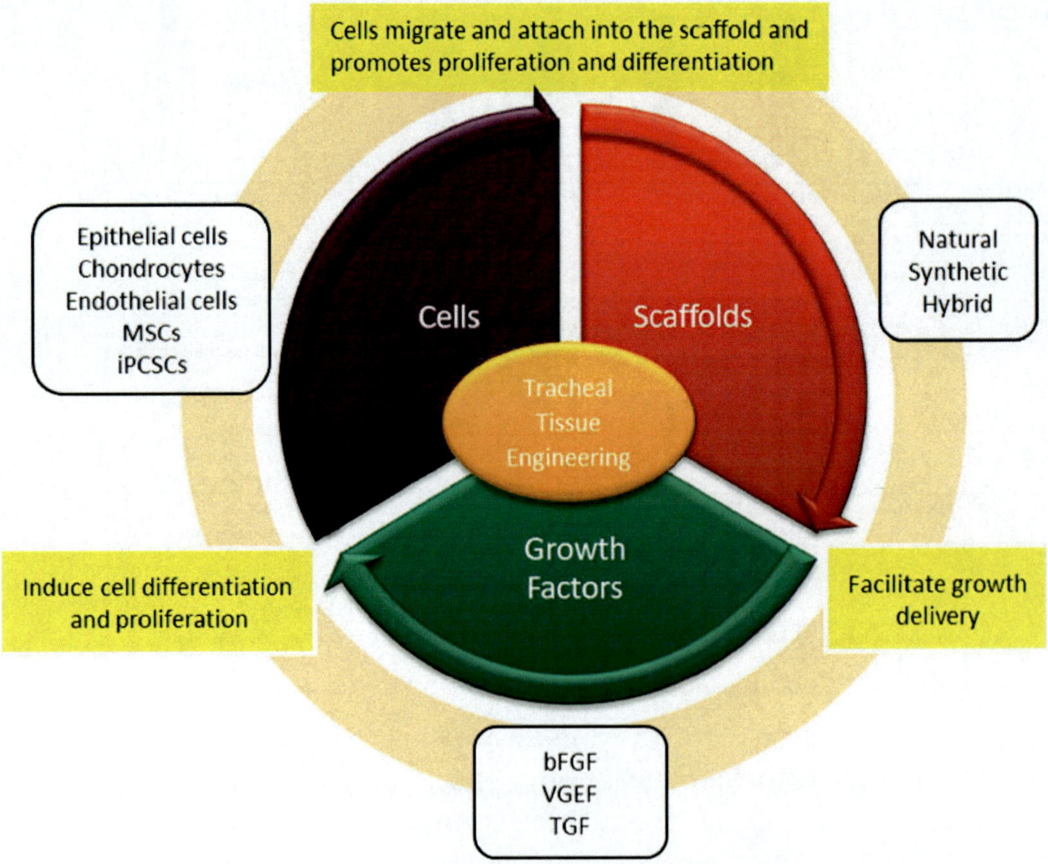

Fig. 2 A schematic diagram of the tissue engineering triad shows the main components of cells, scaffolds, and growth factors

2.1 Scaffolds for Tracheal Tissue Engineering

2.1.1 Natural or Biological Scaffolds

Techniques that utilize biological or natural materials become preferable for tracheal replacement because they preserve the natural ECM composition and the natural 3D structure and biological markers for cell adhesion and development. They are composed of extracellular matrix materials such as collagen, fibrin/hyaluronic acid, and other glycosaminoglycan products (Haykal et al., 2012). The idea of autogenous tissue regeneration focuses on creating a tubularised structure that maintains its viability by using a vascularized pedicle to prevent collapse. Tubularized tissues such as the aorta, radial forearm flap, or omentum are the common tissues used in autologous transplants. However, due to lack of structural integrity, these grafts are usually supported with other prostheses, either patches or stents (Haykal et al., 2014; Mercier et al., 2018; Etienne et al., 2018). The grafts can be assisted by cartilage implantation derived from costal cartilage or other sites (Mercier et al., 2018; Udelsman et al., 2018). Even though this technique does not involve immunosuppressive therapy, the harvested graft retains blood supply, and the procedure is swift. However, the process is technically challenging and may result in mucociliary clearance of the patient. Furthermore, it is known that harvesting grafts are costly, traumatic, and anatomically restricted, with possible morbidity at the donor site because of infection.

Hollow structures such as tracheal and aortic grafts have been used in allograft transplantation. The tissues can be transplanted either fresh or undergo a meticulous process to preserve and eliminate antigenicity. Fresh allografts must be transplanted immediately, and the recipient must be identified and prepared at almost the same time of harvest. Thus, it is usually not available in an emergency setting. In addition, the recipient usually requires lifelong immunosuppression.

On the other hand, the harvested allograft tissue can be frozen or cryopreserved and stored when needed (Rich & Gullane, 2012). The cryopreserved tissues are decellularized through meticulous washing cycles using detergents, salts, enzymes, and physical means to remove the cellular and nuclear components (Conconi et al., 2005; Scarritt et al., 2015). These methods retain the ECM components comprising proteins, collagen subunits, proteoglycans, glycoproteins, and the mechanical properties of the tissues. Additionally, the decellularization method reduces the antigenicity of the tissue matrix and the possibility of immune-rejection risk by eliminating primary histocompatibility complexes class I and II (Bader & Macchiarini, 2010). The decellularized tissues are then repopulated with the patient's cells (autologous) or other cells (epithelial, chondrocytes, MSCs, or combination), as summarized in Fig. 3.

Several studies reported mixed success results from fresh and cryopreserved aortic and tracheal transplantation. Wurtz and colleagues used aortic allografts wrapped with a pectoralis major flap to promote vascularization. Although all grafts developed respiratory epithelial lining, the survived patients require silicone stents to prevent the aortic allografts from collapsing (Wurtz et al., 2006). Tracheal allograft transplantations in paediatric patients carried out by Propst et al. showed that 80% of the patients survived, yet required multiple stenting procedures to prevent the trachea from collapsing (Propst et al., 2011). A similar result was reported by Delaere et al., who heterotopically positioned vascularized tracheal allotransplants in the forearm, with the fascia flap pedicles wrapped around the transplant to create a vascular network to promote revascularization. The trachea was retransplanted orthotopically in the airway and appeared to retain its structural integrity (Delaere et al., 2010; Delaere et al., 2012; Delaere et al., 2014) due to their reduced immuno-rejection capacity and ECM compositions such as collagens.

Allograft transplantation is considered an effective method of replacing the trachea; however, it is limited for several reasons. Vascularization at the heterotopic site requires some time to develop; thus, the graft is almost impossible to be placed in emergency cases. Likewise, shortages

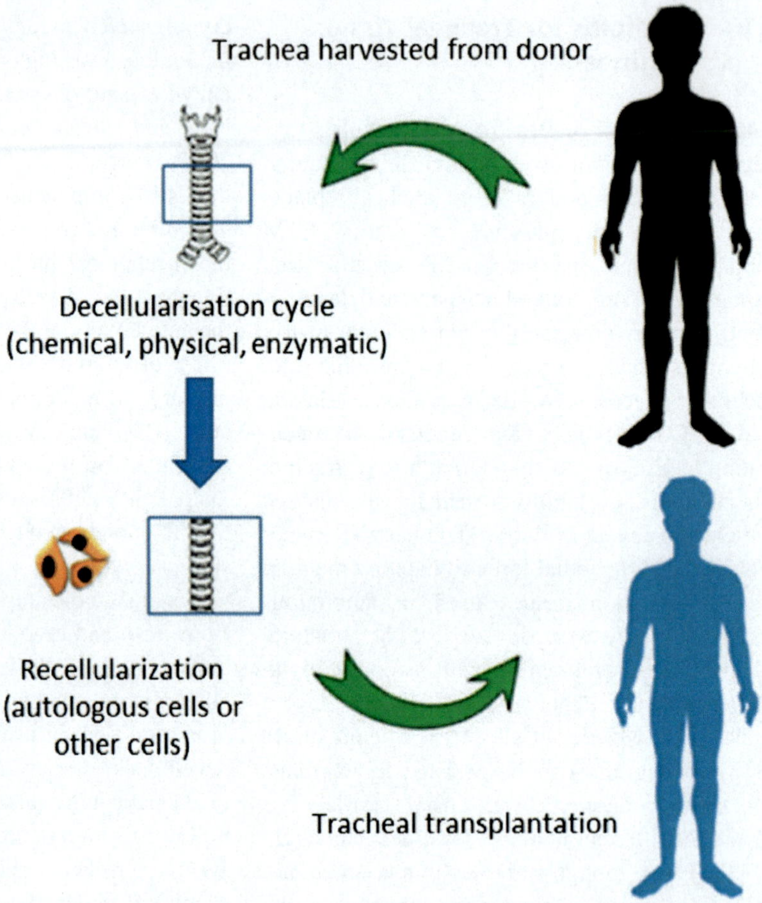

Fig. 3 Schematic diagram shows the decellularization technique in allograft transplant. First, the trachea tissue is harvested from a donor, followed by a complex recellularization method, and transplanted to the recipient

of tissues from suitable donors, unsuitably sized donated trachea, severe immune-rejection risks, and possible complications from infection or disease from the donor-to-patient can be the limiting factors in allografts. Orthotopic tracheal allograft transplantation could be ideal for patients with benign long-segment tracheal stenosis but may not be appropriate for patients who need long-term immunosuppressive therapy. Furthermore, a lack of structural integrity and consistency limit their applications, requiring support from stents and T-tubes.

2.1.2 Synthetic Scaffold

Synthetic materials such as biodegradable polymers are gaining attention as materials in tissue engineering. Expansive processability windows, mechanical properties, and degradation time allow them to be easily manipulated and controlled. Scaffolds are fabricated through various techniques to produce high precision, which suits the application. Some methods used in the fabrication of tracheal scaffolds using biodegradable scaffolds that have been tested on animal models are electrospinning (Townsend et al., 2018), thermally induced phase separation (Jing et al., 2014), and 3D printed technology (additive manufacturing) (Xia et al., 2019; Gao et al., 2017).

The choice of material and design for fabricating tracheal scaffold remains elusive. Multiple criteria must be fulfilled to suit the selection criteria. Scaffolds provide an appropriate 3D niche for the cells to grow, proliferate, and differentiate. Therefore, the scaffold's materials must have specific properties, such as the ability to mimic the ECM to allow cells attachment, proliferation, and differentiation onto the surface of the templates. In addition, it should not trigger marked immune responses that will induce severe inflammatory reactions leading to delayed healing or immune rejection by the host (O'Brien, 2011).

Additionally, the biomaterials should exhibit cell-binding components for attaching, either by resembling native ligands interactions or specifically combining the peptide sequences (Luo et al., 2009; Nomoto et al., 2008). These goals depend highly on the material surface chemistry, which regulates the interaction between cell and substrate (Mooney et al., 1992). Biodegradable polymers that have been used to fabricate tracheal scaffolds include polylactic acid (PLA) (Gao et al., 2019), polyglycolic acid (PGA), polylactide-co-glycolide acid (PLGA) (Grimmer et al., 2004; Komura et al., 2008), polypropylene (Omori et al., 2008), high-density polyethylene (HDPE) (Haliloglu et al., 2000), polycaprolactone (PCL) (Lin et al., 2008; Yin et al., 2017; Chang et al., 2018), and thermoplastic polyurethane (TPU) (Abdul Samat et al., 2021).

PGA, PLA, and their copolymers, PLGA, were reported to have good mechanical properties and are biocompatible with adjustable degradation rates (Dean et al., 2003; Kim et al., 1994). These biopolymers are usually fabricated in combination with other nature-based materials such as alginate (Grimmer et al., 2004) and collagen (Komura et al., 2008). However, each of these elements poses several other issues. For instance, the biodegradable PGA products evoke some local acid environment, which is not conducive for cell growth and development (Luo et al., 2009).

Even though polypropylene was biologically inert, its tendency to shrink is one of the causes of its failure as a tracheal prosthesis. However, it might be suitable in different experimental conditions. Polypropylene has been clinically used as a tracheal stent or as a scaffold material (Omori et al., 2008; Behrend et al., 2006; Baruah et al., 2017). Omori and colleagues used polypropylene mesh or Marlex mesh covered with collagen mesh and transplanted into patients with tracheal defects. A well-epithelialized airway lumen without any obstruction was reported (Omori et al., 2008). Using polytetrafluoroethylene (PTFE), a composite artificial trachea connected on both ends with titanium rings was implanted into dogs. However, the implants could only be used to repair defects of the trachea for a short term (Liu et al., 2013). A better result was demonstrated using a composite tracheal scaffold composed of poly(-ε-caprolactone) stent and a type II collagen sponge onto rabbit model (Lin et al., 2008). Gao and co-workers also produced a similar result, using a 3D-printed PCL scaffold in defective tracheal rabbits (Gao et al., 2017).

After implantation, the scaffold should remain at the site to allow proliferation and secretion of a new matrix of the host's cells to be eventually replaced. Tissue regeneration varies in different implant sites; therefore, it is essential that the degradation rates are controllable and adapted to the needs of a specific application. For example, in hard tissues such as the trachea, the scaffold should provide an adequate provisional framework to endure stresses and forces in vivo until the regenerating cartilage can undertake its structural function to support itself. However, cartilage formation also appears to depend on the methods or types of scaffold and cells used. For instance, the formation of cartilage requires 10 weeks in a scaffold-free cartilage construct (Weidenbecher et al., 2008).

In contrast, only 6 weeks were required in a composite non-woven PGA scaffold seeded with chondrocytes and epithelial cells in subcutaneous implantation of athymic mice (Kojima et al., 2003a). In addition, a 3D-printed mesh structure in a two-layered tubular scaffold from PCL nanofibre and microfibre needs 4 weeks for cartilage regeneration. On the other hand, the neocartilage formation was only seen about 6 to 12 months after implantation in the rabbit model using a 3D-bioprinted multilayered PCL and hydrogel scaffold seeded with nasal epithelial and auricular cartilage cells (Park et al., 2019).

Degradation of the material must not be toxic and should be excreted through the physiologic, metabolic pathways (O'Brien, 2011; Thomson et al., 1995; Gilbert et al., 2007; Sung et al., 2004; Alsberg et al., 2003). Porosity is another essential factor to consider in developing tracheal scaffolds. The cell functions and the regeneration of new tissues depend heavily on pores. The crucial point in developing ideal porous biomaterial and scaffolds is controlling the pores, including the pore size and interconnection. Good scaffold interconnectivity allows the smooth distribution of cells within the scaffolds (Faraj et al., 2007; O'Brien et al., 2005; Silva et al., 2006; Woodfield et al., 2004). The 3D porous structure allows infiltration, adhesion, and proliferation of the cells, thus promoting nutritional and metabolic waste exchange, and stimulating angiogenesis. The pore size ranges 150–300 μm and is reported to encourage chondrocyte growth and encourage the full expression and production of type II collagen and aggrecan with the highest compressive modulus (Zhang et al., 2014; Stenhamre et al., 2011) but maybe improved in 380–405 μm pore scale (Oh et al., 2007). The surface area available for cell attachment is limited in the highly porous scaffold (Law et al., 2016). In general, larger pores facilitate smoother delivery of cells throughout the scaffolds compared to small pores. However, more cells are required to populate the entire space in a highly porous scaffold. A good microenvironment is created when the cells fill the pores and promote chondrogenic differentiation to develop cartilage tissue. In addition, smaller pore sizes may hinder cell seeding and distribution, and thus cartilage regeneration. Inconsistent cartilage regeneration time from multiple studies is reported due to different methods, types of cells, and animal models. Therefore, further research in a precisely controlled condition is necessary, particularly in the long-term development of neo-trachea and its performance after replacement.

2.1.3 Naturally Derived Polymers and Hybrid Polymers

As an ECM, the scaffold plays an important role; however, it is often unable to determine the exact microenvironment during tissue growth to facilitate in vitro or in vivo tissue development. Combining synthetic polymers and naturally derived materials appears to be an appealing strategy to overcome the problem. Attractive characteristics of hybrid materials include a wide range of mechanical and degradation properties, large-scale reproducibility, reasonable processability, and specific biological activity. In their native environment, natural polymers execute diverse sets of functions. For instance, proteins function as structural materials and catalysts, whilst polysaccharides function in membranes and intracellular communication and storage. However, these polymers could not stand alone in a mechanically supported organ like the trachea.

Protein-based polymers resemble specific ECM characteristics, thus controlling cell proliferation, development, and structure during tissue recovery and wound healing (Malafaya et al., 2007). Collagen, gelatin, silk fibroin, fibrin, and hyaluronan acid are the most used protein-based polymers as hybrid scaffolds in tracheal tissue engineering. In addition, collagen has been reported to be used in combination with other synthetic polymers such as polypropylene (Okumuş et al., 2005; Omori et al., 2005; Nomoto et al., 2008), PCL (Lin et al., 2008), PLGA, and PLCL-PLGA (Wu et al., 2018).

Gelatin is a natural polymer derived from collagen. It is commonly utilized in biomedical applications due to its biodegradability, biocompatibility, ease of processing, and gelation properties. It is used in tissue engineering applications to deliver drugs targeted at many tissues such as bone, cartilage, and skin. In addition, gelatin is used as a carrier to provide active biomolecules to improve temporary cell functions (Malafaya et al., 2007). Fibrin-based materials are used in drug distribution to emphasize cell delivery due to their biochemistry, especially cell interactions. The innate ability of fibrin to induce improved cellular interaction and subsequent scaffold remodelling provide an advantage when used in combination with synthetic scaffolds. Furthermore, fibrin can be extracted autologously from an individual, acting as an immunocompatible carrier to supply active biomolecules (Malafaya et al., 2007).

Silk fibroin is derived from silkworms, consisting of H-chain fibroin (FH), L-chain fibroin (FL), and fibrohexamerin at a molar rate of 6:6:1 is another natural polymer that is commonly used in tissue engineering. Fibroin is a fibrous protein built up of silk as the core, surrounded by a glue-like protein called sericin, consists of H-chain fibroin (FH), L-chain fibroin (FL), and fibrohexamerin at a molar rate of 6:6:1 (Inoue et al., 2000). Due to its biocompatibility, degradability, and excellent mechanical properties, silk fibroin has been a desirable biomaterial in tissue engineering (Altman et al., 2003; Dal Pra et al., 2005), degradability (Horan et al., 2005), and excellent mechanical properties.

Polysaccharide is a family of biopolymers containing simple sugar monomers (monosaccharides) (Nishinari & Takahashi, 2003). For example, hyaluronic acid (HA), commonly referred to as hyaluronan. It is a naturally occurring non-sulphated glycosaminoglycan (GAG) and a significant macromolecular part of the intercellular matrix in most connective tissues, such as cartilage, human eye vitreous, umbilical cord, and synovial fluid (Liao et al., 2005). HA is a linear polysaccharide composed of alternating α-1,4-D-glucuronic acid and β-1,3-N-acetyl-D-glucosamine disaccharides, connected by β(13) bonds (Laurent et al., 1995). ECM components are essential building blocks in preparing tissue-engineered scaffolds, mainly because their biological, chemical, and physical characteristics can be regulated.

Chondroitin sulphate is one of the most physiologically essential glycosaminoglycans (GAG). As a component of cartilage, which functions in binding and regulating the growth factors and cytokines, protease inhibition, cell adhesion, migration, proliferation, and differentiation, chondroitin sulphate is an attractive polymer of natural origin cartilage regeneration (Pieper et al., 1999; Lippiello, 2003). However, in tissue engineering, its application is to enhance other polymers' interaction with cells and proteins and improve the cell behaviour of the developed material (Malafaya et al., 2007). Similarly, due to its identical structure to GAGs, biodegradability, and not initiating any immune reaction, chitosan becomes one of the biomaterials used in tissue engineering. However, chitosan per se would not be sufficient for biomechanical strength and thus added to other biomaterials (Nematollahi et al., 2017).

Complete regeneration of the respiratory epithelium at the site of the tracheal defect in a hybrid fibrin/hyaluronan and degradable porous PLGA seeded with chondrocytes was reported by Hong et al. (Hong et al., 2014). Similar results were described with fibrin/HA composite gel seeded with allogenic chondrocytes in rabbits (Kim et al., 2010; Hong et al., 2012). In addition, the successful result was observed using porous silk fibroin co-cultured with primary human tracheobronchial epithelial cells to repair tracheal defects in rabbits (Chen et al., 2018) and silk-fibroin and chitosan blend (Zang et al., 2011). However, even though the biological properties of the naturally derived polymers are similar to the natural tissues and promote cellular migration, proliferation, and differentiation of cells, the scaffold's mechanical strength when used alone is compromised. Therefore, supports from other biomaterials are still required to produce a stable scaffold. Furthermore, in most reported studies, the successful results often come with other essential factors such as types of cells seeded and the potential use of different growth factors.

2.2 Cell Sources

The critical goal in repairing or restoring tracheal tissue engineering is that a scaffold provides mechanical integrity to prevent tracheal failure during respiratory changes. Additionally, the airways play an essential role as the first-line defence against inspired pathogens. This role is carried out by the highly specialized pseudostratified ciliated epithelium composed of innate and adaptive immune systems cells. These goals could not be achieved without sufficient initial vascularization of the scaffolds, followed by a delayed reconstruction of the tracheal defects. In general, tracheal replacement therapies produce structures that mimic native tracheal biomechanics, encourage functioning mucosa, and stimulate rapid revascularization for cell survival. Given the complexity of the anatomical structure of the trachea, it is likely that multiple types of cells are required to be present in an optimized tissue-engineered scaffold.

An optimal cell type for tissue engineering is to develop a maximum number of cells with a universal phenotype that would be non-immunogenic and able to be used in any patient, but with minimal invasive harvesting procedures (Kojima & Vacanti, 2014). In addition, the cells should be able to persist in cultures and orthotopically inducible to differentiate into the functional cell types of any organ system without the risk of adverse effects (Williams, 2004) and express compatible genes to preserve the functionality of the tissue (Rosso et al., 2005). The cells can either be harvested from autologous or allogeneic sources. Autologous cell use is preferable because it eliminates the need for immunosuppression by the recipients.

Chondrocytes are the most common cell type used to promote cartilage regeneration for tracheal scaffold due to easy isolation such as auricle or nasal septum without compromising the functionality of the donor tissue (Liu et al., 2017a). Nasal and trachea-derived chondrocytes were used to construct the human trachea and reported to provide identical mechanical features to natural tissue (Kojima et al., 2003b). However, different results were obtained by Gao et al. when compared cartilage formation in the dorsal subcutaneous spaces of nude mice and rabbit airways. Promising results were observed in the mice due to the compromised immune system and the aseptic condition of the subcutaneous space. Environmental conditions in the airway are not sterile, thus inducing bacteria from the respiratory tract. In addition, local inflammation may be triggered, leading to failure of the cartilage maturation process (Gao et al., 2017). Functionalization of 3D-printed PLCL/gelatin scaffold and gelatin sponge consisting of TGF-β1 also produces a similar result (Park et al., 2015a). However, chondrocytes are proven to be more difficult to harvest and traditionally considered to have low proliferation ability, thus limited proliferation capacity (Bogan et al., 2016).

A complete re-epithelialization is especially important for the long-term performance of a tissue-engineered trachea. However, it is challenging to select a suitable source of epithelial cells since collecting tracheal epithelial tissue is invasive, and the volume of cells collected is limited. In addition, similar to chondrocytes, the phenotype of engineered respiratory epithelium is unstable and readily causes tissue granulation and tracheal stenosis (Walles, 2011). Therefore, different cell origins have been used for trachea re-epithelialization, such as trachea epithelial cells (Kobayashi et al., 2010), nasal (Heikal et al., 2010), abdominal skin (Kim et al., 2004), skin (Kim et al., 2011), and buccal mucosa (Delaere et al., 2010). In addition, a mixture of cells might be harvested during the procedure, including the stem/progenitor cells, which can be further expanded (Ruszymah et al., 2011; Zhao et al., 2012; Li et al., 2004; Liang & Bickenbach, 2002; Fuchs, 2008). Nevertheless, it must be remembered that many issues need to be addressed since the methods to induce differentiation are complex and challenging; the efficacy and reliability of these approaches demand more studies to be done.

Blood supply is usually limited in tracheal tissue engineering due to technical difficulty, which leads to poor results. Currently, the only acceptable way of generating blood supply is to use a vascularized soft tissue flap followed by re-establishing anastomoses between the flap and the trachea adventitia (Delaere & Van Raemdonck, 2016; Law et al., 2016). However, avascular cartilaginous tissue becomes an obstacle for the ingrowth of blood vessels towards the mucosal lining, especially in a long tracheal graft, resulting in ischemic changes. Thus an incision between the cartilaginous ligament is required to enhance the vascularization process (Delaere & Van Raemdonck, 2016). One of the approaches to accelerate neo-vascularization in the bioengineered trachea is to use endothelial cells. A study by Walles et al. used a decellularized porcine jejunum segment seeded with autologous endothelial cells, resulting in uniform reconstitution of endothelial cells in the vascular graft network. However, even though this approach appeared appealing, seeded endothelial cells may not thrive in deprived oxygen conditions as in the implantation site since they are susceptible to oxygen deficiency (Shirai et al., 1998).

As previously mentioned, chondrocytes and epithelial cells have limited differentiation capability; thus, their applicability for tissue regeneration is limited. In addition, chondrocytes become non-viable following implantation into the scaffolds without rapid vascularization. Therefore, co-culturing different cells provide better options for regenerating complete functional tissue implantation to overcome the aforementioned problems. When both chondrocytes and epithelial cells were seeded in decellularized grafts in animals, stenosis and bacterial contamination were reduced. In addition, overall survival was improved compared to animals receiving no-cell or only one-type seeded scaffold (Go et al., 2010). Apart from proliferation and differentiation of other cell types in vitro (Wu et al., 2011), MSCs also support the endogenous epithelium migration towards grafts (Gray et al., 2012). Thus, they are an excellent co-seeding option. Co-transplantation of chondrocytes and MSCs in rabbits by Tsao et al. showed that deposition of type II collagen within biodegradable tracheal grafts was more remarkable than the animal with single-cell or without cells (Tsao et al., 2014). To date, most of the successful reported tissue-engineered tracheal transplantations have utilized the method of co-culturing different cells and the addition of growth factors, which will be discussed in the next section.

Mesenchymal stem cells (MSCs) become a predominant choice of cell in tissue engineering due to their ability to differentiate into multiple lineages, produce wide-ranging trophic and immunomodulatory factors, and easy isolation and expand in vitro (Banas et al., 2007; Bianco et al., 2008; Caplan, 2017; Kern et al., 2006). Integrating MSCs into the damaged site can be achieved by seeding the cells into the engineered tissue or through attracting the wound with the aid of biomaterials and soluble factors such as growth factors, chemokines, and cytokines (Law et al., 2016). Multiple types of MSCs used in tracheal tissue engineering include bone marrow (BM-MSCs) (Park et al., 2016), adipose-derived tissue (ADMSCs) (Kobayashi et al., 2010; Batioglu-Karaaltin et al., 2015), amniotic mesenchymal stem cells (Gray et al., 2012), and human turbinate mesenchymal stromal cell (hTMSC) with good results (Park et al., 2015b).

Co-culturing MSCs and epithelial cells helped cartilage preservation within a decellularized tracheal scaffold in a heterotopic pig model (Haykal et al., 2013). The MSCs elicited an immunomodulatory reaction that increased the number of regulatory T-cells infiltrating relative to the decellularized scaffold alone. Successful transplantation of half-pipe 3D-printed PCL scaffold coated with the MSCs seeded in fibrin into the tracheal defects was reported in a pilot study by Chang et al. (Chang et al., 2014). Analyses of the results revealed a successful integration of MSCs with neocartilage and restoration of ciliated epithelium without any graft rejection. However, the regeneration ability of the BM-MSC was not reported. BM-MSCs and ADMSCs were reported to accelerate vascularization and regeneration of epithelium (Batioglu-Karaaltin et al., 2015; Suzuki et al., 2008).

Similarly, a mature and highly ciliated epithelium was found in the luminal surface of the graft with human turbinate mesenchymal stromal cell (hTMSC) sheets (Park et al., 2015b). In addition, chondrocytes had been reported to stimulate the chondrogenic differentiation of BM-MSCs (Tsao et al., 2014; Kang et al., 2012; Tao et al., 2009; Ko et al., 2016). However, when auricular chondrocytes and BM-MSCs were co-cultured within a PEG/PCL hydrogel, they enhanced expression of chondrogenic genes and increased GAGs and collagen type II accumulation only in vitro conditions but not in the animal model. This condition was presumably due to the disparity in nutritional needs between chondrocytes and BM-MSCs, which are relatively lower but higher in chondrocytes than BM-MSCs. As a result, lower oxygen and nutrition content in tissue fluid limits the growth and chondrogenesis of BMSCs during in vivo cultivation (Chang et al., 2018). Therefore, it is essential to identify the optimal ratio of cells to be co-cultured to obtain the optimal effects of seeded cells.

Induced pluripotent stem cells (iPSCs) are pluripotent stem cells generated directly from an adult somatic cell. It was first developed in 2006 from somatic cells by transfecting four transcription factors in fibroblasts. Compared to multipotent ESCs, iPSCs are reported to have unlimited symmetrical self-renewal properties and generate other stem cells and various immune cells for general clinical requirements (Takahashi & Yamanaka, 2006). Furthermore, the cells offer benefits over embryonic stem cells with limited lineage and ethical considerations (Liang & Zhang, 2013), thus promoting its use in regenerative medicine (Li et al., 2017; Imaizumi et al., 2013). Different types of cells were formed in a study where mixed cells such as human bronchial epithelial cells (hBECs), iPSC-derived mesenchymal stem cells (iPSC-MSCs), and iPSC-derived chondrocytes (iPSC-Chds) were used. The scaffold was cultivated in a bioreactor before transplantation. As a result, the ciliated columnar epithelium was formed in iPSC-MSC groups, whereas in iPSC-Chd groups, neocartilage was observed (Kim et al., 2020). Thus, iPSC may provide a promising approach to long-term functional reconstruction of a segmental tracheal defect. Furthermore, the cells could be used for tissue regenerating in transplants without the possibility of immune rejection as iPSC can autologously be harvested.

The current advanced and innovative technology, which attracts global attention, is the bioprinting technique. It is used in various applications, including regenerative medicine, tissue engineering, and transplantation, drug, and cancer research (Aljohani et al., 2018). The method utilizes a combination of biomaterials and polymers with choices of viable cells being concurrently three-dimensionally printed layer-by-layer in stacking pattern or by incorporating viable cells into 3D-printed biomaterials to fabricate tissues and organs (Ozbolat, 2015). Some advantages of this technique include accurate and regulated deposition of cells, hormones, drugs, growth factors, and the microarchitecture of the tissue constructs. In addition, high-throughput capability can be achieved, possible co-culture ability, and reduced risk of contamination leading to improved tissue regeneration (Peng et al., 2016). Park et al. attempted to fabricate artificial trachea using this method by printing multilayered PCL and hydrogel scaffold incorporated with nasal epithelial and auricular cartilage cells and transplanted into rabbits. Ciliated epithelial cells were identified over 3 months, but no cartilaginous tissue was observed. Neocartilage islets were seen only at 6 and 12 months (Park et al., 2019). Multicellular spheroids composed of chondrocytes, endothelial cells, and bone marrow MSCs within a scaffold-free tubular artificial trachea are cultivated inside the bioreactor with medium perfusion and transplanted into rats by Taniguchi and colleagues. Although chondrogenesis and angiogenesis were observed, the scaffold's structural integrity was compromised (Taniguchi et al., 2018). The bio-3D-printed scaffold-free artificial trachea requires silicone stent to attain sufficient mechanical strength to support the inner lumen during tracheal transplantation to avoid collapse in the acute phase after tracheal transplantation.

There is currently no absolute agreement on the optimal cell–scaffold combination or methods of seeding. Therefore, systematic in vivo experiments are needed to identify inconsistencies between tracheal seeding and types of cells, using reliable outcome assessments, including functionality and immune response.

2.3 Growth Factors

Growth factors are soluble secreted signalling polypeptide molecules that can instruct complex cellular responses during development and can be achieved in adult tissue regeneration. The ability to transmit a specific message to a distinct cell subpopulation is determined by various factors, including its identification and diffusion through ECMs, the target cell number, receptor type, and the intracellular signal transduction following factor binding. Depending on the receptor and cell type that the growth factor binds, the same receptor translates different messages depending on the intracellular transduction pathways, varying from one cell type to another. In addition, the binding ability to ECM, the ECM's degradation property, growth factor concentration, and cell target location are among other factors that govern the ultimate response of a target cell to a specific growth factor (Lee et al., 2011a).

Basic fibroblast growth factor (b-FGF) is one of the most popular growth factors used in tissue engineering due to its effectiveness in inducing angiogenesis and wound healing on smooth muscle cells, endothelial cells, fibroblasts, and epithelial cells (Suzuki et al., 2008; Przybylski, 2009). Meanwhile, bone morphogenetic proteins (BMPs) are from the transforming growth factor-beta (TGF-β) superfamily. BMPs play significant roles in the formation of bone and cartilage, including skeletogenesis, hematopoietic, and epithelial cell differentiation (Deng et al., 2018). In addition, the human recombinant erythropoietin (hrEPO) is a recombinant form of the hormone erythropoietin, characterized as a factor that stimulates differentiation and proliferation of erythroid precursor cells, a protective tissue factor. It is widely used for the treatment of patients with cancer-related anaemia (Engert, 2000).

Igai et al. investigated the effect of b-FGF in a gelatin sponge in tracheal defect dogs. The result showed that slow-releasing b-FGF induced cartilage regeneration after implantation (Igai et al., 2006; Igai et al., 2008). A similar result was reported by Li et al., who observed a significant increase in cartilage regeneration (Li et al., 2009). Furthermore, the combination of b-FGF and BMP-2 resulted in broader and earlier cartilage regeneration in beagle dogs using the controlled release of b-FGF and BMP-2 than that solely induced by the release of each growth factor (Igai et al., 2009). Apart from b-FGF and BMP-2, transforming growth factor-beta 3 (TGF-β3) was shown to cause an increase in angiogenesis (Ott et al., 2015). Other growth factors that promote angiogenesis and re-epithelialization processes are vascular endothelial growth factors (VEGF) (Lee et al., 2011b). The types of scaffold, cells, and growth factors used in the preclinical studies are summarized in Table 1.

3 Culture Systems

3.1 In Vitro Cultivation – Bioreactor

Bioreactors have emerged as an essential role in tissue engineering due to the factors needed for interactions between cells and biomaterials in cultivated mammalian cells. The dynamic system allows fluid flow and controllable conditions like pH, temperature, oxygen tension, cell perfusion, and external stimuli such as mechanical forces to assist the new tissue formation in vitro. Furthermore, the bioreactor system provides the biological and physical regulatory signals crucial for cell proliferation, differentiation, and development of ECM. Additionally, large-scale production can be made available, which cannot be achieved in a conventional static culture system (Takebe et al., 2012; Yoon et al., 2012; Liu et al., 2010).

Multiple studies have been documented using various cell types and scaffolds in bioreactors. Tan et al. and Lin et al., respectively, reported a high cell growth distribution on the scaffolds when cell seeding was performed using a perfusion system and rotational bioreactor. However, only chondrocytes were used in both studies. Thus the epithelial lining was not determined (Tan et al., 2009; Lin et al., 2009). Alternatively, Asnaghi et al. co-cultured autologous respiratory epithelial cells and MSCs onto a decellularized tracheal matrix within the bioreactor and transplanted them into a human body. Angiogenesis was confirmed, viable epithelial cells and chondrocytes were reported.

Similarly, Haykal et al. included epithelial cells and bone marrow–derived mesenchymal stem cells to a decellularized tracheal graft's internal and external surfaces in a bioreactor before transplanting into Yorkshire pigs. While attachment and distribution of the cells were higher and more than static culture, re-epithelialization reduction was observed within the first 2 weeks. Unfortunately, the animals did not survive due to postoperative respiratory distress and tracheal collapse.

The main aim of the bioreactor design is to simulate in vivo tissue formation regeneration efficiently. The cell attachment was greater when co-cultures were used, and cell distribution was even higher than static culture. Despite the inspiring results reported from the use of bioreactors, most bioreactors thus far still have a low-volume performance and require extensive time and effort to form tissues; therefore,

Table 1 Summary of types of scaffold/sources, animal, cell, and growth factors used in the pre-clinical studies

Source/type of scaffold	Type of animal	References	Cell/growth factor	Key findings
PGA mesh and arginine-glycine-aspartic acid (RGD)-modified alginate	Rabbit	(Grimmer et al., 2004)	Autologous chondrocytes	Mature cartilage formation at the tracheal reconstruction site
Composite scaffolds of a collagen sheet, a PGA mesh, and a poly (l-lactide/ε-caprolactone) coarse mesh	Rabbit	(Komura et al., 2008)	Gelatin sponge with b-FGF	Tracheal epithelial regeneration in the internal lumen of this composite scaffold. A gelatin sponge slowly releasing b-FGF might enhance chondrogenesis
Scaffold-free trachea from autologous chondrocytes	Rabbit	(Weidenbecher et al., 2008)	Chondrocytes	The engineered cartilage underwent a remodelling process, forming a solid tracheal framework with better mechanical properties
Poly(ε-caprolactone) (PCL) stent and a type II collagen sponge	Nude mice	(Lin et al., 2008)	Rabbit chondrocytes	Neocartilage is present with abundant proteoglycans and type II collagen
Decellularized trachea	Pigs	(Go et al., 2010)	Epithelial cells and mesenchymal stem cell–derived chondrocytes	Survival of the pigs seeded with both types of cells with no signs of airway collapse or ischemia and healthy epithelium
Fibrin/hyaluronic acid (HA) composite gel	Tracheal defects	(Kim et al., 2010)	Chondrocytes	The tracheal luminal contour and functional epithelial regeneration without graft rejection and inflammation
Collagen sponge with a spiral polypropylene stent and mesh	Rat tracheal defects	(Kobayashi et al., 2010)	Gingival fibroblasts (GFBs) and adipose-derived stem cells (ASCs)	Highly ciliated epithelium in the scaffolds with GFBs, while induction of thick epithelium, and pseudostratified epithelium containing goblet cells formation
The hybrid scaffold of poly(lactic-co-glycolic acid) (PLGA) knitted mesh and collagen sponge	Tracheal defect of rats	(Tatekawa et al., 2010)	b-FGF	Neocartilage was present in the b-FGF group, and limited new cartilage growth and epithelialization were observed at 6 months
Silk fibroin and chitosan blend	Nude mouse – Perichondrium wrapping	(Zang et al., 2011)	Chondrocytes	SFCS-supported chondrogenesis and increased mechanical properties of the regenerated tissue

(continued)

Table 1 (continued)

Source/type of scaffold	Type of animal	References	Cell/growth factor	Key findings
Fibrin/HA composite gel	Rabbit - 1 cm defects	(Hong et al., 2012)	Chondrocytes	Good tracheal luminal contour, functional epithelial regeneration, and preservation of neocartilage without inflammation but lacked adequate mechanical stability
3D-printed scaffold	Nude rat	(Imaizumi et al., 2013)	Induced pluripotent stem cells (iPSCs) in chondrocytes differentiation medium	Cartilage tissue in half of the rats implanted with induced iPSCs
Fibrin/HA hydrogel and degradable porous poly (L-lactic-co-glycolic acid) (PLGA)	Rabbit partial tracheal reconstruction	(Hong et al., 2014)	Chondrocytes	Complete coverage of respiratory epithelium, the good luminal contour of the trachea, and formation of neocartilage with minimal granulation tissue
3D-printed PCL coated with the MSCs seeded in fibrin	Tracheal defect in rabbits (10 × 10 mm)	(Chang et al., 2014)	MSCs	Complete coverage with regenerated respiratory mucosa. The scaffolds were integrated with the adjacent trachea without disruption or granulation tissue formation. Neo-cartilage formation inside the implanted graft was sufficient to maintain the patency of the reconstructed trachea
Decellularized trachea	Rabbit tracheal defects (1.8 cm)	(Batioglu-Karaaltin et al., 2015)	Adipose tissue-derived mesenchymal stem cells (MSCs)	Integration of MSCs seeded and increased angiogenesis
Electrospun poly(D, L-lactide-co-glycolide) (PLGA) (on outer surface) and polycaprolactone (PCL)	Tracheal defects in rabbit	(Ott et al., 2015)	Bone marrow mesenchymal stromal cells (BMSCs), transforming growth factor-beta 3 (TGF-β3)	The BMSC and TGF-β3 group had a greater accumulation of inflammatory cells over the graft, exhibiting normal epithelium, sub-epithelium, and cartilage formation.
Hybrid 3D-printed poly (ε-caprolactone) and human turbinate MSCs (hTMSC) sheets in the luminal surface	Rabbit	(Park et al., 2015b)	hTMSC	A mature and highly ciliated epithelium in grafts with hTMSC sheets
3D-printed PCL	Rabbit	(Gao et al., 2017)	Chondrocytes -cultivated in nude mice	The longer survival rate in the experiment group
Porous 3D silk fibroin	Rabbit tracheal defects	(Chen et al., 2018)	Cocultured of primary human tracheobronchial epithelial cells (hBECs)	Increased epithelium repair and epithelial cell differentiation degree

(continued)

Table 1 (continued)

Source/type of scaffold	Type of animal	References	Cell/growth factor	Key findings
Scaffold-free 3D-bioprinted	Rats	(Taniguchi et al., 2018)	Chondrocytes, endothelial and mesenchymal stem cells – Cultivated in a bioreactor	Chondrogenesis and angiogenesis were present. However, they required support from silicone stents to prevent collapse and until sufficient blood supply was obtained
Polycaprolactone (PCL) and hydrogel	Rabbits	(Park et al., 2019)	3D-bioprinted nasal epithelial and auricular cartilage cells in the printing process	Regeneration of respiratory epithelium present but not in cartilage
3D-printed poly (L-lactic acid) (PLLA)	Rabbits	(Gao et al., 2019)	Chondrocytes – Pre-cultured in vitro for 2 weeks and pre-vascularized in vivo for another 2 weeks. Transplanted with pedicled muscular flaps	Stable blood supply was provided by the pedicled muscular flaps and facilitated the survival of chondrocytes and accelerated epithelialization
PCL nanofibers electrospun (inner) and 3D-printed PCL microfibers (outer).	Rabbit tracheal defect	(Kim et al., 2020)	Human bronchial epithelial cells (hBECs), iPSC-derived mesenchymal stem cells (iPSC-MSCs), and iPSC-derived chondrocytes (iPSC-Chds)	Formation of ciliated columnar epithelium in iPSC-MSC groups and neocartilage in iPSC-Chd groups
Gelatin sponge	Tracheal defects in dogs	(Igai et al., 2006)	b-FGF	Regeneration of cartilage
Gelatin sponge	Dogs	(Igai et al., 2008)	BMP-2	Regeneration of fibrous cartilage and newly formed bone
Gelatin sponge	Dogs	(Igai et al., 2009)	b-FGF and BMP-2	Regeneration of cartilage in the b-FGF + BMP-2 and showed neovascularization

optimization is still needed to fulfil the requirements of clinical practices. Another option is to use 'body' or living bioreactors to counteract the limitation of in vitro bioreactors.

3.2 In Vivo Cultivation

Nude mice were used as living bioreactors before implanting composite biodegradable hydrogel and non-biodegradable HDPE as the internal predetermined shape scaffold seeded with human nasal septum cartilage, noted by Ruszymah et al. (Ruszymah et al., 2005). Likewise, nude mice were used as a natural bioreactor by Kajbafzadeh et al., which implanted a decellularized tracheal graft without any cell seeding. Findings demonstrated a well-formed neo-cartilage, and connective tissue was developed with angiogenesis, but epithelial was not observed. Therefore, cell seeding may be required to expedite re-epithelialization. Another essential intervention to ensure safe regeneration of tubular cartilage tissue and complete restoration of the airways was the in vivo implantation for pre-vascularization. Several studies have documented multiple in vivo implantation sites, such as subcutaneous, omental, and intramuscular implantations (Gao et al., 2019; Luo et al., 2013; Butler et al., 2017). Clinical in vivo implantation was demonstrated by Delaere et al. (Delaere et al., 2010), which involves wrapping the tracheal graft with host tissues, which is made possible. The wrapping allows rapid revascularisation of the graft and further regeneration of cartilage and epithelium, thus reducing graft failure chances. However, contradictory to the principle of using a bioreactor to expedite regenerative tissue formation, this approach has a limitation in terms of size of the implanted scaffold, as well as causing morbidity to the site involved due to multiple surgery requirements and extended time for cultivation.

4 Challenges of Tracheal Replacement

The airway is not a static structure. It moves during physiological movements such as respiration, swallowing, and coughing, which hinders the ingrowth of the foreign materials during healing, thus inhibiting the formation of the neotissues at the site. The internal site of the airway tract is not sterile and directly exposed to the inhaled air. This condition compromises the reparative capacity due to bacterial contamination at the interface site, which will prevent its ingrowth (Delaere et al., 2019). The most frequent failure of trachea replacement results from severe morbidity in the context of a blocked airway, respiratory impairment, and death. The replaced scaffolds should be able to promote epithelial growth to function in mucus secretions and clearing. In addition, the hyaline cartilage is avascular and relies on passive diffusion of nutrients from the vascular plexus in the submucosal layer. This condition provides a low regenerative ability of the cartilage to take over the integrity of the structure. The trachea relies on segmental blood flow, which penetrates through the cartilage rings to nourish the luminal mucosal lining. Thus, any transplanted tracheal scaffold should rapidly develop its blood supply to sustain cellular metabolism (Udelsman et al., 2018). Altogether, these factors must be considered in the design of the engineered tracheal scaffold, where presently, the complexity of these tasks is indicated in both preclinical and clinical studies (Chiang et al., 2016). It is recognized that re-epithelialization of the luminal part and restoration of cartilage tissues on the exterior surface are crucial aspects of effective tissue engineering to restore natural tracheal function and patency of regenerated tracheal tissue (Sekine et al., 2000; Dikina et al., 2015; Crowley et al., 2015).

It is generally accepted that a few challenging requirements are needed to replace the trachea, which is closely related to the anatomical features of the trachea, particularly its proximity to major vessels and segmental blood supply. In addition, the presence of multiple layers of tissues, including respiratory epithelium, submucosa, cartilage, and blood vessels, also contributes to the problems. To prevent chronic inflammation, granular tissue, infection, and erosion, the conduit must be virtually impermeable and become integrated into adjacent tissues. Tracheal substitution materials shall be biocompatible, innocuous, non-immunogenic, non-carcinogenic, and must not dislocate or eradicated with time. It should prevent the accumulation of secretions, resist bacterial colonization, and be permanently located at the site (Grillo, 2002). Despite multiple approaches, an ideal tracheal replacement has not been achieved due to numerous difficulties that hinder the reconstruction of the trachea.

4.1 Regeneration of Tracheal Cartilage

A chondrogenically stable engineered scaffold with excellent biocompatibility must endure and function in the implantation sites in the trachea. There is a lack of a chondro-inductive signal and is usually marked by an aggressive immune response (Liu et al., 2017a). During forced expiration, the hyaline cartilage provides mechanical support in the trachea, whilst the remaining is the trachealis muscle, which allows flexibility to the whole structure. A tracheal scaffold being the main structure in tracheal transplantation must be secured in place. A moderate inflammatory response caused by the material can interfere with its regeneration leading to unstable and collapsed airways (Liu et al., 2017b). Many tracheal transplants using autologous or allogenic trachea have shown compromised framework integrity, where in most cases, patients require initial stenting or post-operation re-stenting procedures (Wurtz et al., 2006; Zhang & Liu, 2015; Martinod et al., 2017; Fabre et al., 2013). Despite being immune to avascularity, cartilage relies on blood supplies from the surrounding tissues (Sykes, 2010). Therefore, vascularisation is crucial to keep the scaffold vitalized. It is characterized by the initial angiogenic response that promotes the proliferation of endothelial cells and the development of a capillary network.

4.2 Re-epithelialization of the Tracheal Lumen

One of the challenges in the fabrication and implantation of tracheal scaffold is to reproduce characteristic ciliated epithelium resembling the typical structure of the trachea to be physiologically functional (Park et al., 2019). However, the regenerative capability of the epithelium is limited, with the average half-life of ciliated cells being approximately 6 months, and it is susceptible to multiple stimuli and causes (Rawlins & Hogan, 2006; Rawlins & Hogan, 2008). The epithelial regeneration is accomplished by a relatively slow process involving stages in normal conditions, including dedifferentiation, migration, proliferation, and subsequent redifferentiation of the cells. Many soluble factors are secreted by the resident cells and infiltrating leukocytes which mediate the proliferation process. The member of the epidermal growth factor (EGF) and transforming growth factor (TGF) families were reported to produce a significant impact on the healing rate of the airway epithelium (Erjefält et al., 1995; Ryan et al., 1994). In the case of the only epithelial surface being damaged, adjacent cells immediately migrate and spread around the injured areas, proliferating and differentiating to completely seal the wound, generating a pseudostratified epithelium that may preserve its regular function. However, in more extensive injury, airway remodelling occurs where collagen is deposited, fibrosis and scarring may result in tracheal stenosis and obstruction (Zhang et al., 2015). Any delayed process leads to an interruption of this mucociliary ladder, local inflammation, and infection leading to stenosis (Soleas et al., 2012).

4.3 Revascularization of the Trachea

Sufficient blood flow is crucial for cartilage survival and epithelial reconstruction. Cartilage is nonvascular and depends on indirect nourishment from the adjacent tissues (Sykes, 2010). Due to its anatomical structure and intricate blood supply, restoring blood supply following a tracheal transplant is considered one of the most challenging organs to be transplanted. The mucosal lining of the trachea receives blood supply from tiny branches. Therefore, vascularization is necessary for adequate scaffold vitalization. It is marked by an initial angiogenic reaction, which induces proliferation and formation of endothelial cells and capillaries. This reaction initiates neoangiogenesis from pre-existing vessels needed for the scaffold's effective grafting in the neighbouring host tissues. Therefore, pre-vascularization is one of the prerequisites that must be fulfilled before successful implantation (Delaere et al., 1995; Schumann et al., 2009).

Furthermore, the effects from surgical preparation of the host induce a local inflammatory response and cause infiltration of leucocytes. This condition promotes angiogenesis through several angiogenic factors (Sung et al., 2004; Laschke et al., 2005; Möhle et al., 1997; Shaw et al., 2003). Nevertheless, an excessive inflammatory reaction can negatively affect the angiogenesis process and integrate cells onto the implanted scaffolds.

5 Perspective

While numerous reports have been published in recent decades concerning the regeneration of tracheal tissue in preclinical research, the findings appeared inconsistent. Multiple variables could influence the analysis, and comparison between different studies is heterogeneous and becomes extremely difficult. Therefore, standard study designs and outcome parameters such as histological, biochemical, and biomechanical evaluations are essential (Ahern et al., 2009). Another possible affecting variable is the duration of implantation in animals. The longer the pre-cultivation time in vitro or in vivo will eliminate post-implantation of an inflammatory reaction and, therefore, stable and stronger cartilage and epithelial can be regenerated. Prolonged implantation may simulate the actual condition in situ, allowing further investigation on cellular activities and reactions towards the implanted scaffold. In addition, the material's mechanical and biodegradability properties can be further analysed. These factors may affect the choice of materials, especially for paediatric patients. Finally, variations in the animal model may pose problems in extrapolating the accuracy of findings. Larger animals are more challenging to handle than small animals in their housing and post-operative care; thus, results are difficult to reproduce.

Furthermore, goats have more acute immune systems than smaller animals, affecting their regenerative capacity (Ahern et al., 2009). The wound size influences the healing time; however, the scale of tracheal defects produced in animal models is not standardized. In addition, smaller animals have more excellent intrinsic healing capability than larger animals (Xing et al., 2016); therefore, healing time would differ to a certain degree. All the strategies in tracheal tissue engineering are summarized in Fig. 4.

The clinical studies of tracheal replacement using different methods have been reported with mixed results. Notwithstanding the prerequisite for numerous stenting and bronchoscopic procedures, particularly within the early year of implantation, the tissue-engineered tracheal replacement using autologous and allograft sources was the most chosen clinical method. However, it is a great challenge to obtain a perfect match due to a shortage of donors. On the other hand, the successful clinical application of biodegradable scaffolds has not yet been proven. Furthermore, long-term post-operative results were not available. Therefore, a valid conclusion on the technique approach could not be verified.

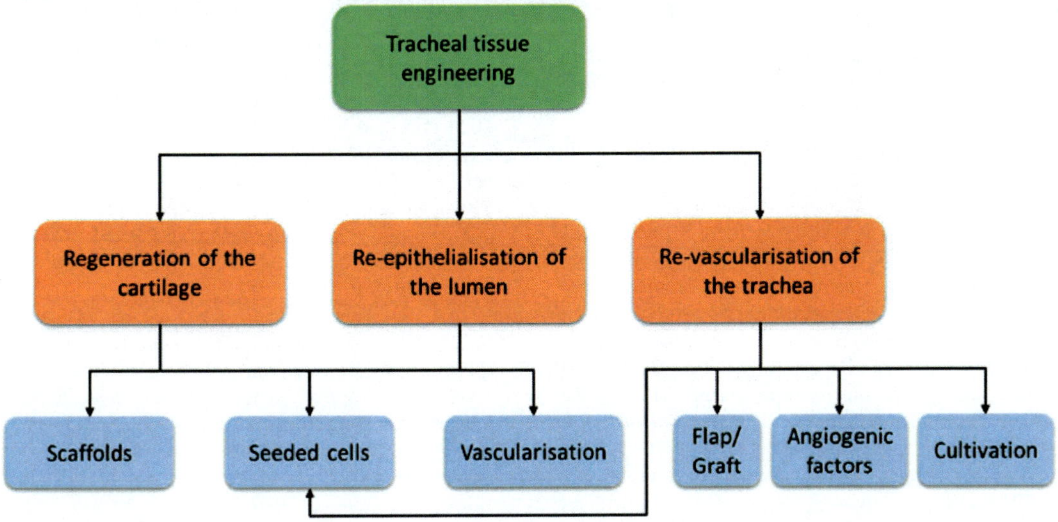

Fig. 4 Summary of the strategies and challenges in reconstructing tissue engineered trachea for trachea replacement

6 Conclusion

The human trachea requires a stable, rigid, but flexible framework to withstand the forces during respiration; a functional ciliated luminal epithelium as the first defence mechanism in the respiratory system; and adequate blood supply to nourish the whole structure. Despite being revolutionary and lifesaving, multiple obstacles need to be overcome to reconstruct large defects of the trachea. All problems are interrelated and mutually linked. Various approaches to transplantation have been studied; however, an optimal scaffold remains elusive. With advanced tissue engineering technologies, it is hoped that the airway tissue engineering solution will be successful in the immediate future.

Acknowledgement This research was funded by the Ministry of Higher Education, Malaysia, for the Prototype Research Grant Scheme (PRGS) with Project Code PRGS/1/2021/SKK07/USM/02/1 and Universiti Sains Malaysia (USM) Research University Grant (1001/CIPPT/8012203). Additionally, the author would like to thank the Ministry of Higher Education and the International Islamic University, Malaysia, for their support through the Academic Staff Training Scheme fellowship (ASTS) for Asmak Abdul Samat.

Conflict of Interest All authors declare no conflict of interest. The authors alone are responsible for the content and writing of the article.

References

Abdul Samat, A., Abdul Hamid, Z. A., Jaafar, M., & Yahaya, B. H.. (2021 September 13). Mechanical Properties and In Vitro Evaluation of Thermoplastic Polyurethane and Polylactic Acid Blend for Fabrication of 3D Filaments for Tracheal Tissue Engineering. *Polymers (Basel)* [Internet]. [cited 2021 Sep 20];13(18):3087. Available from: https://www.mdpi.com/2073-4360/13/18/3087/htm

Ahern, B. J., Parvizi, J., Boston, R., & Schaer, T. P. (2009, June). Preclinical animal models in single site cartilage defect testing: A systematic review. *Osteoarthritis and Cartilage* [Internet]. [cited 2020 Aug 3];17(6):705–13. Available from: https://pubmed.ncbi.nlm.nih.gov/19101179/

Aljohani, W., Ullah, M. W., Zhang, X., & Yang, G. (2018). Bioprinting and its applications in tissue engineering and regenerative medicine [Internet]. Vol. 107, *International Journal of Biological Macromolecules*. Elsevier B.V.; [cited 2020 Feb 9]. p. 261–75. Available from: https://doi.org/10.1016/j.ijbiomac.2017.08.171.

Alsberg, E., Kong, H. J., Hirano, Y., Smith, M. K., Albeiruti, A., & Mooney, D. J. (2003). Regulating bone formation via controlled scaffold degradation. *Journal of Dental Research, 82*(11), 903–908.

Altman, G. H., Diaz, F., Jakuba, C., Calabro, T., Horan, R. L., Chen, J., et al. (2003). Silk-based biomaterials. *Biomaterials, 24*(3), 401–416.

Bader, A., & Macchiarini, P. (2010). Moving towards in situ tracheal regeneration: The bionic tissue engineered transplantation approach. *Journal of Cellular and Molecular Medicine, 14*(7), 1877–1889.

Bailey, A. M., Mendicino, M., & Au, P. (2014). An FDA perspective on preclinical development of cell-based regenerative medicine products. *Nature Biotechnology*. Nature Publishing Group, *32*, 721–723.

Banas, A., Teratani, T., Yamamoto, Y., Tokuhara, M., Takeshita, F., & Quinn, G., et al. (2007, July). Adipose tissue-derived mesenchymal stem cells as a source of human hepatocytes. *Hepatology* [Internet]. [cited 2020 Jul 13];46(1):219–28. Available from: https://pubmed.ncbi.nlm.nih.gov/17596885/

Baruah, B., Sarawgi, M., Sahu, P., Dubey, K. P., Gupta, A., & Kumar, A. (2017). Polypropylene in endoscopic Dacryocystorhinostomy: A novel stent. *Indian Journal of Otolaryngology and Head & Neck Surgery, 70*(2), 1–4.

Batioglu-Karaaltin, A., Karaaltin, M. V., Ovali, E., Yigit, O., Kongur, M., & Inan, O., et al. (2015, April 1). In Vivo Tissue-Engineered Allogenic Trachea Transplantation in Rabbits: A Preliminary Report. *Stem Cell Review Reports* [Internet]. [cited 2020 Jul 11];11(2):347–56. Available from: https://link.springer.com/article/10.1007/s12015-014-9570-8

Behrend, M., Kluge, E., & Schüttler, W. (2006). On the use of unsealed polypropylene mesh as tracheal replacement. *ASAIO Journal, 52*(3), 328–333.

Bianco, P., Robey, P. G., & Simmons, P. J. (2008, April 10). Mesenchymal stem cells: Revisiting history, concepts, and assays. *Cell Stem Cell* [Internet]. [cited 2020 Jul 13];2(4):313–9. Available from: /pmc/articles/PMC2613570/?report=abstract.

Bogan, S. L., Teoh, G. Z., & Birchall, M. A. (2016 July 1). Tissue engineered airways: A prospects article. *Journal of Cellular Biochemistry* [Internet]. [cited 2020 Jan 15];117(7):1497–505. Available from: http://doi.wiley.com/10.1002/jcb.25512

Butler, C. R., Hynds, R. E., Crowley, C., Gowers, K. H. C., Partington, L., Hamilton, N. J., et al. (2017). Vacuum-assisted decellularization: An accelerated protocol to generate tissue-engineered human tracheal scaffolds. *Biomaterials, 124*, 95–105.

Caplan, A. I. (2017). Mesenchymal stem cells: Time to change the name! *Stem Cells Translational Medicine* [Internet]. [cited 2020 Jul 13];6(6):1445–51. Available from: https://pubmed.ncbi.nlm.nih.gov/28452204/

Chang, J. W., Park, S. A., Park, J. K., Choi, J. W., Kim, Y. S., & Shin, Y. S., et al. (2014). Tissue-engineered tracheal reconstruction using three-dimensionally printed artificial tracheal graft: Preliminary report. *Artificial Organs* [Internet]. [cited 2020 Jan 11];38(6). Available from: https://pubmed.ncbi.nlm.nih.gov/24750044/

Chang, C. S., Yang, C. Y., Hsiao, H. Y., Chen, L., Chu, I. M., Cheng, M. H., et al. (2018). Cultivation of auricular chondrocytes in poly(ethylene glycol)/poly (ε-caprolactone) hydrogel for tracheal cartilage tissue engineering in a rabbit model. *European Cells & Material, 35*, 350–364.

Chen, Q. (2010). A novel technique of tracheal reconstruction with autologous bronchial pedicle flap. *The Thoracic and Cardiovascular Surgeon, 58*(7), 427–428.

Chen, Z., Zhong, N., Wen, J., Jia, M., Guo, Y., Shao, Z., et al. (2018). Porous three-dimensional silk fibroin scaffolds for tracheal epithelial regeneration in vitro and in vivo. *ACS Biomaterials Science & Engineering, 4*(8), 2977–2985.

Chiang, T., Pepper, V., Best, C., Onwuka, E., & Breuer, C. K. (2016 Nov 1). Clinical translation of tissue engineered trachea grafts. *The Annals of Otology, Rhinology, and Laryngology* [Internet]. [cited 2019 Nov 4];125(11):873–85. Available from: /pmc/articles/PMC5800778/?report=abstract

Conconi, M. T., De Coppi, P., Di Liddo, R., Vigolo, S., Zanon, G. F., Parnigotto, P. P., et al. (2005). Tracheal matrices, obtained by a detergent-enzymatic method, support in vitro the adhesion of chondrocytes and tracheal epithelial cells. *Transplant International, 18*(6), 727–734.

Crowley, C., Birchall, M., & Seifalian, A. M. (2015, April). Trachea transplantation: From laboratory to patient. *Journal of Tissue Engineering and Regenerative Medicine* [Internet]. [cited 2019 Dec 24];9(4): 357–67. Available from: http://doi.wiley.com/10.1002/term.1847

Dal Pra, I., Freddi, G., Minic, J., Chiarini, A., & Armato, U. (2005). De novo engineering of reticular connective tissue in vivo by silk fibroin nonwoven materials. *Biomaterials, 26*(14), 1987–1999.

Dean, D., Topham, N. S., Meneghetti, S. C., Wolfe, M. S., Jepsen, K., He, S., et al. (2003). Poly(propylene fumarate) and poly(DL-lactic-co-glycolic acid) as scaffold materials for solid and foam-coated composite tissue-engineered constructs for cranial reconstruction. *Tissue Engineering*, 495–504.

Delaere, P., & Van Raemdonck, D. (2016). Tracheal replacement. *Journal of Thoracic Disease, 8*, S186–S196.

Delaere, P. R., Liu, Z. Y., Hermans, R., Sciot, R., & Feenstra, L. (1995). Experimental tracheal allograft revascularization and transplantation. *The Journal of Thoracic and Cardiovascular Surgery* [Internet]. [cited 2020 Jun 27];110(3):728–37. Available from: https://pubmed.ncbi.nlm.nih.gov/7564440/

Delaere, P., Vranckx, J., Verleden, G., De Leyn, P., Van Raemdonck, D. (2010, January 14). Tracheal allotransplantation after withdrawal of immunosuppressive therapy. *The New England Journal of Medicine* [Internet]. [cited 2020 Aug 3];362(2):138–45. Available from: https://pubmed.ncbi.nlm.nih.gov/20071703/

Delaere, P. R., Vranckx, J. J., Meulemans, J., Vander Poorten, V., Segers, K., Van Raemdonck, D., et al. (2012). Learning curve in tracheal allotransplantation. *American Journal of Transplantation, 12*(9), 2538–2545.

Delaere, P. R., Vranckx, J. J., & Hondt, M. Den. (2014). Tracheal allograft after withdrawal of immunosuppressive therapy [Internet]. Vol. 370, *New England Journal of Medicine*. Massachussetts Medical Society; [cited 2020 Aug 3]. pp. 1568–70. Available from: http://www.nejm.org/doi/10.1056/NEJMc1315273

Delaere, P., Van Raemdonck, D., & Vranckx, J. (2019). *Tracheal transplantation* (Intensive Care Medicine) (Vol. 45, pp. 391–393). Springer.

Deng, Z. H., Li, Y. S., Gao, X., Lei, G. H., & Huard, J. (2018). *Bone morphogenetic proteins for articular cartilage regeneration* (Osteoarthritis and Cartilage) (Vol. 26, pp. 1153–1161). W.B. Saunders Ltd.

Dikina, A. D., Strobel, H. A., Lai, B. P., Rolle, M. W., & Alsberg, E. (2015). Engineered cartilaginous tubes for tracheal tissue replacement via self-assembly and fusion of human mesenchymal stem cell constructs. *Biomaterials, 52*(1), 452–462.

Elliott, M. J., De Coppi, P., Speggiorin, S., Roebuck, D., Butler, C. R., Samuel, E., et al. (2012). Stem-cell-based, tissue engineered tracheal replacement in a child: A 2-year follow-up study. *Lancet* [Internet]. [cited 2020 Aug 31];380(9846):994–1000. Available from: https://pubmed.ncbi.nlm.nih.gov/22841419/

Engert, A. (2000). Recombinant human erythropoietin as an alternative to blood transfusion in cancer-related anaemia. *Disease Management & Health Outcomes, 8*(5), 259–272.

Erjefält, J. S., Erjefält, I., Sundler, F., & Persson, C. G. A. (1995, August). In vivo restitution of airway epithelium. *Cell Tissue Research* [Internet]. [cited 2020 Jul 12];281(2):305–16. Available from: https://link.springer.com/article/10.1007/BF00583399

Etienne, H., Fabre, D., Caro, A. G., Kolb, F., Mussot, S., Mercier, O., et al. (2018). Tracheal replacement. *European Respiratory Journal, 51*.

Fabre, D., Kolb, F., Fadel, E., Mercier, O., Mussot, S., Le Chevalier, T., et al. (2013). Successful tracheal replacement in humans using autologous tissues: An 8-year experience. *Annals of Thoracic Surgery*, 1146–1155.

Faraj, K. A., Van Kuppevelt, T. H., & Daamen, W. F. (2007). Construction of collagen scaffolds that mimic the three-dimensional architecture of specific tissues. *Tissue Engineering, 13*(10), 2387–2394.

Fuchs, E. (2008). Skin stem cells: Rising to the surface. *Journal of Cell Biology, 180*, 273–284.

Gao, M., Zhang, H., Dong, W., Bai, J., Gao, B., Xia, D., et al. (2017). Tissue-engineered trachea from a 3D-printed scaffold enhances whole-segment tracheal repair. *Scientific Reports, 7*(1).

Gao, B., Jing, H., Gao, M., Wang, S., Fu, W., Zhang, X., et al. (2019). Long-segmental tracheal reconstruction in rabbits with pedicled tissue-engineered trachea based on a 3D-printed scaffold. *Acta Biomaterialia, 97*, 177–186.

George, M., Lang, F., Pasche, P., & Monnier, P. (2005). Surgical management of laryngotracheal stenosis in adults. *European Arch Oto-Rhino-Laryngology, 262*(8), 609–615.

Gilbert, T. W., Stewart-Akers, A. M., & Badylak, S. F. (2007). A quantitative method for evaluating the degradation of biologic scaffold materials. *Biomaterials, 28*(2), 147–150.

Giwa, S., Lewis, J. K., Alvarez, L., Langer, R., Roth, A. E., Church, G. M., et al. (2017). The promise of organ and tissue preservation to transform medicine. *Nature Biotechnology, 35*(6), 530–542.

Go, T., Jungebluth, P., Baiguero, S., Asnaghi, A., Martorell, J., & Ostertag, H., et al. (2010, February). Both epithelial cells and mesenchymal stem cell-derived chondrocytes contribute to the survival of tissue-engineered airway transplants in pigs. *The Journal of Thoracic and Cardiovascular Surgery* [Internet]. [cited 2020 Jun 26];139(2):437–43. Available from: http://www.ncbi.nlm.nih.gov/pubmed/19995663

GonfiOtti, A., Jaus, M. O., Barale, D., Baiguera, S., Comin, C., Lavorini, F., et al. (2014). The fi rst tissue-engineered airway transplantation: 5-year follow-up results. *Lancet, 383*(9913), 238–244.

Gray, F. L., Turner, C. G., Ahmed, A., Calvert, C. E., Zurakowski, D., & Fauza, D. O. (2012). Prenatal tracheal reconstruction with a hybrid amniotic mesenchymal stem cells-engineered construct derived from decellularized airway. *Journal of Pediatric Surgery*, 1072–1079.

Grillo, H. C. (2002). Tracheal replacement: A critical review. *The Annals of Thoracic Surgery* [Internet]. [cited 2020 Sep 4];73(6):1995–2004. Available from: https://pubmed.ncbi.nlm.nih.gov/12078821/

Grimmer, J. F., Gunnlaugsson, C. B., Alsberg, E., Murphy, H. S., Kong, H. J., Mooney, D. J., et al. (2004). Tracheal reconstruction using tissue-engineered cartilage. *Archives of Otorhinolaryngology-Head & Neck Surgery, 130*(10), 1191–1196.

Haliloglu, T., Onar, V., Yildirim, G., Sapci, T., Savci, N., Kahvecioglu, O., et al. (2000). Tracheal reconstruction with porous high-density polyethylene tracheal prosthesis. *The Annals of Otology, Rhinology, and Laryngology, 109*(10 I), 981–987.

Hamilton, N. J., Kanani, M., Roebuck, D. J., Hewitt, R. J., Cetto, R., Culme-Seymour, E. J., et al. (2015). Tissue-engineered tracheal replacement in a child: A 4-year follow-up study. *American Journal of Transplantation, 15*(10), 2750–2757.

Haykal, S., Soleas, J. P., Salna, M., Hofer, S. O. P., & Waddell, T. K. (2012). Evaluation of the structural integrity and extracellular matrix components of tracheal allografts following cyclical decellularization techniques: Comparison of three protocols. *Tissue Engineering - Part C Methods, 18*(8), 614–623.

Haykal, S., Zhou, Y., Marcus, P., Salna, M., Machuca, T., Hofer, S. O. P., et al. (2013). The effect of decellularization of tracheal allografts on leukocyte infiltration and of recellularization on regulatory T cell recruitment. *Biomaterials, 34*(23), 5821–5832.

Haykal, S., Salna, M., Waddell, T. K., & Hofer, S. O. (2014). Advances in tracheal reconstruction. *Plastic and Reconstructive Surgery, 2*(7).

Heikal, M. Y. M., Aminuddin, B. S., Jeevanan, J., Chen, H. C., Sharifah, S. H., & Ruszymah, B. H. I. (2010 October). Autologous implantation of bilayered tissue-engineered respiratory epithelium for tracheal mucosal regenesis in a sheep model. *Cells Tissues Organs* [Internet]. [cited 2020 Feb 3];192(5): 292–302. Available from: https://pubmed.ncbi.nlm.nih.gov/20616535/

Hong, H. J., Lee, J. S., Choi, J. W., Min, B-H., Lee, H-B., Kim, C-H. (2012, November 1). Transplantation of autologous chondrocytes seeded on a fibrin/hyaluronan composite gel into tracheal cartilage defects in rabbits: Preliminary results. *Artificial Organs* [Internet]. [cited 2020 Jan 15];36(11):998–1006. Available from: http://doi.wiley.com/10.1111/j.1525-1594.2012.01486.x

Hong, H. J., Chang, J. W., Park, J. K., Choi, J. W., Kim, Y. S., Shin, Y. S., et al. (2014, November 1). Tracheal reconstruction using chondrocytes seeded on a poly (L-lactic-co-glycolic acid)-fibrin/hyaluronan. *Journal of Biomedical Materials Research - Part A* [Internet]. [cited 2020 Jan 31];102(11):4142–50. Available from: https://pubmed.ncbi.nlm.nih.gov/24443290/

Horan, R. L., Antle, K., Collette, A. L., Wang, Y., Huang, J., Moreau, J. E., et al. (2005). In vitro degradation of silk fibroin. *Biomaterials, 26*(17), 3385–3393.

Igai, H., Chang, S. S., Gotoh, M., Yamamoto, Y., Misaki, N., Okamoto, T., et al. (2006). Regeneration of canine tracheal cartilage by slow release of basic fibroblast growth factor from gelatin sponge. *ASAIO Journal, 52*(1), 86–91.

Igai, H., Chang, S. S., Gotoh, M., Yamamoto, Y., Yamamoto, M., Tabata, Y., et al. (2008). Tracheal cartilage regeneration and new bone formation by slow release of bone morphogenetic protein (BMP)-2. *ASAIO Journal, 54*(1), 104–108.

Igai, H., Chang, S. S., Gotoh, M., Yamamoto, Y., Yamamoto, M., & Tabata, Y., et al. (2009, May). Widespread and early tracheal cartilage regeneration by synchronous slow release of b-FGF and BMP-2. *ASAIO Journal* [Internet]. [cited 2020 Aug 2];55(3):266–70. Available from: https://pubmed.ncbi.nlm.nih.gov/19318920/

Imaizumi, M., Nomoto, Y., Sato, Y., Sugino, T., Miyake, M., Wada, I., et al. (2013). Evaluation of the use of induced pluripotent stem cells (iPSCs) for the regeneration of tracheal cartilage. *Cell Transplantation, 22*(2), 341–353.

Inoue, S., Tanaka, K., Arisaka, F., Kimura, S., Ohtomo, K., & Mizuno, S. (2000). Silk fibroin of Bombyx mori is secreted, assembling a high molecular mass elementary unit consisting of H-chain, L-chain, and P25, with a 6:6:1 molar ratio. *The Journal of Biological Chemistry, 275*(51), 40517–40528.

Jing, X., Mi, H. Y., Salick, M. R., Cordie, T., Crone, W. C., & Peng, X. F., et al. (2014, July 6). Morphology, mechanical properties, and shape memory effects of poly(lactic acid)/ thermoplastic polyurethane blend scaffolds prepared by thermally induced phase separation. Journal of Cellular Plastics [Internet]. [cited 2020 Jan 15];50(4):361–79. Available from: http://journals.sagepub.com/doi/10.1177/0021955X14525959

Jones, B., & Bes, M. (2012). Keeping kidneys. *Bulletin of the World Health Organization, 90*, 718–719.

Kang, N., Liu, X., Guan, Y., Wang, J., Gong, F., Yang, X., et al. (2012). Effects of co-culturing BMSCs and auricular chondrocytes on the elastic modulus and hypertrophy of tissue engineered cartilage. *Biomaterials, 33*(18), 4535–4544.

Keogh, M. B., O'Brien, F. J., & Daly, J. S. (2010). Substrate stiffness and contractile behaviour modulate the functional maturation of osteoblasts on a collagen-GAG scaffold. *Acta Biomaterialia, 6*(11), 4305–4313.

Kern, S., Eichler, H., Stoeve, J., Klüter, H., & Bieback, K. (2006, May). Comparative analysis of mesenchymal stem cells from bone marrow, umbilical cord blood, or adipose tissue. *Stem Cells* [Internet]. [cited 2020 Jul 13];24(5):1294–301. Available from: https://pubmed.ncbi.nlm.nih.gov/16410387/

Kim, W. S., Vacanti, J. P., Cima, L., Mooney, D., Upton, J., Puelacher, W. C., et al. (1994). Cartilage engineered in predetermined shapes employing cell transplantation on synthetic biodegradable polymers. *Plastic and Reconstructive Surgery, 94*(2), 233–237.

Kim, J. H. J., Suh, S. W., Shin, J. Y., Kim, J. H. J., Choi, Y. S., & Kim, H. (2004 July). Replacement of a tracheal defect with a tissue-engineered prosthesis: Early results from animal experiments. The Journal of Thoracic and Cardiovascular Surgery [Internet]. [cited 2020 Jun 28];128(1):124–9. Available from: https://pubmed.ncbi.nlm.nih.gov/15224031/

Kim, D. Y., Pyun, J. H., Choi, J. W., Kim, J. H., Lee, J. S., Shin, H. A., et al. (2010). Tissue-engineered allograft tracheal cartilage using fibrin/hyaluronan composite gel and its in vivo implantation. *The Laryngoscope, 120*(1), 30–38.

Kim, J. H., Kong, W. H., Kim, J. G., Kim, H. J., Seo, S. W. (2011, February). Possibility of Skin Epithelial Cell Transdifferentiation in Tracheal Reconstruction. *Artificial Organs* [Internet]. [cited 2020 Jun 28];35(2): 122–30. Available from: https://pubmed.ncbi.nlm.nih.gov/20946312/

Kim, I. G., Park, S. A., Lee, S. H., Choi, J. S., Cho, H., & Lee, S. J., et al. (2020, December 1). Transplantation of a 3D-printed tracheal graft combined with iPS cell-derived MSCs and chondrocytes. Sci Rep [Internet]. [cited 2020 Jul 11];10(1):1–14. Available from: https://doi.org/10.1038/s41598-020-61405-4.

Klimek, K., & Ginalska, G. (2020). Proteins and peptides as important modifiers of the polymer scaffolds for tissue engineering applications-A review. *Polymers (Basel), 12*(4).

Knoepfler, P. S. (2015). From bench to FDA to bedside: US regulatory trends for new stem cell therapies. *Advanced Drug Delivery Reviews*. Elsevier, *82*, 192–196.

Ko, C. Y., Ku, K. L., Yang, S. R., Lin, T. Y., Peng, S., Peng, Y. S., et al. (2016). In vitro and in vivo co-culture of chondrocytes and bone marrow stem cells in photocrosslinked PCL–PEG–PCL hydrogels enhances cartilage formation. *Journal of Tissue Engineering and Regenerative Medicine, 10*(10), E485–E496.

Kobayashi, K., Suzuki, T., Nomoto, Y., Tada, Y., Miyake, M., & Hazama, A., et al. (2010 June). A tissue-engineered trachea derived from a framed collagen scaffold, gingival fibroblasts and adipose-derived stem cells. *Biomaterials [Internet]*. [cited 2020 Jun 28];31(18):4855–63. Available from: https://pubmed.ncbi.nlm.nih.gov/20347137/

Kojima, K., & Vacanti, C. A. (2014). Tissue engineering in the trachea. *The Anatomical Record, 297*(1), 44–50.

Kojima, K., Bonassar, L. J., Roy, A. K., Mizuno, H., Cortiella, J., & Vacanti, C. A. (2003a). A composite tissue-engineered trachea using sheep nasal chondrocyte and epithelial cells. *The FASEB Journal, 17*(8), 823–828.

Kojima, K., Bonassar, L. J., Ignotz, R. A., Syed, K., Cortiella, J., & Vacanti, C. A. (2003b). Comparison of tracheal and nasal chondrocytes for tissue engineering of the trachea. *The Annals of Thoracic Surgery, 76*(6), 1884–1888.

Komura, M., Komura, H., Kanamori, Y., Tanaka, Y., Suzuki, K., Sugiyama, M., et al. (2008). An animal model study for tissue-engineered trachea fabricated from a biodegradable scaffold using chondrocytes to augment repair of tracheal stenosis. *Journal of Pediatric Surgery, 43*(12), 2141–2146.

Kucera, K. A., Doss, A. E., Dunn, S. S., Clemson, L. A., & Zwischenberger, J. B. (2007). Tracheal replacements: Part 1. *ASAIO Journal, 53*, 497–505.

Laschke, M. W., Häufel, J. M., Thorlacius, H., & Menger, M. D. (2005, September 15). New experimental approach to study host tissue response to surgical mesh materials in vivo. *Journal of Biomedical Materials Research - Part A* [Internet]. [cited 2020 Jun 27];74(4):696–704. Available from: https://pubmed.ncbi.nlm.nih.gov/16037956/

Laurent, T. C., Laurent, U. B. G., & Fraser, J. R. (1995). Functions of hyaluronan. *Annals of the Rheumatic Diseases, 54*(5), 429–432.

Law, J. X., Liau, L. L., & Aminuddin, B. S. (2016, December 1). Ruszymah BHI. Tissue-engineered trachea: A review. *International Journal of Pediatric Otorhinolaryngology* [Internet]. [cited 2019 Dec 17];91:55–63. Available from: https://pubmed.ncbi.nlm.nih.gov/27863642/

Lee, K., Silva, E. A., & Mooney, D. J. (2011a). Growth factor delivery-based tissue engineering: General approaches and a review of recent developments. *Journal of the Royal Society Interface, 8*(55), 153–170.

Lee, Y. C., Hung, M. H., Liu, L. Y., Chang, K. T., Chou, T. Y., Wang, Y. C., et al. (2011b). The roles of transforming growth factor-β 1 and vascular endothelial

growth factor in the tracheal granulation formation. *Pulmonary Pharmacology & Therapeutics, 24*(1), 23–31.
Lee, J. Y., Park, J. H., & Cho, D. W. (2018). Comparison of tracheal reconstruction with allograft, fresh xenograft and artificial trachea scaffold in a rabbit model. *Journal of Artificial Organs, 21*(3), 325–331.
Li, A., Pouliot, N., Redvers, R., & Kaur, P. (2004). Extensive tissue-regenerative capacity of neonatal human keratinocyte stem cells and their progeny. *The Journal of Clinical Investigation, 113*(3), 390–400.
Li, X., Wang, J., Ni, Y., Yan, X., Lu, Q., Xu, H., et al. (2009). Bone morphogenetic Protein-2 stimulation of cartilage regeneration in canine tracheal graft. *The Journal of Heart and Lung Transplantation, 28*(3), 285–289.
Li, Y. C., Zhu, K., & Young, T. H. (2017). Induced pluripotent stem cells, form in vitro tissue engineering to in vivo allogeneic transplantation [Internet]. Vol. 9, *Journal of Thoracic Disease*. AME Publishing Company; [cited 2020 Jul 13]. pp. 455–9. Available from: https://www.ncbi.nlm.nih.gov/pmc/articles/PMC5394022/
Liang, L., & Bickenbach, J. R. (2002). Somatic epidermal stem cells can produce multiple cell lineages during development. *Stem Cells, 20*(1), 21–31.
Liang, G., & Zhang, Y. (2013). Embryonic stem cell and induced pluripotent stem cell: An epigenetic perspective [Internet]. Vol. 23, *Cell Research*. [cited 2020 Jul 13]. p. 49–69. Available from: https://pubmed.ncbi.nlm.nih.gov/23247625/
Liao, Y. H., Jones, S. A., Forbes, B., Martin, G. P., & Brown, M. B. (2005). Hyaluronan: Pharmaceutical characterization and drug delivery. *Drug Delivery, 12*(6), 327–342.
Lin, C. H., Su, J. M., Hsu, S. H. (2008, March 1). Evaluation of type II collagen scaffolds reinforced by poly(ε-caprolactone) as tissue-engineered trachea. *Tissue Engineering - Part C Methods* [Internet]. [cited 2020 Jan 15];14(1):69–77. Available from: https://pubmed.ncbi.nlm.nih.gov/18454647/
Lin, C. H., Hui, H. S., Huang, C. E., Cheng, W. T., & Su, J. M. (2009). A scaffold-bioreactor system for a tissue-engineered trachea. *Biomaterials, 30*(25), 4117–4126.
Lippiello, L. (2003). Glucosamine and chondroitin sulfate: Biological response modifiers of chondrocytes under simulated conditions of joint stress. *Osteoarthritis and Cartilage, 11*(5), 335–342.
Liu, L., Wu, W., Tuo, X., Geng, W., Zhao, J., & Wei, J, et al. (2010, May). Novel strategy to engineer trachea cartilage graft with marrow mesenchymal stem cell macroaggregate and hydrolyzable scaffold. Artif Organs [Internet]. [cited 2020 Feb 19];34(5):426–33. Available from: https://pubmed.ncbi.nlm.nih.gov/20633157/
Liu, W., Xiao, P., Liang, H., An, R., Cheng, G., & Yu, F. (2013). Trachea repair and reconstruction with new composite artificial trachea transplantation. *Chinese Journal of Reparative and Reconstructive Surgery, 27*(3), 330–334.
Liu, Y., Zhou, G., & Cao, Y. (2017a). Recent Progress in cartilage tissue engineering—Our experience and future directions. *Engineering, 3*(1), 28–35.
Liu, Y. Y., Li, D., Yin, Z., Luo, X., Liu, W., Zhang, W., et al. (2017b, February 1). Prolonged in vitro precultivation alleviates post-implantation inflammation and promotes stable subcutaneous cartilage formation in a goat model. *Biomedical Materials* [Internet]. [cited 2020 Jan 11];12(1):015006. Available from: https://iopscience.iop.org/article/10.1088/1748-605X/12/1/015006
Luo, X., Zhou, G., Liu, W., Zhang, W. J., Cen, L., Cui, L., et al. (2009). In vitro precultivation alleviates post-implantation inflammation and enhances development of tissue-engineered tubular cartilage. *Biomedical Materials, 4*(2).
Luo, X., Liu, Y., Zhang, Z., Tao, R., Liu, Y., He, A., et al. (2013). Long-term functional reconstruction of segmental tracheal defect by pedicled tissue-engineered trachea in rabbits. *Biomaterials, 34*(13), 3336–3344.
Malafaya, P. B., Silva, G. A., & Reis, R. L. (2007). Natural-origin polymers as carriers and scaffolds for biomolecules and cell delivery in tissue engineering applications. *Advanced Drug Delivery Reviews, 59*(4–5), 207–233.
Martinod, E., Paquet, J., Dutau, H., Radu, D. M., Bensidhoum, M., Abad, S., et al. (2017). In vivo tissue engineering of Human airways. *The Annals of Thoracic Surgery, 103*(5), 1631–1640.
Mercier, O., Kolb, F., & Dartevelle, P. G. (2018). Autologous tracheal replacement: Surgical technique and outcomes. *Thoracic Surgery Clinics, 28*(3), 347–355.
Mhanna, R., & Hasan, A. (2017). Introduction to tissue engineering. In: *Tissue engineering for artificial organs* [Internet]. Weinheim, Germany: Wiley-VCH Verlag GmbH & Co. KGaA; [cited 2019 Dec 12]. pp. 1–34. Available from: http://doi.wiley.com/10.1002/9783527689934.ch1
Möhle, R., Green, D., Moore, M. A. S., Nachman, R. L, & Rafii, S. (1997, January 21). Constitutive production and thrombin-induced release of vascular endothelial growth factor by human megakaryocytes and platelets. *Proceedings of the National Academy of Sciences of the United States of America* [Internet]. [cited 2020 Jun 27];94(2):663–8. Available from: https://pubmed.ncbi.nlm.nih.gov/9012841/
Mooney, D., Hansen, L., Vacanti, J., Langer, R., Farmer, S., & Ingber, D. (1992). Switching from differentiation to growth in hepatocytes: Control by extracellular matrix. *Journal of Cellular Physiology, 151*(3), 497–505.
Mueller, D. K. (2018). Tracheal resection: Background, indications, Contraindications. *Medscape Med Journal* [Internet]. [cited 2019 Nov 26]; Available from: https://emedicine.medscape.com/article/1969880-overview
Murphy, C. M., O'Brien, F. J., Little, D. G., & Schindeler, A. (2013). Cell-scaffold interactions in the bone tissue engineering triad. *European Cells & Materials, 26*, 120–132.

Nematollahi, Z., Tafazzoli-Shadpour, M., Zamanian, A., Seyedsalehi, A., Mohammad-Behgam, S., Ghorbani, F., et al. (2017). Fabrication of chitosan silk-based tracheal scaffold using freeze-casting method. *Iranian Biomedical Journal, 21*(4), 228–239.

Nishinari, K., & Takahashi, R. (2003). Interaction in polysaccharide solutions and gels. *Current Opinion in Colloid & Interface Science, 8*(4–5), 396–400.

Nomoto, Y., Kobayashi, K., Tada, Y., Wada, I., Nakamura, T., & Omori, K. (2008). Effect of fibroblasts on epithelial regeneration on the surface of a bioengineered trachea. *The Annals of Otology, Rhinology, and Laryngology, 117*(1), 59–64.

O'Brien, F. J. (2011). Biomaterials & scaffolds for tissue engineering. *Materials Today, 14*(3), 88–95.

O'Brien, F. J., Harley, B. A., Yannas, I. V., & Gibson, L. J. (2005). The effect of pore size on cell adhesion in collagen-GAG scaffolds. *Biomaterials, 26*(4), 433–441.

Oh, S. H., Park, I. K., Kim, J. M., & Lee, J. H. (2007). In vitro and in vivo characteristics of PCL scaffolds with pore size gradient fabricated by a centrifugation method. *Biomaterials, 28*(9), 1664–1671.

Okumuş, A., Çizmeci, O., Kabakas, F., Kuvat, S. V., Bilir, A., & Aydin, A. (2005). Circumferential trachea reconstruction with a prefabricated axial bio-synthetic flap: Experimental study. *International Journal of Pediatric Otorhinolaryngology, 69*(3), 335–344.

Omori, K., Nakamura, T., Kanemaru, S., Asato, R., Yamashita, M., & Tanaka, S., et al. (2005 June). Regenerative medicine of the trachea: the first human case. *The Annals of Otology, Rhinology, and Laryngology* [Internet]. [cited 2020 Jun 27];114(6):429–33. Available from: http://www.ncbi.nlm.nih.gov/pubmed/16042099

Omori, K., Tada, Y., Suzuki, T., Nomoto, Y., Matsuzuka, T., Kobayashi, K., et al. (2008). Clinical application of in situ tissue engineering using a scaffolding technique for reconstruction of the larynx and trachea. *The Annals of Otology, Rhinology, and Laryngology, 117*(9), 673–678.

Ott, L. M., Vu, C. H., Farris, A. L., Fox, K. D., Galbraith, R. A., Weiss, M. L., et al. (2015, September 1). Functional reconstruction of tracheal defects by protein-loaded, cell-seeded, fibrous constructs in rabbits. *Tissue Engineering - Part A* [Internet]. [cited 2020 Aug 2];21(17–18):2390–403. Available from: https://pubmed.ncbi.nlm.nih.gov/26094554/

Ozbolat, I. T. (2015). *Bioprinting scale-up tissue and organ constructs for transplantation* (Trends in Biotechnology) (Vol. 33, pp. 395–400). Elsevier Ltd.

Park, J. H., Hong, J. M., Ju, Y. M., Jung, J. W., Kang, H. W., & Lee, S. J., et al. (2015a, September 1). A novel tissue-engineered trachea with a mechanical behavior similar to native trachea. *Biomaterials [Internet].* [cited 2020 Jul 11];62:106–15. Available from: https://pubmed.ncbi.nlm.nih.gov/26041482/

Park JH, Park JY, Nam IC, Hwang SH, Kim CS, Jung JW, et al. (2015b, October 1). Human turbinate mesenchymal stromal cell sheets with bellows graft for rapid tracheal epithelial regeneration. *Acta Biomaterials* [Internet]. [cited 2020 Aug 3];25:56–64. Available from: https://doi.org/10.1016/j.actbio.2015.07.014

Park, S. Y., Choi, J. W., Park, J. K., Song, E. H., Park, S. A., Kim, Y. S., et al. (2016). Tissue-engineered artificial oesophagus patch using three-dimensionally printed polycaprolactone with mesenchymal stem cells: A preliminary report. *Interactive Cardiovascular and Thoracic Surgery, 22*(6), 712–717.

Park, J. H., Yoon, J. K., Lee, J. B., Shin, Y. M., Lee, K. W., Bae, S. W., et al. (2019). Experimental tracheal replacement using 3-dimensional bioprinted artificial trachea with autologous epithelial cells and chondrocytes. *Scientific Reports, 9*(1), 1–11.

Peng, W., Unutmaz, D., & Ozbolat, I. T. (2016). Bioprinting towards physiologically relevant tissue models for pharmaceutics. *Trends in Biotechnology, 34*(9), 722–732.

Pieper, J. S., Oosterhof, A., Dijkstra, P. J., Veerkamp, J. H., & Van Kuppevelt, T. H. (1999). Preparation and characterization of porous crosslinked collagenous matrices containing bioavailable chondroitin sulphate. *Biomaterials, 20*(9), 847–858.

Propst, E. J., Prager, J. D., Meinzen-Derr, J., Clark, S. L., Cotton, R. T., Rutter, M. J.. (2011, June 20). Pediatric tracheal reconstruction using cadaveric homograft. *Archives of Otorhinolaryngology-Head & Neck Surgery* [Internet]. [cited 2020 Sep 4];137(6):583–90. Available from: https://jamanetwork.com/

Przybylski, M. (2009). A review of the current research on the role of bFGF and VEGF in angiogenesis. [Internet]. Vol. 18, *Journal of Wound Care*. [cited 2020 Jul 19]. p. 516–9. Available from: https://pubmed.ncbi.nlm.nih.gov/20081577/

Ravindran, S., Song, Y., & George, A. (2010, January). Development of three-dimensional biomimetic scaffold to study epithelial-mesenchymal interactions. *Tissue Engineering Part A* [Internet]. [cited 2020 Feb 3];16(1):327–42. Available from: http://www.ncbi.nlm.nih.gov/pubmed/19712044

Rawlins, E. L., & Hogan, B. L. M. (2006, July). Epithelial stem cells of the lung: Privileged few or opportunities for many? *Development* [Internet]. [cited 2020 Jun 27];133(13):2455–65. Available from: https://pubmed.ncbi.nlm.nih.gov/16735479/

Rawlins, E. L., & Hogan, B. L. M. (2008, July). Ciliated epithelial cell lifespan in the mouse trachea and lung. *The American Journal of Physiology-Lung Cellular and Molecular Physiology* [Internet]. [cited 2020 Jun 27];295(1):L231. Available from: /pmc/articles/PMC2494792/?report=abstract.

Rich, J. T., & Gullane, P. J. (2012, August). Current concepts in tracheal reconstruction. *Current Opinion in Otolaryngology & Head and Neck Surgery* [Internet]. [cited 2020 Sep 4];20(4):246–53. Available from: https://pubmed.ncbi.nlm.nih.gov/22894992/

Ris, H. B., Krueger, T., Cheng, C., Pasche, P., Monnier, P., & Magnusson, L. (2008). Tracheo-carinal reconstructions using extrathoracic muscle flaps. *European Journal of Cardio-Thoracic Surgery, 33*(2), 276–283.

Rosso, F., Marino, G., Giordano, A., Barbarisi, M., Parmeggiani, D., & Barbarisi, A. (2005 June). Smart materials as scaffolds for tissue engineering. *Journal of Cellular Physiology* [Internet]. [cited 2019 Nov 22];203(3):465–70. Available from: http://doi.wiley.com/10.1002/jcp.20270

Ruszymah, B. H. I., Chua, K., Latif, M. A., Nor Hussein, F., & Bin, S. A. (2005). Formation of in vivo tissue engineered human hyaline cartilage in the shape of a trachea with internal support. *International Journal of Pediatric Otorhinolaryngology, 69*(11), 1489–1495.

Ruszymah, B. H. I., Azrul Izham, B. A., Mohd Heikal, M. Y., Khor, S. F., Fauzi, M. B., & Aminuddin, B. S. (2011). Human respiratory epithelial cells from nasal turbinate expressed stem cell genes even after serial passaging. *The Medical Journal of Malaysia, 66*(5), 440–442.

Ryan, R. M, Mineo-Kuhn, M. M., Kramer, C. M., & Finkelstein, J. N. (1994). Growth factors alter neonatal type II alveolar epithelial cell proliferation. *The American Journal of Physiology-Lung Cellular and Molecular Physiology* [Internet]. [cited 2020 Jul 12];266 (1 10–1). Available from: https://pubmed.ncbi.nlm.nih.gov/8304465/

Scarritt, M. E., Pashos, N. C., & Bunnell, B. A. (2015). A review of cellularization strategies for tissue engineering of whole organs. *Frontiers in Bioengineering and Biotechnology. Frontiers Media S.A, 3*, Article 43.

Schumann, P., Tavassol, F., Lindhorst, D., Stuehmer, C., Bormann, K. H., & Kampmann, A., et al. (2009, September). Consequences of seeded cell type on vascularization of tissue engineering constructs in vivo. *Microvascular Research* [Internet]. [cited 2020 Jan 12];78(2):180–90. Available from: https://pubmed.ncbi.nlm.nih.gov/19540853/

Sekine, T., Nakamura, T., Matsumoto, K., Liu, Y., Ueda, H., Tamura, N., et al. (2000). Carinal reconstruction with a Y-shaped collagen-conjugated prosthesis. *The Journal of Thoracic and Cardiovascular Surgery, 119*(6), 1162–1168.

Shaw, J. P., Chuang, N., Yee, H., & Shamamian, P. (2003). Polymorphonuclear neutrophils promote rFGF-2-induced angiogenesis in vivo. *The Journal of Surgical Research, 109*(1), 37–42.

Shirai, T., Rao, V., Weisel, R. D., Ikonomidis, J. S., Li, R. K., Tumiati, L. C., et al. (1998). Preconditioning human cardiomyocytes and endothelial cells. *The Journal of Thoracic and Cardiovascular Surgery, 115*(1), 210–219.

Silva, M. M. C. G., Cyster, L. A., Barry, J. J. A., Yang, X. B., Oreffo, R. O. C., Grant, D. M., et al. (2006). The effect of anisotropic architecture on cell and tissue infiltration into tissue engineering scaffolds. *Biomaterials, 27*(35), 5909–5917.

Soleas, J. P., Paz, A., Marcus, P., McGuigan, A., & Waddell, T. K. (2012). Engineering airway epithelium. *Journal of Biomedicine & Biotechnology, 2012*.

Stenhamre, H., Nannmark, U., Lindahl, A., Gatenholm, P., & Brittberg, M. (2011). Influence of pore size on the redifferentiation potential of human articular chondrocytes in poly(urethane urea) scaffolds. *Journal of Tissue Engineering and Regenerative Medicine, 5*(7), 578–588.

Sung, H. J., Meredith, C., Johnson, C., & Galis, Z. S. (2004 November). The effect of scaffold degradation rate on three-dimensional cell growth and angiogenesis. *Biomaterials [Internet].* [cited 2019 Nov 21];25 (26):5735–42. Available from: https://pubmed.ncbi.nlm.nih.gov/15147819/

Suzuki, T., Kobayashi, K., Tada, Y., Suzuki, Y., Wada, I., Nakamura, T., et al. (2008). Regeneration of the trachea using a bioengineered scaffold with adipose-derived stem cells. *The Annals of Otology, Rhinology, and Laryngology, 117*(6), 453–463.

Sykes, M. (2010, January 14). Immune evasion by chimeric trachea. *The New England Journal of Medicine* [Internet]. [cited 2019 Dec 4];362(2):172–4. Available from: https://www.ncbi.nlm.nih.gov/pmc/articles/PMC4045014/

Takahashi, K., & Yamanaka, S. (2006, August 25). Induction of pluripotent stem cells from mouse embryonic and adult fibroblast cultures by defined factors. *Cell* [Internet]. [cited 2020 Jul 13];126(4):663–76. Available from: https://pubmed.ncbi.nlm.nih.gov/16904174/

Takebe, T., Kobayashi, S., Kan, H., Suzuki, H., Yabuki, Y., & Mizuno, M., et al. (2012). Human elastic cartilage engineering from cartilage progenitor cells using rotating wall vessel bioreactor. In: *Transplantation Proceedings* [Internet]. [cited 2020 Aug 3]. p. 1158–61. Available from: https://pubmed.ncbi.nlm.nih.gov/22564652/

Tan, Q., Hillinger, S., Van Blitterswijk, C. A., & Weder, W. (2009, January). Intra-scaffold continuous medium flow combines chondrocyte seeding and culture systems for tissue engineered trachea construction. *Interactive CardioVascular and Thoracic Surgery* [Internet]. [cited 2020 Aug 3];8(1):27–30. Available from: https://pubmed.ncbi.nlm.nih.gov/18550604/

Tang, X. L., Li, Q., Rokosh, G., Sanganalmath, S. K., Chen, N., Ou, Q., et al. (2016). Long-term outcome of administration of c-kitPOS cardiac progenitor cells after acute myocardial infarction: Transplanted cells do not become cardiomyocytes, but structural and functional improvement and proliferation of endogenous cells persist for at lea. *Circulation Research, 118*(7), 1091–1105.

Taniguchi, D., Matsumoto, K., Tsuchiya, T., Machino, R., Takeoka, Y., & Elgalad, A., et al. (2018, May 1). Scaffold-free trachea regeneration by tissue engineering with bio-3D printing†. *Interactive CardioVascular and Thoracic Surgery* [Internet]. [cited 2019 Nov 26];26(5):745–52. Available from: https://academic.oup.com/icvts/article/26/5/745/4810383

Tao, M. X., Chun, G. S., Qi, X. H., Deng, L., Zhi, W., Xiang, Z., et al. (2009). Variations in the ratios of co-cultured mesenchymal stem cells and chondrocytes regulate the expression of cartilaginous and osseous phenotype in alginate constructs. *Bone, 45*(1), 42–51.

Tatekawa, Y., Kawazoe, N., Chen, G., Shirasaki, Y., Komuro, H., & Kaneko, M. (2010). Tracheal defect repair using a PLGA-collagen hybrid scaffold reinforced by a copolymer stent with bFGF-impregnated gelatin hydrogel. *Pediatric Surgery International, 26*(6), 575–580.

Thomson, R. C., Wake, M. C., Yaszemski, M. J., & Mikos, A. G. (1995). Biodegradable polymer scaffolds to regenerate organs. *Advances in Polymer Science, 122*, 218–274.

Townsend, J. M., Ott, L. M., Salash, J. R., Fung, K. M., Easley, J. T., Seim, H. B., et al. (2018). Reinforced electrospun Polycaprolactone nanofibers for tracheal repair in an in vivo ovine model. *Tissue Engineering - Part A, 24*(17–18), 1301–1308.

Tsao, C. K., Ko, C. Y., Yang, S. R., Yang, C. Y., Brey, E. M., Huang, S., et al. (2014). An ectopic approach for engineering a vascularized tracheal substitute. *Biomaterials, 35*(4), 1163–1175.

Udelsman, B., Mathisen, D. J., & Ott, H. C.. (2018). A reassessment of tracheal substitutes-A systematic review [Internet]. Vol. 7, *Annals of Cardiothoracic Surgery*. AME Publishing Company; [cited 2019 Dec 23]. p. 175–82. Available from: www.annalscts.com

US Health Resources and Services Administration. Organ Donation Statistics [Internet]. July. 2019 [cited 2019 Dec 2]. Available from: https://www.organdonor.gov/statistics-stories/statistics.html#waiting-list

Viola, J., Lal, B., & Grad, O. (2003) The emergence of tissue engineering as a research field contract # EEC-9815425 task order A-07 Acknowledgements.

Walles, T. (2011). Tracheobronchial bio-engineering: Biotechnology fulfilling unmet medical needs. *Advanced Drug Delivery Reviews, 63*, 367–374.

Weidenbecher, M., Tucker, H. M., Awadallah, A., & Dennis, J. E. (2008). Fabrication of a neotrachea using engineered cartilage. *The Laryngoscope, 118*(4), 593–598.

Williams, D. (2004). Benefit and risk in tissue engineering. *Materials Today, 7*(5), 24–29.

Woodfield, T. B. F., Malda, J., De Wijn, J., Péters, F., Riesle, J., & Van Blitterswijk, C. A. (2004). Design of porous scaffolds for cartilage tissue engineering using a three-dimensional fiber-deposition technique. *Biomaterials, 25*(18), 4149–4161.

Wu, L., Leijten, J. C. H., Georgi, N., Post, J. N., Van Blitterswijk, C. A., & Karperien, M. (2011). Trophic effects of mesenchymal stem cells increase chondrocyte proliferation and matrix formation. *Tissue Engineering - Part A, 17*(9–10), 1425–1436.

Wu, T., Zhang, J., Wang, Y., Li, D., Sun, B., El-Hamshary, H., et al. (2018). Fabrication and preliminary study of a biomimetic tri-layer tubular graft based on fibers and fiber yarns for vascular tissue engineering. *Materials Science and Engineering: C, 82*, 121–129.

Wurtz, A., Porte, H., Conti, M., Desbordes, J., Copin, M. C., Azorin, J., et al. (2006). Tracheal replacement with aortic allografts. *The New England Journal of Medicine, 355*(18), 1938–1940.

Xia, D., Jin, D., Wang, Q., Gao, M., Zhang, J., Zhang, H., et al. (2019). Tissue-engineered trachea from a 3D-printed scaffold enhances whole-segment tracheal repair in a goat model. *Journal of Tissue Engineering and Regenerative Medicine, 13*(4), 694–703.

Xing, D., Chen, J., Yang, J., Heng, B. C., Ge, Z., & Lin, J. (2016). Perspectives on animal models utilized for the Research and Development of regenerative therapies for articular cartilage. *Current Molecular Biology Reports, 2*(2), 90–100.

Yin, H., Wang, J., Gu, Z., Feng, W., Gao, M., Wu, Y., et al. (2017). Evaluation of the potential of kartogenin encapsulated poly(L-lactic acid-co-caprolactone)/collagen nanofibers for tracheal cartilage regeneration. *Journal of Biomaterials Applications, 32*(3), 331–341.

Yoon, H. H., Bhang, S. H., Shin, J. Y., Shin, J., & Kim, B. S. (2012, October 1). Enhanced cartilage formation via three-dimensional cell engineering of human adipose-derived stem cells. *Tissue Engineering - Part A* [Internet]. [cited 2020 Aug 3];18(19–20):1949–56. Available from: https://pubmed.ncbi.nlm.nih.gov/22881427/

Zang, M., Zhang, Q., Davis, G., Huang, G., Jaffari, M., Ríos, C. N., et al. (2011). Perichondrium directed cartilage formation in silk fibroin and chitosan blend scaffolds for tracheal transplantation. *Acta Biomaterialia, 7*(9), 3422–3431.

Zeltinger, J., Sherwood, J. K., Graham, D. A., Müeller, R., & Griffith, L. G. (2001). Effect of pore size and void fraction on cellular adhesion, proliferation, and matrix deposition. *Tissue Engineering, 7*(5), 557–572.

Zhang, S., & Liu, Z. (2015). Airway reconstruction with autologous pulmonary tissue flap and an elastic metallic stent. *World Journal of Surgery, 39*(8), 1981–1985.

Zhang, Q., Lu, H., Kawazoe, N., & Chen, G. (2014). Pore size effect of collagen scaffolds on cartilage regeneration. *Acta Biomaterialia, 10*(5), 2005–2013.

Zhang, H., Fu, W., & Xu, Z. (2015, November 1). Re-epithelialization: A key element in tracheal tissue engineering. *Regenerative Medicine* [Internet]. [cited 2020 Jan 14];10(8):1005–23. Available from: https://pubmed.ncbi.nlm.nih.gov/26388452/

Zhao, X., Yu, F., Li, C., Li, Y., Chao, S. S., Loh, W. S., et al. (2012). The use of nasal epithelial stem/progenitor cells to produce functioning ciliated cells in vitro. *American Journal of Rhinology & Allergy, 26*(5), 345–350.

Adv Exp Med Biol - Innovations in Cancer Research and Regenerative Medicine (2023) 4: 165–179
https://doi.org/10.1007/5584_2023_767

Published online: 30 March 2023

The Rapid Development of Airway Organoids: A Direct Culture Strategy

Syahidatulamali Che Shaffi, Norashikin Zakaria, Nur Shuhaidatul Sarmiza Abdul Halim, Anan A. Ishtiah, Azim Ab Patar, and Badrul Hisham Yahaya

Abstract

Introduction The lung is a complex organ composed of numerous cell types. Exposure to air pollutants, cigarette smoke, bacteria, viruses, and many others may cause injury to the epithelial cells that line the conducting airways and alveoli. Organoids are the 3D self-organising structures grown from stem cells and generated from adult stem and progenitor cells. Lung organoids are fascinating tools to investigate human lung development in vitro. The objective of this study was to establish a rapid method for generating lung organoids with a direct culture strategy.

Methods Trachea and lung organoids were derived from mixed cell populations of mice primary airway epithelial cells, fibroblasts, and lung microvascular endothelial cells and directly digested from the whole cell population in the distal lung.

Results The formation of spheres appeared as early as 3 days and continued to proliferate until day 5. The generation of trachea and lung organoids self-organised into discrete epithelial structures was formed within less than 10 days.

Conclusion We conclude that researchers will be able to examine cellular involvement during organ formation and molecular networks because organoids come in a variety of morphologies and stages of development, and this organoid protocol may be used for modelling lung diseases as a platform for therapeutic purposes and suitable for personalised medicine for respiratory diseases.

Keywords

3D culture · Direct culture · Lung organoid · Trachea organoid

S. C. Shaffi, N. Zakaria, N. S. S. A. Halim, A. A. Ishtiah, and B. H. Yahaya (✉)
Lung Stem Cell and Gene Therapy Group, Department of Biomedical Sciences, Advanced Medical and Dental Institute (IPPT), SAINS@BERTAM, Universiti Sains Malaysia, Penang, Malaysia
e-mail: badrul@usm.my

A. A. Patar
Department of Neurosciences, School of Medical Sciences, Universiti Sains Malaysia, Kubang Kerian, Kelantan, Malaysia

1 Introduction

The lung is a complex organ composed of interconnected epithelial stromal cells, including vascular cells, immune cells, and extracellular matrix (ECM), synergising to maintain lung integrity. Exposure to air pollutants, cigarette smoke, bacteria, viruses, and many others in the lung may cause injury to the epithelial cells that line the conducting airways and alveoli.

Repetitive exposure to these insults may result in an inflammatory storm contributing to disease progression and respiratory failure. Chronic obstructive pulmonary disease (COPD), idiopathic pulmonary fibrosis (IPF), bronchial asthma, and respiratory infections like the current COVID-19 pandemic and others are among the respiratory illnesses that cause severe damage to the lungs and leads to an increase in mortality rate (Kong et al., 2021; Zhu et al., 2020). Therefore, establishing suitable models to study human respiratory diseases is critically important in advancing our understanding of tissue homeostasis mechanisms and pathological changes in the airways. Recently, advances in the understanding of stem cell niches and the role of crucial signalling modulators in controlling stem cell maintenance and differentiation have fuelled the development of new in vitro three-dimensional (3D) culture technologies that sustain the stem cell-driven formation of near-physiological, self-organising structures called organoids (Wörsdörfer et al., 2020).

Lung organoids are fascinating tools to investigate human lung development in vitro. They raise hopes of understanding the mechanism of diseases affecting the human lung. Pluripotent stem cell-based organoid cultures are state-of-the-art in vitro platforms that recapitulate fundamental aspects of organogenesis, allowing researchers to model and investigate human lung development and diseases (Sahabian et al., 2019). Moreover, they represent promising tools for drug discovery and toxicity testing. However, using induced pluripotent stem cells (iPSCs) involves an extra step, as somatic cells must first be transformed into iPSCs by expressing transcriptional factors such as *OCT4*, *KLF4*, *SOX2*, and *MYC*. Following that, iPSCs are subjected to germ layer and tissue-specific patterning factors followed by embedding in Matrigel to aid 3D architecture development and treatment with differentiation factors to create the desired organoids. These complex stepwise differentiation protocols require many growth factors and supplements that are not cost-effective and time-consuming. This is because pluripotent stem cell tissues generated through directed differentiation are immature; they can be used to study human diseases that are immature or premature. For example, Snoeck and co-workers generated lung bud organoids from human iPSCs that mimic foetal lung development (Chen et al., 2017). However, since iPSC-derived organoids do not cause the mature cell types observed in adult tissues, they may not accurately replicate adult disease presentations in many cases (Sachs et al., 2019). This is especially true in lung cancer, where it looks like a waste of effort to generate iPSCs from the tumour rather than directly producing cancer organoids. Organoids produced from adult stem cells (ASC) mimic the more advanced stages of the human lung.

In contrast to iPSC, organoids generated from the adult stem and progenitor cells reliably preserve their in vivo regeneration activity in vitro, allowing for comprehensive pictures of tissue repair following damage. Moreover, organoids maintain their organ identity and are genetically stable throughout time (Lu et al., 2021). Organ development, tissue homeostasis, and illnesses may all be studied because of the capacity to generate organoids from patient-derived healthy and sick tissue.

Recent studies on liver organoids suggest that intercellular signalling between mesenchymal cells, endothelial cells, and hepatocytes is required for proper organoid maturation, and similar interactions likely play a role in other tissues and organ systems as well (Asai et al., 2017). Another study has found that a culture protocol for lung stem/progenitor cells that includes Matrigel, feeder cells, and lung fibroblast to provide a cocktail of growth factors is less supportive for distal lung stem/progenitor cell organoid culture, most likely due to insufficient concentrations of essential growth factors (Barkauskas et al., 2013). When stromal cells are replaced with large amounts of FGF10 and hepatocyte growth factor, lung stem/progenitor cells form organoids with limited colony-forming capacity, implying that additional growth factors are required for alveolar organoid formation (Lu et al., 2021). Other studies also found that incorporating mesenchymal stem cells into lung organoids improved alveolar differentiation (Leeman et al.,

2019). These results demonstrate that multi-lineage cell communication drives tissue differentiation and maturation in organoid culture. Hence, the objective of this study was to introduce a direct co-culture system that is numerous, feasible, rapid, and robust in the development of lung organoids. A mixed population of airway epithelial cells, lung fibroblast, and microvascular endothelial cells was randomly seeded in Matrigel™ and cultured with an organoid culture medium containing minimal supplements to generate the lung organoid. The mixed cell population has generated proximal airway and distal lung organoids that self-organised into a discrete epithelial structure within less than 10 days. Our findings suggest that the supporting cells within the lung and tracheal cells secret growth factor cocktails that are critical for the rapid and robust generation of lung organoids. Hence, our improved method in this study showed that the rapid generation of organoid culture would give the ability to generate organoids from individual patients, enabling the personalised testing of chemotherapies and targeted therapies. In addition, the advantages of the lung organoid culture are propagation ability, amenability for high throughput screening, and the feasibility of establishing a proximal or distal lung model.

2 Materials and Methods

2.1 Ethics Statement

This study was approved by the Universiti Sains Malaysia Animal Ethics Committee [USM/IACUC/2020/(122)(1050)]. Mice were housed and maintained in an air-conditioned room, kept on a 12-h light/dark cycle, and had access to clean food and water.

2.2 Animals

A total of six Balb/c mice (male, 8 weeks old, weight 19–25 grammes) were used in this study. The animals were grouped into two groups ($n = 3$); (1) Group A, without ROCK inhibitor, TGF-β inhibitor, R-spondin, and noggin, and (2) Group B – with the addition of ROCK inhibitor, TGF-b inhibitor, R-spondin, and noggin group (Table 1). The laboratory mice were euthanised by intravenous injection of sodium pentobarbital (200 mg/ml) (Virbac, Carros, France).

2.3 Culture Medium

The trachea and lung organoids were grown in a complete culture medium consisting of the basal medium of DMEM/F12 (Gibco, Thermo Scientific, MA, USA) supplemented with 5% foetal bovine serum (FBS) (Gibco), 2 mM GlutaMAX™ (Gibco), 25 ng/μL epidermal growth factors (Gibco), 10 μg/mL), insulin (Elabscience, MD, Texas, USA), 5 μg/mL transferrin (Gibco), and 15 mg protein/mL bovine pituitary extract (Gibco). Proliferation medium was prepared by adding 10 μM of Rho-associated protein kinase (ROCK) inhibitor (Cayman Chemical, Michigan, USA), 10 μM of transforming growth factor-β (TGF-β) type I receptor inhibitor (Cayman Chemical) to the complete culture media with the addition of recombinant mouse noggin protein (25ug), and

Table 1 Comparison of supplements used in group A and group B organoid culture for the trachea and lung

	Trachea		Lung	
Supplements	Group A	Group B	Group A	Group B
[a]ROCK inhibitor	–	+	–	+
[a]TGF-β inhibitor	–	+	–	+
R-spondin			–	+
Noggin			–	+

[a]ROCK inhibitor and TGF-β were removed in differentiation medium

recombinant mouse R-spondin (25ug) to the lung organoid media. On the other hand, organoid differentiation media consist of similar complete culture media with reduced FBS to 2% and removed ROCK inhibitor and TGF-β inhibitor.

2.4 Isolation of Trachea and Lung Epithelial Cells

The isolation of airway and lung epithelial cells was performed as described by Barkauskas et al. (2013), with some modifications. The mice were euthanised by injection of sodium pentobarbital (200 mg/kg) and immediately followed by dissecting using sterile equipment to collect the lung and trachea. Briefly, the mouse body was sprayed with 70% alcohol, followed by an incision to expose the thoracic cavity. The trachea and lung organs were collected and placed in a 15 ml tube containing ice-cold sterile phosphate-buffered saline (PBS) supplemented with 100 μl/ml antibiotic/antimycotic solution containing 10,000 units/ml of penicillin, 10 mg/ml of streptomycin, and 25 μg/ml amphotericin B (Sigma-Aldrich, Missouri, USA) and transport to cell culture lab. Under the aseptic condition, excess fat surrounding the tissue was removed. The trachea tissue was opened lengthwise and cut into 1 cm^2 piece, and the lung tissues were minced until they became slurry. Tissue from both samples was digested overnight at 4 °C (continuous shaking at 200 rpm) in a digestion medium containing 1 mg/ml Pronase (Gibco) and 100 μl/ml antibiotic/antimycotic solutions prepared in Modified Hanks' Balanced Salt Solution (HBSS) (Sigma-Aldrich). After overnight digestion, a cell scraper was used to remove the epithelium from the trachea and the lung. The epithelial cells were transferred into a 15 ml tube containing 0.25% trypsin/EDTA (Gibco), incubated for 5 min at 37 °C, and passed through a 70 μm cell strainer. The digestion was stopped by adding 10% FBS, and the cells were pelleted by centrifugation at approximately 600 × g for 5 min at 4 °C (Eppendorf 5804 R, Hamburg, Germany). After resuspension, red blood cells were removed by incubation in ammonium chloride (red blood cell lysing buffer; Sigma-Aldrich) for 10 min, followed by centrifugation at approximately 600 × g for 5 min at 4 °C to remove the lysis buffer. The epithelial cell pellet was washed twice with PBS, centrifuged at 600 × g for 5 min at 4 °C, and resuspended complete culture medium, followed by cell counting.

2.5 Lung and Trachea Organoid Culture

The organoid culture was established from the isolated lung and trachea epithelial cells by culturing the cells in growth factor reduced Matrigel™ (Corning, NY, USA). Briefly, 25×10^3 cells and 50 × 103 cells from the trachea and lung were mixed with thawed Matrigel (on ice). Then, two 15 μl drops of the Matrigel-cell suspension were pipetted into wells of a Nunc™ 24-well-Nunclon Delta-treated plate (Thermo Fisher Scientific) and allowed to solidify for 15 min at 37 °C. Then 400 μl of trachea or lung organoid proliferation medium respective added into the well. The cultures were monitored daily to assess the formation of organoids. The cells were grown for 6 days with medium change every 3 days. After day 6, the cell will form a spheroid and to promote the maturation and differentiation of the spheroids into organoids, the medium was changed to a medium containing 2% FBS (instead of 5% in the proliferation medium) and with/without ROCK and TGF-β inhibitors.

2.6 Passaging and Cryopreservation of Trachea and Lung Organoid Culture

Once the necrotic cells were visible, organoid cultures were subcultured with a fresh proliferation medium. Briefly, the culture medium was removed without disturbing the matrix dome, and the dome was carefully washed using the basal medium. The matrix dome was dissolved, and the organoids were dissociated by adding TrypLE™ Express Enzyme (Gibco). The dome was broken up by gently pipetting the enzyme solution up and down until the organoids were released from the matrix. Then, the organoids were transferred into a new 15 ml tube and incubated for 10–15 min at 37 °C. The digestion was stopped by adding 10% FBS, and the cells were pelleted by centrifugation at about 100 × g for 5 min at 4 °C. The cells were counted and resuspended in thawed Matrigel™ and re-cultured as described above. The remaining cells were cryopreserved in a freezing medium containing a complete culture medium containing 10 μM ROCK inhibitor and 10% DMSO (Sigma-Aldrich).

2.7 Organoid Fixation and Processing

Mature organoids were harvested on day 10 of culture for histological examination, such as H&E staining, immunofluorescence, and immunohistochemistry staining. Briefly, the medium was discarded, the matrix dome containing the organoids was washed with PBS, and the matrix was gently broken up by pipetting through 1000 μl pipette tips with their end cut to increase the bore size. The organoids were then transferred to a 15 ml tube, pelleted by centrifugation at approximately 100 × g for 5 min at 4 °C, and fixed in 10% buffered formalin overnight at 4 °C. Then, the cell block was prepared by mixing the organoid with in-house prepared CytoQuick gel solution and centrifuged for 10 min at 600 × g, and the gel was allowed to solidify for 10 min at 4 °C. Finally, the cell block containing organoids was processed in a standard automated tissue histology processor, embedded in paraffin, and sectioned into 3 μm thick slices using a semi-automated rotary microtome (Leica Biosystems, Wetzlar, Germany) and mounted on frosted-edge slides and poly-L-lysine-coated slides for histological staining and immunostaining, respectively.

2.8 Haematoxylin and Eosin (H&E) Staining

A standard H&E staining procedure was performed to evaluate the structure and composition of the spheroids and organoids. The procedure involved a 2-min deparaffinisation step with xylene followed by a rehydration step with graded ethanol solutions (100%, 90%, 80%, and 70%) and washed in tap water (2 min for each step). Next, samples were stained with Harris' haematoxylin for 5 min, followed by incubation with 0.5% acid alcohol for 2 s, tap water for 2 min, 0.2% ammonia water for 20 s, and tap water for 2 min. Next, samples were counterstained with eosin for 3 min and dehydrated using graded ethanol solutions (70%, 80%, 90%, and 100%; each for 10 s), followed by a final clearing step with xylene for 2 min before the slides were mounted using dibutyl phthalate polystyrene xylene (DPX; Sigma-Aldrich). Unless otherwise stated, all reagents used are from Sigma-Aldrich.

2.9 Immunofluorescence

Organoid sections on poly-L-lysine-coated glass slides were heated to 60 °C and dipped in xylene for 3 min to remove the surrounding paraffin. The sections were rehydrated with graded ethanol solutions (100%, 90%, 80%, and 70%) and PBS (for 3 min each). The antigen retrieval was performed in 10 mM citrate buffer and heated for 20 min at 95 °C. The slides were allowed to cool down and washed with PBS for 5 min. Next, the tissue was permeabilised with 0.25% Triton X-100 (Sigma-Aldrich) in PBS for 10 min at room temperature, washed with PBS for 5 min, and blocked with 5% bovine serum albumin (BSA; Sigma-Aldrich) in PBS-Tween 20 (Sigma-Aldrich) for 20 min. The following antibodies were used: anti-Cytokeratin 14, anti-Cytokeratin 5 + 8 (25 μg/ml each; Thermo Fisher Scientific), anti-CD44, anti-Mucin 5 ac, and anti-pan-cytokeratin (10 μg/ml each; Abcam). Meanwhile, the antibodies used for lung organoids were SFTPC (1 mg/ml; Elabscience), AQP5 (200 μg/ml; Santa Cruz), and CC10 (10 μg/ml; Santa Cruz). All primary antibodies were diluted in PBS-Tween 20 containing 1% BSA, placed into a humidified chamber, and incubated overnight at 4 °C. Slides were then washed two times with PBS for 5 min, and then appropriate secondary antibodies, diluted in PBS with 1% BSA, were added. The slides were incubated at room temperature for another 40 min, washed two times with PBS, and counterstained with 4′,6-diamidino-2-phenylindole (DAPI) (Sigma-Aldrich) for 5 min at room temperature and mounted with Fluoroshield™ (Sigma-Aldrich). All staining was examined under a bright-field phase-contrast/fluorescence-inverted TI-U microscope (Nikon, Tokyo, Japan). Image analysis and processing were performed with NIS-Elements software (Nikon).

2.10 Imaging Analysis

Both trachea and lung organoid cultures were observed daily, and the culture images were captured using an inverted microscope. Ten random images were captured for each sample. In addition, semi-quantitative analysis was used to count the number of organoid formations, and the organoid sizes for both trachea and lung organoids were measured.

Figure 1 shows the schematic diagram of the steps involved in the development of organoids from trachea and lung tissues.

2.11 Statistical Analysis

Statistical analysis was performed using GraphPad Prism 9 version 9.0 (GraphPad Software, California, USA). The number and size of trachea and lung organoid cultures were tested using a two-tailed Student's *t*-test analysis to assess the statistical significance of differences between the two experimental groups. Differences were considered statistically significant when $p \leq 0.05$, whereas $p \leq 0.0001$ was considered highly effective.

3 Results

3.1 ROCK Inhibitor Increases the Number and Size of Organoids

In this study, organoids from a whole-cell population were isolated and developed from the lung and trachea tissue of mice. The initial stage of organoid development starts with the formation of spheres which begin to appear as early as 3 days in culture and continue to proliferate and

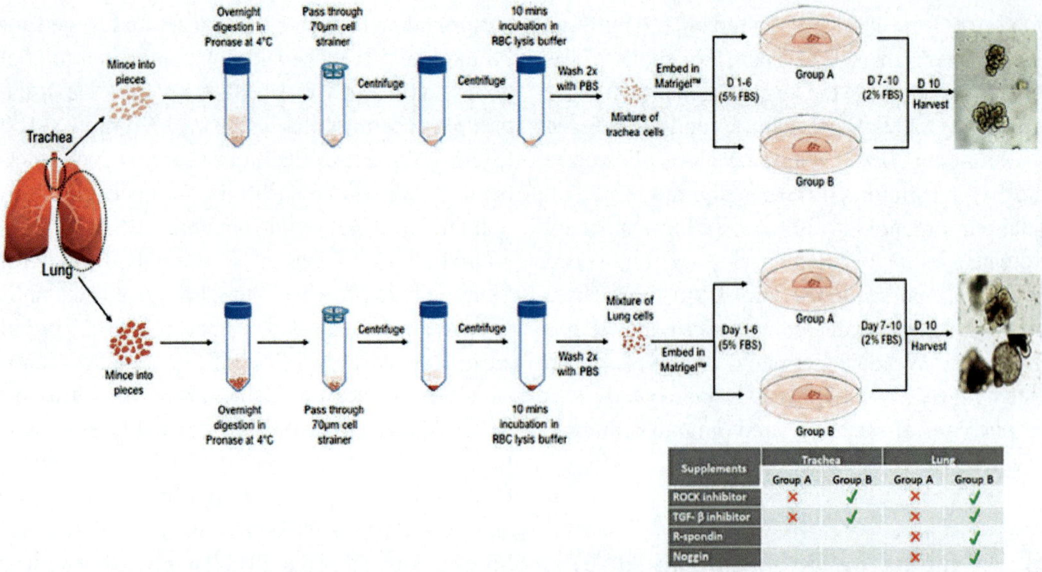

Supplements	Trachea		Lung	
	Group A	Group B	Group A	Group B
ROCK inhibitor	✗	✓	✗	✓
TGF-β inhibitor	✗	✓	✗	✓
R-spondin			✗	✓
Noggin			✗	✓

Fig. 1 Illustration of the overview methodology for the development of organoids from the trachea and lung. Trachea and lung organ were collected and continued with overnight digestion before proceeding to organoid culture procedure. The culture was divided into two groups, Group A, without ROCK inhibitor, TGF-β inhibitor, R-spondin, and noggin, and (2) Group B – with the addition of ROCK inhibitor, TGF-b inhibitor, R-spondin, and noggin group

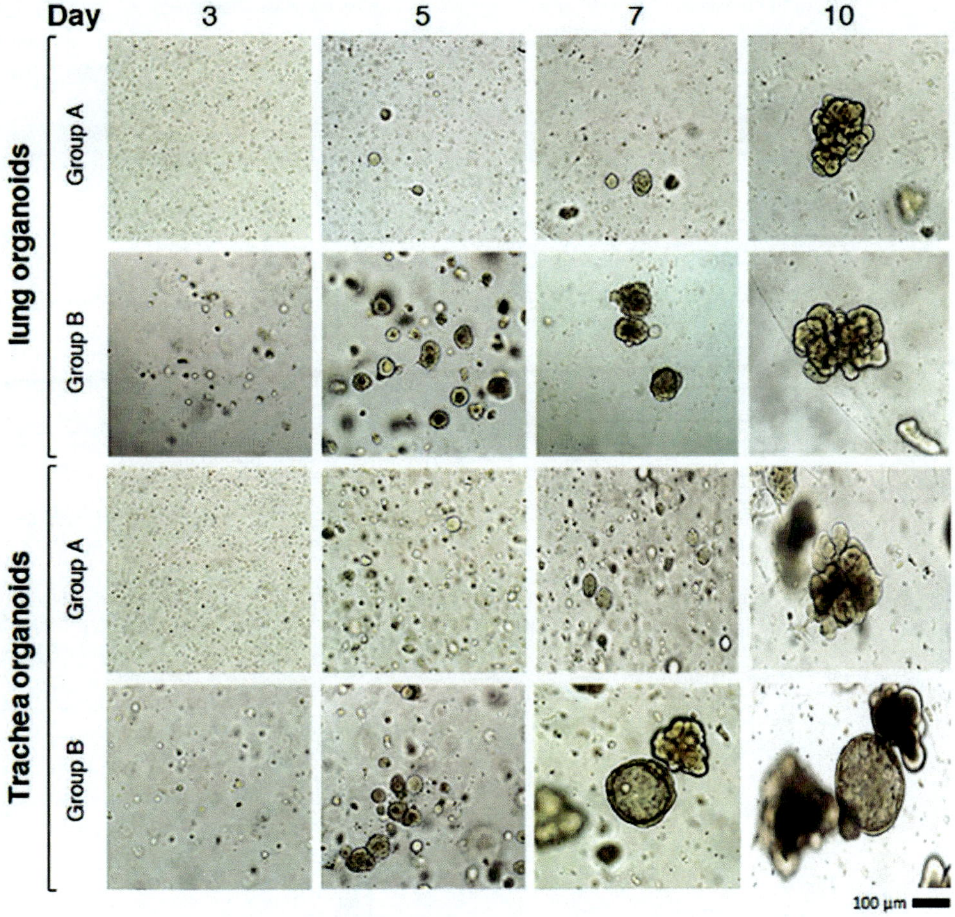

Fig. 2 The culture images for trachea and lung organoid on day 3, day 5, day 7, and day 10. The images showed the comparison between Group A and Group B at different time points. Group B showed the growth of spheres begin to develop as early as 3 days in culture and continue to proliferate until day 5 and the size spheres continued to increase gradually up to day 10

increase in size until day 5 (Fig. 2). Then, on day 6, the spheres were induced to differentiate into the organoid structure. At this stage, the size of the trachea and lung spheres continued to increase gradually up to day 10, where the mature organoid was formed (Fig. 2). On the contrary, the number of organoids/spheres gradually decreased from day 3 until day 10 as the spheres underwent differentiation to develop organoids (Fig. 2). This indicates that not all spheres undergo differentiation and form mature organoids.

Two different culture conditions were tested to increase the number and size of organoids: (1) Group A without ROCK inhibitor and TGF-β inhibitor and Group B with ROCK inhibitor and TGF-β inhibitor. The addition of ROCK inhibitor and TGF-β inhibitor causes a significant ($p < 0.0001$) increase in the sizes (Fig. 3) and number (Fig. 4) of the trachea and lung organoid/spheres. In addition, the presence of ROCK inhibitor has caused an exponential increase in the size of the trachea and lung organoid/spheres (Fig. 3a). The result also shows that trachea tissue generates

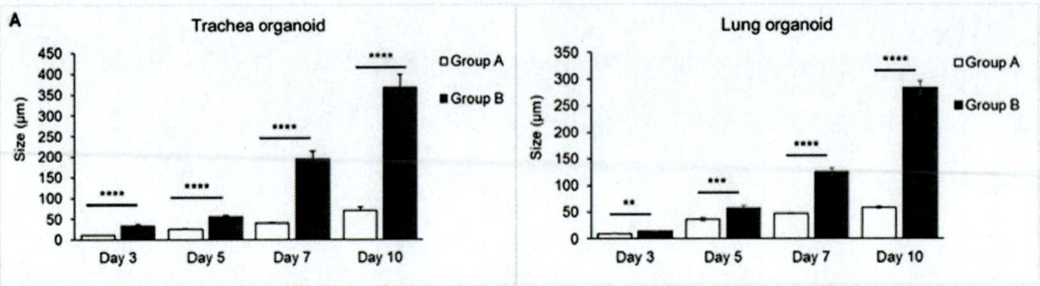

B

Mean Size (μm)	Trachea organoid		Lung organoid	
	Group A	Group B	Group A	Group B
Day 3	11.39 ± 0.31	35.09 ± 2.32	9.86 ± 0.38	14.63 ± 0.73
Day 5	25.61 ± 2.04	56.12 ± 2.95	36.80 ± 3.38	57.84 ± 4.47
Day 7	39.93 ± 1.30	195.21 ± 18.73	47.73 ± 1.55	126.53 ± 6.66
Day 10	70.78 ± 7.83	367.91 ± 29.97	59.24 ± 2.49	283.95 ± 12.32

C

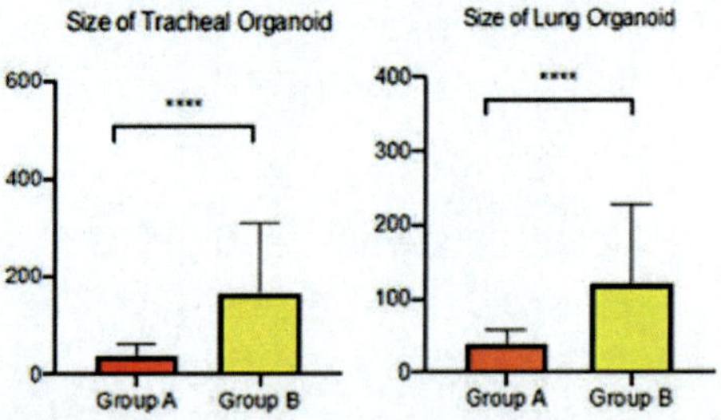

Fig. 3 The trachea and lung organoid size in group A and B culture on days 3, 5, 7, and 10. (**a**) Bar chart of trachea and lung sizes on day 3–day 10 of culture (p-value). (**b**) Table showing the mean ± standard error of trachea and lung organoid on day 3–day 10. (**c**) Bar chart showing organoid size comparison between groups A and B ($p < 0.0001$)

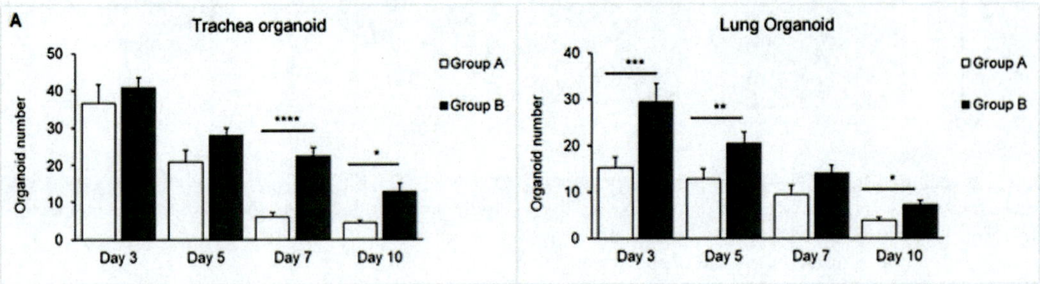

B

Mean number	Trachea organoid		Lung organoid	
	Group A	Group B	Group A	Group B
Day 3	36.7 ± 4.86	40.8 ± 2.82	15.2 ± 2.32	29.4 ± 3.84
Day 5	20.90 ± 3.16	28.1 ± 1.91	12.8 ± 2.21	20.5 ± 2.42
Day 7	6.1 ± 1.05	22.4 ± 2.36	9.5 ± 1.88	14.1 ± 1.69
Day 10	4.4 ± 0.64	12.8 ± 2.29	3.8 ± 0.7	7.3 ± 0.88

C

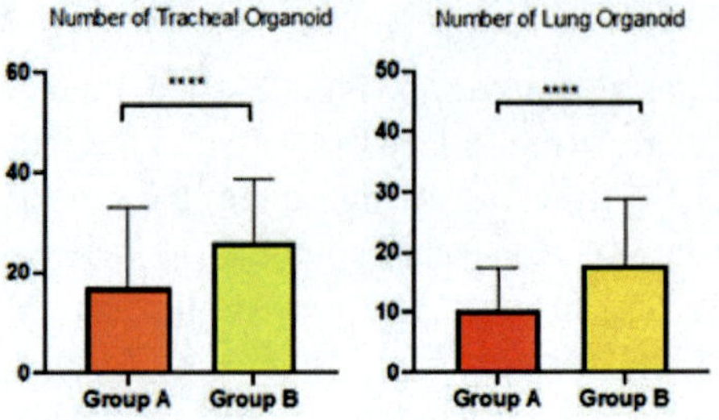

Fig. 4 The number of the trachea and lung organoids developed in group A and B culture on days 3, 5, 7, and 10. (**a**) Bar chart of the trachea and lung number on day 3–day 10 of culture. (**b**) Table showing the mean ± standard error of trachea and lung number on day 3–day 10. (**c**) Bar chart showing a comparison in the number of organoids formed between groups A and B ($p < 0.0001$)

more organoids than lung tissue (Fig. 4a), but the lung forms larger organoids (Fig. 3a). Since group B culture condition produces more and larger organoids in a short time, only this group was selected for further investigation.

3.2 Luminal and Multilobular Morphology Resembles Trachea and Lung Organoid

The developed trachea and lung spheres were induced to differentiate to organoid structure from day 6 up to day 10. One day after the initiation of the differentiation process, some of the spheres showed changes in the cellular morphology that resembled the trachea and lung. For example, the lumen at the middle of the organoid was pronounced, which denotes the trachea maturation and multilobular morphology resembling the lung region (Fig. 5a). To confirm our hypothesis that some of the spheres have entered the organoid differentiation stage, the harvested organoids proceeded with basic H&E, immunofluorescence, and immunohistochemistry staining to examine the expression of specific markers related to the trachea and lung. Several cell-specific markers were used, such as CK14, pan-cytokeratin, CK5/8, and alpha-tubulin antibodies, to observe stem cell differentiation in the trachea organoids. In contrast, the AQP5, SFTPC, and CC10 antibodies were used to examine the expression of alveolar epithelial cell type I (AEC I), alveolar epithelial cell type I (AEC II), and club cells in lung organoids.

The H&E staining showed that these organoids comprised a monolayer of epithelial cells surrounding a liquid-filled lumen in culture as early as day 7. The presence of pseudostratified epithelium was profound on days 7 and 10 in both organoids (Fig. 5b). Matured trachea organoids showed the typical hallmarks of differentiated trachea epithelial lineages based on immunofluorescence staining. The study also showed that organoids had differentiated into several proximal airway epithelial cell types, including epithelial cells, ciliated cells, and luminal cells, by day 7 of culture. The study also observed tracheospheres with diameters ranging from <100 μm to >400 μm, suggesting heterogeneity exists within the organoid populations, as shown in Fig. 2. These tracheospheres consisted of a pseudostratified epithelium with $CK14^{+}$ basal cells peripheral to luminal $CK5/8^{+}$ cells (Fig. 6). By day 10, ciliated cells were observed. This was confirmed by antibody staining for acetylated tubulin (Fig. 5c). Meanwhile, for the lung region, the cells expressing AQP5 for AEC I, SFTPC for AEC II, and the expression of Clara cell (CC10) were observed in the lung organoid (Fig. 7).

4 Discussions

This study has developed a rapid method to establish an organoid culture from the whole trachea and lung tissues of mice with a complex airway cellular structure. The current study has reported that airway organoids can be cultured and grown from small amounts of trachea and lung tissues

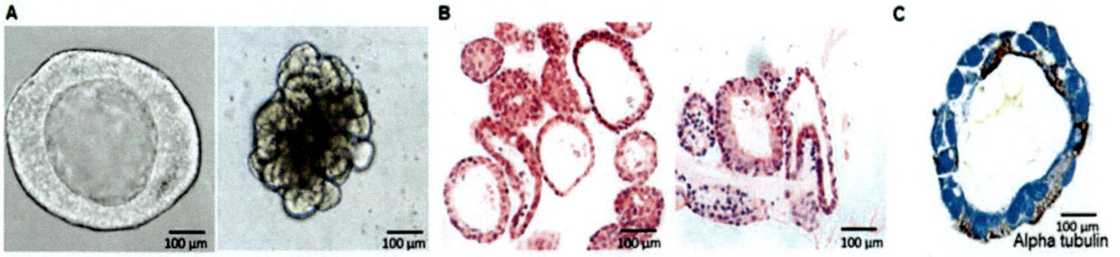

Fig. 5 Morphology of differentiated mice trachea organoids. (**a**) Organoids formed after 4 days in a culture medium were observed to have either a cystic or multilobular morphology. (**b**) H&E staining for trachea organoid. (**c**) Immunostaining of alpha tubulin marker in trachea organoid

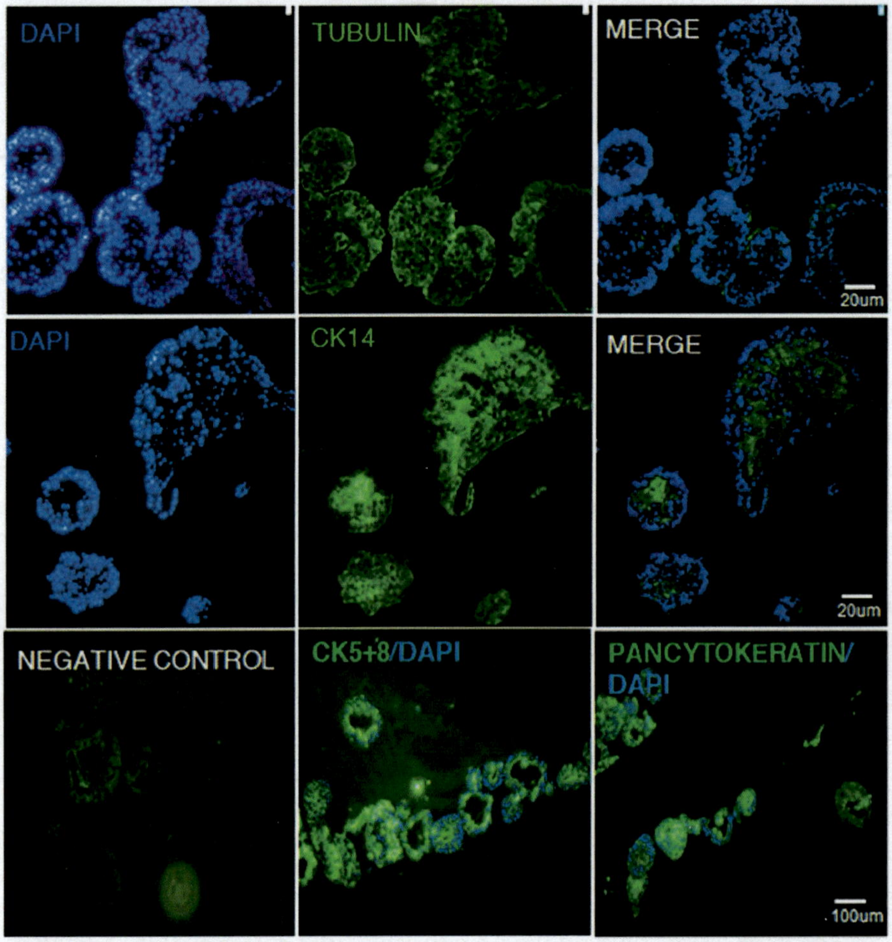

Fig. 6 Identification of different cell types in mice trachea organoids. Mice trachea organoids were immunostained with an epithelial cell marker tubulin, cytokeratin 14, CK5+8, and pan-cytokeratin. Nuclei were counterstained with DAPI (blue). Nuclei were counterstained with DAPI (blue)

within a short time. Furthermore, only a small number of cells were required to generate airway organoids, where the initial seeding cell was 25×10^3 and 50×10^3 in a 24-well plate for trachea and lung organoids, respectively, in which significantly lower cell numbers were required as compared to a previous study reported by Leeman et al. (2019) (the initial seeding cell was 1×10^5/well in a 24-well plate).

The organoid morphologies observed in the current study vary for the trachea and lung. Different morphologies and stages of organoid development were observed even at the same time point in both cultures. This indicates that the development and maturation of cells depend on stem cell behaviour. The addition of several growth factors in the organoid culture, including ROCK inhibitor, TGF-β, R-spondin, and noggin, contributed to the significant ($p < 0.05$) increase in sizes and organoid number, which enhanced the organoid formation with pronounced morphology. These effects are different from the group with no addition of growth factors. R-spondin and noggin growth factors facilitate organoid growth, allowing stem cells to create their niche, self-organise, and form organoids with larger sizes (Sato & Clevers, 2015; Lugli et al., 2016). Moreover, Wnt, R-spondin, and

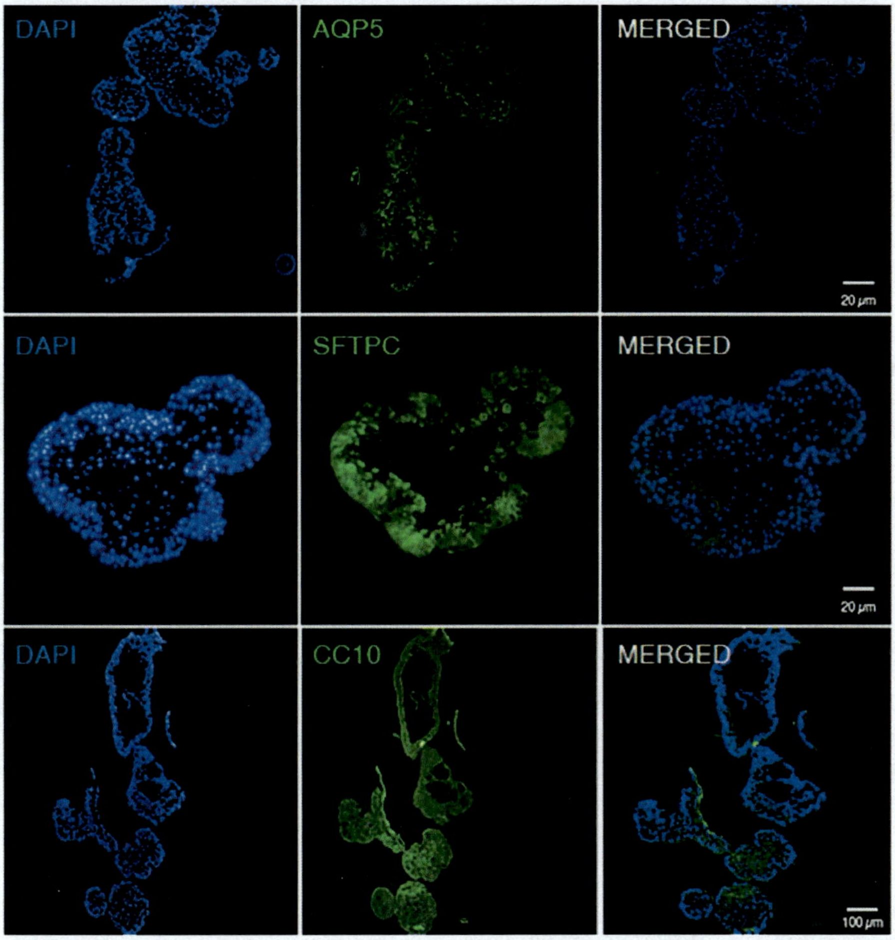

Fig. 7 Identification of different cell types in mice lungs. Lung organoids were immunostained with alveolar type I, alveolar type II, and club cell markers; AQP5, SFTPC, and CC10. Nuclei were counterstained with DAPI (blue)

noggin are the primary growth factors required to maintain a stem cell population in mouse or human spheroid cultures (Miyoshi & Stappenbeck, 2013).

In previous studies, L-WRN-conditioned media that were supplemented with ROCK and a TGF-β inhibitor were used to generate intestinal spheroids from human and mouse cells (Miyoshi & Stappenbeck, 2013; Powell & Behnke, 2017). The BMP/TGF-β signalling pathway is responsible for epithelial cell differentiation, and the addition of TGF-β antagonists such as SB431542 in the culture medium prevents spontaneous differentiation (Mishra et al., 2005). While testing different culture media, the current study has found that adding ROCK and TGF-β inhibitors significantly improved the organoids' growth in terms of numbers and sizes. This finding agrees with previous studies by Ida et al. (2021) and Koslow et al. (2019), where the ROCK inhibitor enhances the formation of 3D organoids, increases proliferation, and reduces apoptosis of progenitor cells grown in vitro. Several ROCK substrates are involved in the execution of organoid formation and possibly initiating apoptosis. ROCK inhibitor (Y-27632 substrate) has been used to control stress conditions and enhance cell recovery in primary cell isolation and cryopreservation (Emre et al., 2010) and reduce dissociation-induced apoptosis in human embryonic stem cell culture and primary primate corneal endothelial cells (Okumura et al., 2009;

Watanabe et al., 2007). However, the response to ROCK substrates is cell type-specific and depends on the apoptotic stimulus, which may explain why using ROCK substrate is only beneficial in some organoid cultures.

Cells' ability to form organoids depends on the number of stem cells present in a specific region of the airways. In this study, trachea tissue generated more spheroids that later mature into organoids than the lung. This indicates that tissue from two different airway regions has various organoid formation capacities. Furthermore, the abundance of basal cells in the trachea has contributed significantly to organoid formation capability. The trachea and main stem bronchi are lined with pseudostratified columnar epithelium organised into basal and luminal cells. These basal cells can proliferate and undergo multipotent differentiation to give rise to secretory cells, ciliated cells, ionocytes, and tuff cells (Montoro et al., 2018; Plasschaert et al., 2018).

Moreover, basal cells are the predominant cell type responsible for large airway regeneration, making them the most widely accepted proximal airway stem cells (Rock et al., 2010; Alysandratos et al., 2020). Meanwhile, the number of stem cells in the alveolar epithelium depends on AEC II capable of proliferating with subsequent AEC I. While the AEC II remains the most accepted stem cell of the alveolus, some reports suggest it's still depending on the extent and type of injury to the alveolar epithelium that will serve as facultative progenitors that contribute to the regeneration of the alveolar epithelium (Alysandratos et al., 2020; Zacharias et al., 2018).

The generation of differentiated trachea and lung organoids was reported in the current study as early as 7 days in culture. Our newly established trachea and lung organoid cultures resemble the original parental tissue characteristics. For example, the presence of multiple cell types, such as basal cells, epithelial cells, and ciliated cells, was profound in this study. Evidence also suggests that the AEC II and club cells are essential as stem cells in forming lung organoids (Kong et al., 2021). Gas-exchanging AEC I cells and surfactant-secreting AEC II cells make up the alveoli. Since both cell types arise from the same progenitor, a rare self-renewing type-II cell works as the stem cell to regenerate the alveolar epithelium. Both AEC I and AEC II cells are derived directly from a bipotent progenitor; however, new AEC I cells arise from long-lived, self-renewing mature AEC II cells that produce slowly expanding clonal foci of alveolar renewal, and AEC widely activates the function of this stem cell I injury (Alysandratos et al., 2020). Meanwhile, AEC II self-renewal is selectively induced by EGFR ligands and oncogenic *Kras* that are efficiently generated (Desai et al., 2014). Through the ability to proliferate, these unique self-renewing AEC II cells in lung epithelial cell lineages can re-enter the cell cycle, replace lost cells, and repopulate the lost cells. The lung could respond to injury and stress by activating stem cell populations and re-entering the cell cycle (Kotton & Morrisey 2014). Hence, the higher expression marker of AEC II cells in this current study showed the ability of AEC II cells in our organoid to proliferate, enter the cell cycle, and undergo the differentiated state. This was in line with previous reports on airway organoids derived from iPSCs. The organoids possess cell types like lung mesenchyme, expressing alveolar cell markers, and the structure resembles the bronchi or bronchioles human airway (Miller et al., 2019; Sachs et al., 2019).

Our findings highlight a direct effect of supporting cells, including fibroblast, mesenchymal stem cells, endothelial cells, and many other cells, found in mice lungs cultured together with the basal cells on the development of an organoid. Fibroblast activation is crucial for normal repair in various tissues. A previous study mentioned that organoids did not form in the absence of fibroblast and found that fibroblast is essential for the epithelial organoid formation and organoids form only if fibroblasts are in the Matrigel™ itself (Güney et al., 2021). This indicates the involvement of direct contact of supporting cells secreted factors in their supportive role in organoid formation.

We also found a multilobular structure on lung organoids as early as day 7. This might be due to the use of the mix-cell population of the trachea and lung organs to develop organoids. The use of a whole mixture of cell populations creates intercellular signalling between these supporting cells and induces secretion of growth factor cocktails responsible for organoids' rapid development and maturation. This contrasts with a previous study that grew organoids from pure AEC II cells, and the lung organoids started to develop buds and peripheral alveolar-like structures by day 21 (Vazquez-Armendariz et al., 2020). Moreover, most of the previous studies used to generate multicellular airway organoids from undifferentiated pluripotent stem cells such as iPSCs and briefly mentioned that the airway epithelial cell culture was dissociated on days 14 and 21 (Demchenko et al., 2022; Kim et al., 2021). Meanwhile, in our study, the whole cell population's use will enhance the rapid development of organoids within 7 days, and multiple morphologies were found. Our finding was in line with a previous study in liver organoids where the organoid was more matured and functional if the hepatoblasts were co-cultured with mesenchymal stem cells and endothelial cells (Asai et al., 2017; Ayabe et al., 2018; Camp et al., 2017). Furthermore, it was reported that multi-lineage communication between endothelial cells and mesenchymal cells was mediated by vascular endothelial growth factor (VEGF), stimulating endothelial network formation (Schmidt et al., 2022; Vazquez-Armendariz et al., 2020; Ergün & Wörsdörfer, 2022).

5 Conclusion

In conclusion, our study provides a methodology for developing short-time and rapid organoid cultures that can demonstrate the utility of the establishment for drug testing and biomarker validation. As reported by our study, the advantage of using whole lung and tracheal cells to develop organoids is that it gives the opportunities to harvest different morphologies and stages during organoid development at the same time point and in the same culture conditions. One can study the development and maturation of trachea and lung organoids during the early stage towards fully differentiated organoids. Having multiple morphologies and phases of organoid development will allow researchers to study cellular involvement during organ development and molecular networks. This organoid culture can be widely used for therapeutic purposes, disease modelling, and suitable methods for personalised medicine. However, the limitations of the empirical findings presented here should be considered. The inability of either mouse or human AEC II to be expanded effectively in culture before seeding in Matrigel represents a significant barrier to using organoids to examine gene function in the alveolar epithelium. Other than that, current organoid technology is still imperfect as in the beginning, organoids just had epithelial layers and lacked tissue microenvironments like the immune and vascularisation systems. Additionally, the culture medium needs further improvement for some organoids to grow over an extended period. Lastly, the culture medium's growth factors or molecular inhibitors may impact how well organoids respond to drugs. Despite its current limitations, the promising organoid technology has enormous potential for properly simulating organ development and human diseases. Other than that, it can also provide an excellent platform for the discovery of drugs, customised medicine, and the prediction of treatment outcomes. Additionally, organoid technology is a fantastic substitute for transplant use. However, future work will undoubtedly enhance this innovative clinical tool.

Acknowledgement This research was funded by the Ministry of Higher Education Malaysia for the Fundamental Research Grant Scheme (FRGS) with Project Code FRGS/1/2019/STG03/USM/02/2.

Conflict of Interest All authors declare no conflict of interest. The authors alone are responsible for the content and writing of the article.

References

Alysandratos, K.-D., Herriges, M. J., & Kotton, D. N. (2020). Epithelial stem and progenitor cells in lung repair and regeneration. *The Annual Review of Physiology, 83*, 529–550. https://doi.org/10.1146/annurev-physiol-041520

Asai, A., Aihara, E., Watson, C., Mourya, R., Mizuochi, T., Shivakumar, P., Phelan, K., Mayhew, C., Helmrath, M., Takebe, T., Wells, J., & Bezerra, J. A. (2017). Paracrine signals regulate human liver organoid maturation from induced pluripotent stem cells. *Development (Cambridge)*. https://doi.org/10.1242/dev.142794

Ayabe, H., Anada, T., Kamoya, T., Sato, T., Kimura, M., Yoshizawa, E., Kikuchi, S., Ueno, Y., Sekine, K., Camp, J. G., Treutlein, B., Ferguson, A., Suzuki, O., Takebe, T., & Taniguchi, H. (2018). Optimal hypoxia regulates human iPSC-derived liver bud differentiation through intercellular TGFB Signaling. *Stem Cell Reports*. https://doi.org/10.1016/j.stemcr.2018.06.015

Barkauskas, C. E., Cronce, M. J., Rackley, C. R., Bowie, E. J., Keene, D. R., Stripp, B. R., Randell, S. H., Noble, P. W., & Hogan, B. L. M. (2013). Type 2 alveolar cells are stem cells in adult lungs. *Journal of Clinical Investigation*. https://doi.org/10.1172/JCI68782

Camp, J. G., Sekine, K., Gerber, T., Loeffler-Wirth, H., Binder, H., Gac, M., Kanton, S., Kageyama, J., Damm, G., Seehofer, D., Belicova, L., Bickle, M., Barsacchi, R., Okuda, R., Yoshizawa, E., Kimura, M., Ayabe, H., Taniguchi, H., Takebe, T., & Treutlein, B. (2017). Multi-lineage communication regulates human liver bud development from pluripotency. *Nature*. https://doi.org/10.1038/nature22796

Chen, Y. W., Huang, S. X., De Carvalho, A. L. R. T., Ho, S. H., Islam, M. N., Volpi, S., Notarangelo, L. D., Ciancanelli, M., Casanova, J. L., Bhattacharya, J., Liang, A. F., Palermo, L. M., Porotto, M., Moscona, A., & Snoeck, H. W. (2017). A three-dimensional model of human lung development and disease from pluripotent stem cells. *Nature Cell Biology*. https://doi.org/10.1038/ncb3510

Demchenko, A., Lavrov, A., & Smirnikhina, S. (2022). Lung organoids: Current strategies for generation and transplantation. *Cell and Tissue Research, 390*(3), 317–333. Springer. https://doi.org/10.1007/s00441-022-03686-x

Desai, T. J., Brownfield, D. G., & Krasnow, M. A. (2014). Alveolar progenitor and stem cells in lung development, renewal and cancer. *Nature, 507*(7491), 190–194. https://doi.org/10.1038/nature12930

Emre, N., Vidal, J. G., Elia, J., O'Connor, E. D., Paramban, R. I., Hefferan, M. P., Navarro, R., Goldberg, D. S., Varki, N. M., Marsala, M., & Carson, C. T. (2010). The ROCK inhibitor Y-27632 improves recovery of human embryonic stem cells after fluorescence-activated cell sorting with multiple cell surface markers. *PLoS One*. https://doi.org/10.1371/journal.pone.0012148

Ergün, S., & Wörsdörfer, P. (2022). Organoids, assembloids and embryoids: New avenues for developmental biology, disease modeling, drug testing and toxicity assessment without animal experimentation. *Organoids, 1*(1), 37–40. https://doi.org/10.3390/organoids1010004

Güney, T. G., Herranz, A. M., Mumby, S., Dunlop, I. E., & Adcock, I. M. (2021). Epithelial–stromal cell interactions and extracellular matrix mechanics drive the formation of airway-mimetic tubular morphology in lung organoids. *iScience, 24*(9). https://doi.org/10.1016/j.isci.2021

Ida, Y., Hikage, F., & Ohguro, H. (2021). ROCK inhibitors enhance the production of large lipid-enriched 3D organoids of 3T3-L1 cells. *Scientific Reports, 11*(1). https://doi.org/10.1038/s41598-021-84955

Kim, J. H., An, G. H., Kim, J. Y., Rasaei, R., Kim, W. J., Jin, X., Woo, D. H., Han, C., Yang, S. R., Kim, J. H., & Hong, S. H. (2021). Human pluripotent stem-cell-derived alveolar organoids for modeling pulmonary fibrosis and drug testing. *Cell Death Discovery, 7*(1). https://doi.org/10.1038/s41420-021-00439-7

Kong, J., Wen, S., Cao, W., Yue, P., Xu, X., Zhang, Y., Luo, L., Chen, T., Li, L., Wang, F., Tao, J., Zhou, G., Luo, S., Liu, A., & Bao, F. (2021). Lung organoids, useful tools for investigating epithelial repair after lung injury. *Stem Cell Research and Therapy*. https://doi.org/10.1186/s13287-021-02172-5

Koslow, M., O'Keefe, K. J., Hosseini, Z. F., Nelson, D. A., & Larsen, M. (2019). ROCK inhibitor increases proacinar cells in adult salivary gland organoids. *Stem Cell Research*. https://doi.org/10.1016/j.scr.2019.101608

Kotton, D. N., & Morrisey, E. E. (2014). Lung regeneration: Mechanisms, applications and emerging stem cell populations. *Nature Medicine, 20*(8), 822–832. Nature Publishing Group. https://doi.org/10.1038/nm.3642

Leeman, K. T., Pessina, P., Lee, J. H., & Kim, C. F. (2019). Mesenchymal stem cells increase alveolar differentiation in lung progenitor organoid cultures. *Scientific Reports*. https://doi.org/10.1038/s41598-019-42819-1

Lu, T., Cao, Y., Zhao, P., Shen, S., & Xi, Y. (2021). Organoid: A powerful tool to study lung regeneration and disease. *Cell Regeneration*. https://doi.org/10.1186/s13619-021-00082-8

Lugli, N., Kamileri, I., Keogh, A., Malinka, T., Sarris, M. E., Talianidis, I., Schaad, O., Candinas, D., Stroka, D., & Halazonetis, T. D. (2016). R-spondin 1 and noggin facilitate expansion of resident stem cells from non-damaged gallbladders. *EMBO Reports, 17*(5), 769–779. https://doi.org/10.15252/embr.201642169

Miller, A. J., Dye, B. R., Ferrer-Torres, D., Hill, D. R., Overeem, A. W., Shea, L. D., & Spence, J. R. (2019). Generation of lung organoids from human pluripotent stem cells in vitro. *Nature Protocols*. https://doi.org/10.1038/s41596-018-0104-8

Mishra, L., Derynck, R., & Mishra, B. (2005). Transforming growth factor-b signaling in stem cells and cancer roles of TGF-b family members in neural stem cells. *Science 310*(5745), 68–71. https://doi.org/10.1126/science.1118389

Miyoshi, H., & Stappenbeck, T. S. (2013). In vitro expansion and genetic modification of gastrointestinal stem cells in spheroid culture. *Nature Protocols*. https://doi.org/10.1038/nprot.2013.153

Montoro, D. T., Haber, A. L., Biton, M., Vinarsky, V., Lin, B., Birket, S. E., Yuan, F., Chen, S., Leung, H. M., Villoria, J., Rogel, N., Burgin, G., Tsankov, A. M., Waghray, A., Slyper, M., Waldman, J., Nguyen, L., Dionne, D., Rozenblatt-Rosen, O., … Rajagopal, J. (2018). A revised airway epithelial hierarchy includes CFTR-expressing ionocytes. *Nature, 560*(7718), 319–324. https://doi.org/10.1038/s41586-018-0393-7

Okumura, N., Ueno, M., Koizumi, N., Sakamoto, Y., Hirata, K., Hamuro, J., & Kinoshita, S. (2009). Enhancement on primate corneal endothelial cell survival in vitro by a rock inhibitor. *Investigative Ophthalmology and Visual Science*. https://doi.org/10.1167/iovs.08-2634

Plasschaert, L. W., Žilionis, R., Choo-Wing, R., Savova, V., Knehr, J., Roma, G., Klein, A. M., & Jaffe, A. B. (2018). A single-cell atlas of the airway epithelium reveals the CFTR-rich pulmonary ionocyte. *Nature, 560*(7718), 377–381. https://doi.org/10.1038/s41586-018-0394-6

Powell, R. H., & Behnke, M. S. (2017). WRN conditioned media is sufficient for in vitro propagation of intestinal organoids from large farm and small companion animals. *Biology Open*. https://doi.org/10.1242/bio.021717

Rock, J. R., Randell, S. H., & Hogan, B. L. M. (2010). Airway basal stem cells: A perspective on their roles in epithelial homeostasis and remodeling. *DMM Disease Models and Mechanisms, 3*(9–10), 545–556. https://doi.org/10.1242/dmm.006031

Sachs, N., Papaspyropoulos, A., Zomer-van Ommen, D. D., Heo, I., Böttinger, L., Klay, D., Weeber, F., Huelsz-Prince, G., Iakobachvili, N., Amatngalim, G. D., Ligt, J., Hoeck, A., Proost, N., Viveen, M. C., Lyubimova, A., Teeven, L., Derakhshan, S., Korving, J., Begthel, H., et al. (2019). Long-term expanding human airway organoids for disease modeling. *The EMBO Journal*. https://doi.org/10.15252/embj.2018100300

Sahabian, A., Sgodda, M., Naujok, O., Dettmer, R., Dahlmann, J., Manstein, F., Cantz, T., Zweigerdt, R., Martin, U., & Olmer, R. (2019). Chemically-defined, Xeno-free, scalable production of hPSC-derived definitive endoderm aggregates with multi-lineage differentiation potential. *Cell*. https://doi.org/10.3390/cells8121571

Sato, T., & Clevers, H. (2015). Snap shot: Growing organoids from stem cells. *Cell, 161*(7), 1700–1700.e1. Cell Press. https://doi.org/10.1016/j.cell.2015.06.028

Schmidt, S., Alt, Y., Deoghare, N., Krüger, S., Kern, A., Rockel, A. F., Wagner, N., Ergün, S., & Wörsdörfer, P. (2022). A blood vessel organoid model recapitulating aspects of vasculogenesis, angiogenesis and vessel wall maturation. *Organoids, 1*(1), 41–53. https://doi.org/10.3390/organoids1010005

Vazquez-Armendariz, A. I., Heiner, M., el Agha, E., Salwig, I., Hoek, A., Hessler, M. C., Shalashova, I., Shrestha, A., Carraro, G., Mengel, J. P., Günther, A., Morty, R. E., Vadász, I., Schwemmle, M., Kummer, W., Hain, T., Goesmann, A., Bellusci, S., Seeger, W., et al. (2020). Multilineage murine stem cells generate complex organoids to model distal lung development and disease. *The EMBO Journal, 39*(21). https://doi.org/10.15252/embj.2019103476

Watanabe, K., Ueno, M., Kamiya, D., Nishiyama, A., Matsumura, M., Wataya, T., Takahashi, J. B., Nishikawa, S., Nishikawa, S. I., Muguruma, K., & Sasai, Y. (2007). A ROCK inhibitor permits survival of dissociated human embryonic stem cells. *Nature Biotechnology*. https://doi.org/10.1038/nbt1310

Wörsdörfer, P., Takashi, I., Asahina, I., Sumita, Y., & Ergün, S. (2020). Do not keep it simple: Recent advances in the generation of complex organoids. *Journal of Neural Transmission*. https://doi.org/10.1007/s00702-020-02198-8

Zacharias, W. J., Frank, D. B., Zepp, J. A., Morley, M. P., Alkhaleel, F. A., Kong, J., Zhou, S., Cantu, E., & Morrisey, E. E. (2018). Regeneration of the lung alveolus by an evolutionarily conserved epithelial progenitor. *Nature, 555*(7695), 251–255. https://doi.org/10.1038/nature25786

Zhu, N., Zhang, D., Wang, W., Li, X., Yang, B., Song, J., Zhao, X., Huang, B., Shi, W., Lu, R., Niu, P., Zhan, F., Ma, X., Wang, D., Xu, W., Wu, G., Gao, G. F., & Tan, W. (2020). A novel coronavirus from patients with pneumonia in China, 2019. *New England Journal of Medicine*. https://doi.org/10.1056/nejmoa2001017

Adv Exp Med Biol - Innovations in Cancer Research and Regenerative Medicine (2023) 4: 181–191
https://doi.org/10.1007/5584_2021_669

Published online: 6 November 2021

A Simple Method to Produce Engineered Cartilage from Human Adipose-Derived Mesenchymal Stem Cells and Poly ε-Caprolactone Scaffolds

Hue Thi-Ngoc Nguyen and Ngoc Bich Vu

Abstract

Introduction The damaged articular cartilage has limited self-regeneration capacity because of the absence of blood vessels, lymphatics, and nerves. Cartilage transplantation is, hence, a popular method used to treat this disease. However, sources of autograft and allogenic cartilage for transplantation are limited. Therefore, this study aims to suggest a simple method to produce engineered cartilage from human adipose-derived mesenchymal stem cells (ADSCs) and poly (ε-caprolactone) (PCL) scaffolds.

Methods ADSCs were isolated and expanded from fat tissues according to published protocols. PCL-porous scaffolds were produced from PCL with $5 \times 5 \times 0.6$ mm^3 with 200–400 μ m pore sizes. ADSCs were seeded on the PCL scaffolds at three different densities (10^4, 10^5, 10^6 cells per scaffold). The adherence of ADSCs on the surface of PCL scaffolds was evaluated based on an immunostaining assay to determine the presence of ADSCs. The cell proliferation on PCL scaffolds was determined by MTT assay. The complexity in ADSCs and PCL scaffolds was induced to cartilage using a chondrogenesis medium. The engineered cartilage was characterized by the accumulation of proteoglycan and aggrecan by Safranin O staining assay. Their structures were evaluated using an H-E staining assay. Finally, these engineered cartilage tissues were transplanted into mice to assess cartilage maturation when compared to natural cartilage.

Results The results showed that the engineered cartilage tissues could be successfully produced by cultures of ADSCs on poly ε-caprolactone scaffolds in combination with chondrogenesis medium. The suitable density of ADSCs was 10^6 cells/per scaffold of $5 \times 6 \times 0.6$ mm^3 with pore size from 200 to 400 μ m.

Conclusion The results showed that an in vitro cartilage tissue was created from ADSCs and PCL scaffold. The cartilage tissue exists in the mice for 6 months.

Keywords

Adipose-derived stem cells · Cartilage · Engineered cartilage · PCL scaffold · Tissue engineering

H. T.-N. Nguyen and N. B. Vu (✉)
Stem Cell Institute, University of Science, Ho Chi Minh City, Vietnam

Vietnam National University Ho Chi Minh City, Ho Chi Minh City, Vietnam
e-mail: ngocvu@sci.edu.vn; vbngoc@hcmus.edu.vn

Abbreviations

ADSC	Adipose-derived stem cells
CD	Cluster of differentiation
ECM	Extracellular matrix
MSC	Mesenchymal stem cell
PBS	Phosphate buffer saline
PCL	Poly (ε-caprolactone)

1 Introduction

Cartilaginous tissue is a special connective tissue without blood vessels, lymphatics, or nerves. Therefore, this tissue is considered as a low-level regeneration tissue (Sophia Fox et al., 2009; Tuan et al., 2013a). Some methods used in current treatment, such as microfracture, osteochondral autograft transplantation, and autologous chondrocyte implantation, can only improve certain symptoms. In the large-volume defect, joint replacement surgery or large osteochondral allograft must be performed (Tuan et al., 2013a; Pisanu et al., 2018). However, limitations exist for these treatments including acute graft-versus-host disease, joint replacement lifespan, lifetime risks of revision surgery, or the recovery process after transplantation (Toonstra et al., 2016; Bayliss et al., 2017).

Recently, tissue engineering proved to be the novel strategy for injured cartilage regeneration. Stem cells combined with growth factors and scaffolds can produce engineered tissues that are expected to bring new applications in regenerative medicine, specifically cartilaginous tissue (Iwamoto et al., 2013; Tuan et al., 2013b; Oldershaw, 2012). Cell sources for tissue engineering are usually differentiated cells or stem cells (Colnot, 2011; Khademhosseini et al., 2014). Some cell sources such as fibroblasts or stem cells have been used in cartilage tissue engineering (Chung & Burdick, 2008; Vinatier & Guicheux, 2016). In there, mesenchymal stem cells (MSCs) showed many advantages for cartilage tissue engineering because they are easily differentiated into chondrocytes (Vinatier & Guicheux, 2016; de Isla et al., 2010).

Besides stem cells, a scaffold is one of three components of tissue engineering. Some kinds of scaffolds used in tissue engineering are poly (glycerol sebacate) (PGS), poly(L-lactic acid) (PLLA) or poly (ϵ-caprolactone) (PCL), poly (lactide-co-glycolide acid) (PLGA), poly(lactide acid) (PLA) (Oh et al., 2007; Alaribe Franca et al., 2016; Frydrych et al., 2015; Mondal et al., 2016; Mano et al., 2007; He & Lu, 2016), or natural materials, such as collagen, matrigel, chitosan, and fibrin (Mano et al., 2007; Hubbell, 2003; Tuzlakoglu et al., 2004; Tabata et al., 2000). In these kinds of scaffolds, PCL is most commonly used in tissue engineering. PCL was successfully synthesized by Carothers's research group in 1930, and was commercialized, and is in use even today (Woodruff & Hutmacher, 2010). A PCL scaffold is a resilient polyester made with more structures that can be synthesized with other polymers, such as gelatin and chitosan, depending on application (Fu et al., 2014). PCL scaffold applications have been developed extensively, especially in drug delivery systems, medical devices, or tissue engineering (Woodruff & Hutmacher, 2010). In tissue engineering, they were used to create artificial cartilage or bone tissue (Choong et al., 2004).

Chondrogenesis of stem cells in scaffolds can be induced by some growth factors and chemicals. The most common growth factors used in cartilage tissue engineering are TGF, FGF, BMP, and IGF (Kawakami et al., 2006). TGF- β 1, TGF- β 3, and BMP-2 enhance the action of the cartilage ECM production (Valcourt et al., 1999). BMP-4 supports the mature chondrocytes during differentiation (Miljkovic et al., 2008). FGF-2 joins to produce proteoglycan in cartilage ECM. Other factors, such as BMP-4, BMP-7, and IGF-1, also support the creation of artificial cartilage (Danisovic et al., 2012; Vinatier et al., 2009).

2 Materials and Methods

2.1 Human Adipose-Derived Stem Cells Isolation and Expansion

ADSCs were isolated and expanded using published protocols (Van Pham et al., 2014; Truong et al., 2018). Briefly, human belly adipose tissues were collected from the hospital after cosmetic surgery with a consent form. These samples were transferred to the laboratory for stromal vascular fraction (SVF) extraction and then ADSC isolation by culture. SVF was extracted from the adipose tissues using the Cell Extraction Kit (Regenmedlab, HCMC, VN). The extracted SVFs were washed twice with Washing Buffer before being plated on the T-75 flasks in ADSCCult I primary (Regenmedlab). The flasks were incubated at 37 °C, 5% CO_2, for 5 days without refreshing the media. After this period, the media in the flasks were refreshed by ADSCCult I medium every 3 days until 70–80% confluence and then sub-cultured using Deattachment reagent (Regenmedlab). The ADSCs at fourth were used for further experiments.

2.2 Characterization of Human Adipose-Derived Stem Cells

ADSC candidates were characterized as mesenchymal stem cells based on the minimal criteria for mesenchymal stem cells as suggested by the International Society for Cellular Therapy (ISCT) (Dominici et al., 2006). ADSCs were also stained with CD14-FITC, CD34-FITC, CD73-FITC, CD105-PE (Santa Cruz Biotechnology, USA), CD44-PE, and HLA-DR (FITC-BD Biosciences) to determine immunophenotype.

To evaluate the in vitro differentiation potential, candidates were cultured in StemPro® Adipogenesis Differentiation Media, StemPro® Osteogenesis Differentiation Media, and StemPro® Chondrogenesis Differentiation Media (Gibco/Thermo Fisher Scientific, USA) to differentiate into adipocytes, osteoblasts, and chondroblasts, respectively. After 8–14 days, the cells were stained with Oil Red O to observe the appearance of adipocytes. After 28 days, the cells were stained with Alizarin Red to determine the accumulation of Ca^{2+} and Mg^{2+} in osteoblasts, as well as with Safranin O to determine the synthesis and storage of proteoglycan and collagens in chondroblasts.

2.3 Poly ε-Caprolactone (PCL) Scaffold Preparation

The PCL scaffolds with 5 × 5 × 0.6 mm^3 (length x width x thickness) (produced by the Institute of Applied Informatics and Mechanics, Vietnam Academy of Science and Technology) were used in this study. The pore size of the scaffolds is 200–400 μ m. PCL scaffolds were sterilized by soaking in 96% alcohol for 24 h, followed by washing twice in phosphate-buffered saline (PBS). The PCL scaffolds were then allowed to dry and stored in sterile tubes naturally.

2.4 ADSC Adhesion and Proliferation on PCL Scaffolds

ADSCs were loaded on three pieces of PCL scaffolds at three densities 10^4, 10^5, and 10^6 cells/ml in a 15-ml falcon. The medium was changed every 3 days. The adherence of ADSCs on the PCL scaffold was observed under the light microscope and scanning electron microscope (SEM).

ADSC proliferation on PCL was recorded based on MTT assay at 3, 7, and 14 days after loading. Briefly, the culture medium was removed and 10 μ L MTT solution was added. The samples were incubated at 37 °C for 4 h. Finally, 100 μ L DMSO was added to dissolve the formazan crystal. The optical density (OD) values were recorded at 595 nm.

2.5 Inducing PCL-ADSCs into Cartilage Tissue

PCL-ADSCs were induced into cartilage tissue by culturing them in StemPro® Chondrogenesis Differentiation Media (Thermo Fisher). The medium was changed every 5 days for 28 days.

To confirm that the cartilage was forming, structures of PCL-ADSCs were fixed in 4% PFA for 30 min. The samples were then incubated with Safranin O for 10 min and washed with 96% alcohol or stained with Alcian blue for 30 min, before being washed by PBS. The accumulation of proteoglycan was showed by red color after Safranin O staining. The expression of aggrecan was showed by blue color after Alcian blue staining.

2.6 Statistical Analysis

Data were analyzed by the GraphPad Prism 8 software. Statistical analysis was performed by t-test. P-value <0.05 was considered significant. All data were presented as mean ± SD.

3 Results

3.1 Characteristics of Mesenchymal Stem Cell Candidates

ADSCs were characterized as MSCs using the minimal criteria for MSCs suggested by the ISCT (Dominici et al., 2006). The results showed that ADSCs could adhere and spread on the plastic flask surface and be displayed like fibroblast morphology (Fig. 1). After being induced by adipogenesis medium for 14 days, tiny lipid droplets were formed in the cytosol of hADSCs; they were stained red when the cells were stained with Oil red O dye (Fig. 1). In the in vitro differentiation assays of ADSCs toward osteoblasts and chondrocytes, after 28 days of induction, ADSCs were positive with Alizarin Red staining (Fig. 1) and positive with Safranin O staining (Fig. 1), respectively. These results confirmed that ADSCs exhibited the differentiation potential of MSCs. These ADSCs also displayed the MSC particular marker profile and were positive with CD44, CD73, and CD105 (Fig. 2), but negative with CD14, CD34, and HLA-DR (Fig. 2).

3.2 ADSC Adhesion and Proliferation on PCL Scaffolds

The results showed some differences in the adhesion ability of ADSCs on PCL surfaces at various loading densities of ADSCs. On day 3, at the loading density of 10^4 cells/ml, a few ADSCs appeared and adhered on the surface of scaffolds (Fig. 3a), while at the loading density of 10^5 cells/ml, cells could cover and fill some pores of PCL scaffolds (Fig. 3b). At a density of 10^6 cells/ml, all of the pores of PCL scaffolds were filled and covered by ADSCs (Fig. 3c).

On days 7 and 14, the density and coverage percentage of loading cells in pores of scaffolds gradually increased compared to day 3 in all groups. Notably, on day 14, 100% pores of PCL scaffolds were filled by ADSCs in the group of 10^6 cells/ml. The MIT assay supported these observations. The number of ADSCs significantly increased from day 3 to day 14 in all groups (Fig. 4).

Finally, the adhesion of ADSCs on the PCL scaffold was confirmed by observing the PCL's surface under an SEM microscope. The results demonstrated that ADSCs existed and filled the pores of PCL scaffolds (Fig. 5).

3.3 ADSC-PCL Complexes Formed Engineered Cartilage Tissues After Induced in Chondrogenesis Medium In Vitro and Implanted into Mice

ADSCs on the complex of ADSC-PCL were stained red with Safranin O (Fig. 6) and blue with Alcian blue (Fig. 6) when cultured in osteogenic-induced medium at day 31. These showed that there was an accumulation of proteoglycan and aggrecan.

The PCL cross sections showed that the PCL fibers were round and that the gaps surrounding the fibers were the pores (Fig. 7a). After subcutaneous transplantation of experimental animal, the PCL pores include for PCL alone (Fig. 7b, e) or

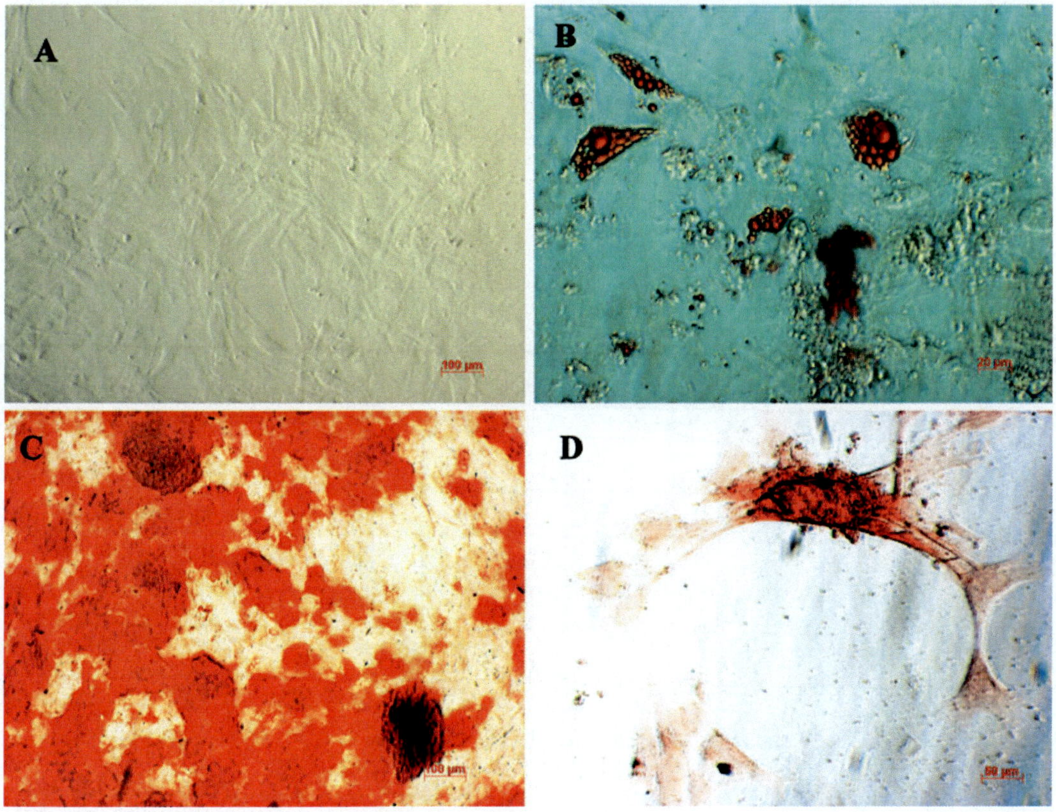

Fig. 1 The morphology and differentiation potential of ADSC in vitro. ADSCs well adhered on a plastic flask (**a**). When cultured in a differentiation-induced medium, ADSCs could be differentiated into adipocytes (**b**), osteoblasts (**c**), and chondrocytes (**d**)

PCL with chondrogenic-induced ADSC (Fig. 7c, f) had the presence of cytoplasm, cell nucleus, and blood vessels. There were some changes in the PCL scaffold after being transplanted under mice skin for 9 months. We found the migration and invasion of mouse cells into the ADSC-PCL complex. The cell surrounded the PCL fibers and filled up the gaps of the pores (Fig. 7b, e).

On the other hand, after 9 months of grafting, mice did not express any physiological damage including inflammation or tumor formation at the graft site. Significantly, there was the presentation of chondrocyte-like cells inside ADSC-PCL (Fig. 7f) after 6 months of mouse subcutaneous transplantation, similar to in vivo cartilage structure (Hoemann et al., 2012) (Fig. 7d). However, the size of the fibers was not significantly decreased both in the case of PCL implantation and ADSC-PCL implantation.

Finally, the expression of HLA-DR antigen in chondrocytes in engineered cartilage tissue after 6 months of implantation was used to confirm the engineered cartilage tissues containing implanted ADSCs. The results showed that ADSCs before differentiation into chondrocytes were negative with HLA-DR (Fig. 8a–d). However, they become positive with this antigen after differentiated into chondrocytes in vitro (Fig. 8e–h). Furthermore, after 6 months of implantation in mice, the cross sections of engineered cartilage tissues were also positive with HLA-DR (Fig. 8i–m).

4 Discussion

This study aims to create in vitro engineered cartilage tissues from ADSC and PCL scaffolds. In the first experiment, we evaluated and

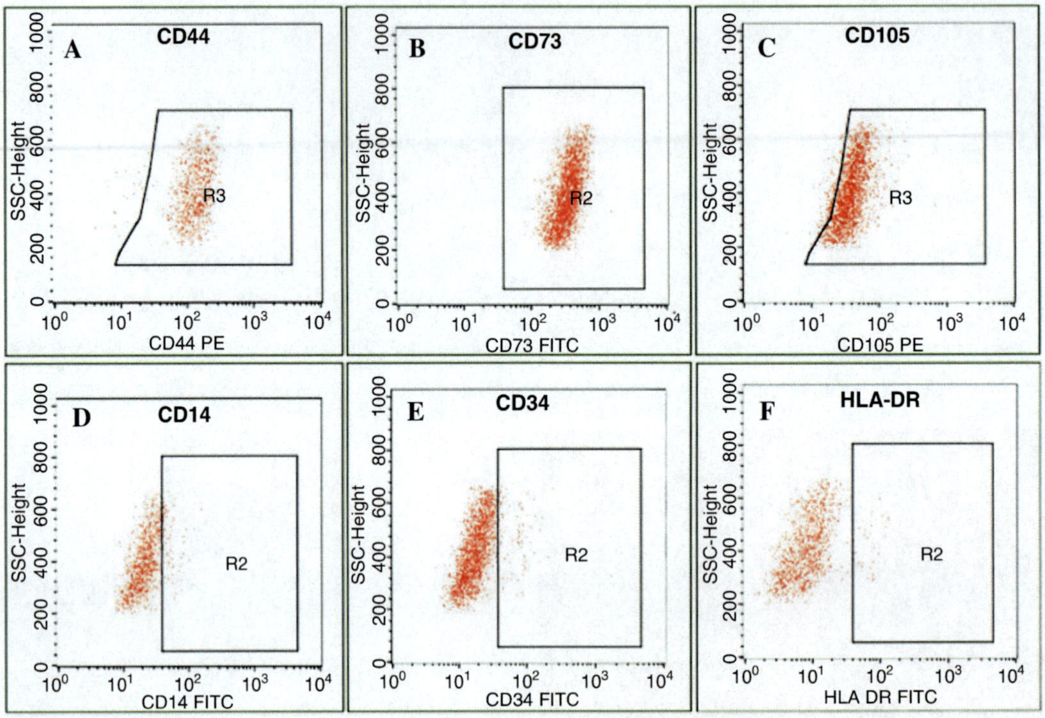

Fig. 2 **The surface markers expression of ADSCs.** ADSCs were positive with CD44-PE (**a**), CD73-FITC (**b**), and CD105-PE (**c**) and negative with CD14-FITC (**d**), CD34-FITC (**e**), and HLA-DR-FITC (**f**)

confirmed ADSCs as MSCs according to the suggestion of ISCT about minimal criteria of MSCs. The results confirmed that isolated and expanded ADSCs agreed with these criteria.

In the experiment, we investigated the suitable density of loading cells on the PCL scaffolds. Based on the proliferation rate and the filling capacity of scaffold pores, the loading density of 10^6 cells/ml was more suitable than others. These data also showed that ADSCs were well adhered to and proliferated on the surface of the scaffold. This was similar to the published study by Hutmacher et al. (Woodruff & Hutmacher, 2010). The cellular adhesion to the scaffold surface is essential for cell-cell and cell-extracellular matrix (Murphy et al., 2013). There are two discovered mechanisms for cell adhesion on artificial materials: direct non-receptor-mediated cell-material binding or receptor-mediated binding through ECM molecules or their parts (Bacakova et al., 2004). ADSC adherence on PCL may be via weak chemical bonds between various molecules on the polymers' cell membrane and functional chemical groups. In addition, cell-cell interactions and cell culture media are also important factors for cell adhesion and proliferation on PCL scaffold. Some molecules, such as fibrinogen (Tang et al., 1998), vitronectin, collagen, or laminin (Groth et al., 1999) secreted by cells or from culture media, can be absorbed on the PCL surface (Bacakova et al., 2004). Thus, cell adhesion and proliferation on PCL are enhanced.

In the next step, the complexes of ADSC-PCLs were incubated in the chondrogenesis medium to induce ADSCs toward chondrocytes. The results confirmed that, after 31 days, proteoglycan and aggrecan successfully accumulated in the ADSCs on PCL. Indeed, Safranin O and Alcian blue staining are used to detect chondrogenesis. Safranin O is a cationic dye that stains glycosaminoglycan (GAG) in proteoglycan (Wall & Board, 2014) while aggrecan is a major proteoglycan present in cartilage tissues, and Alcian blue is

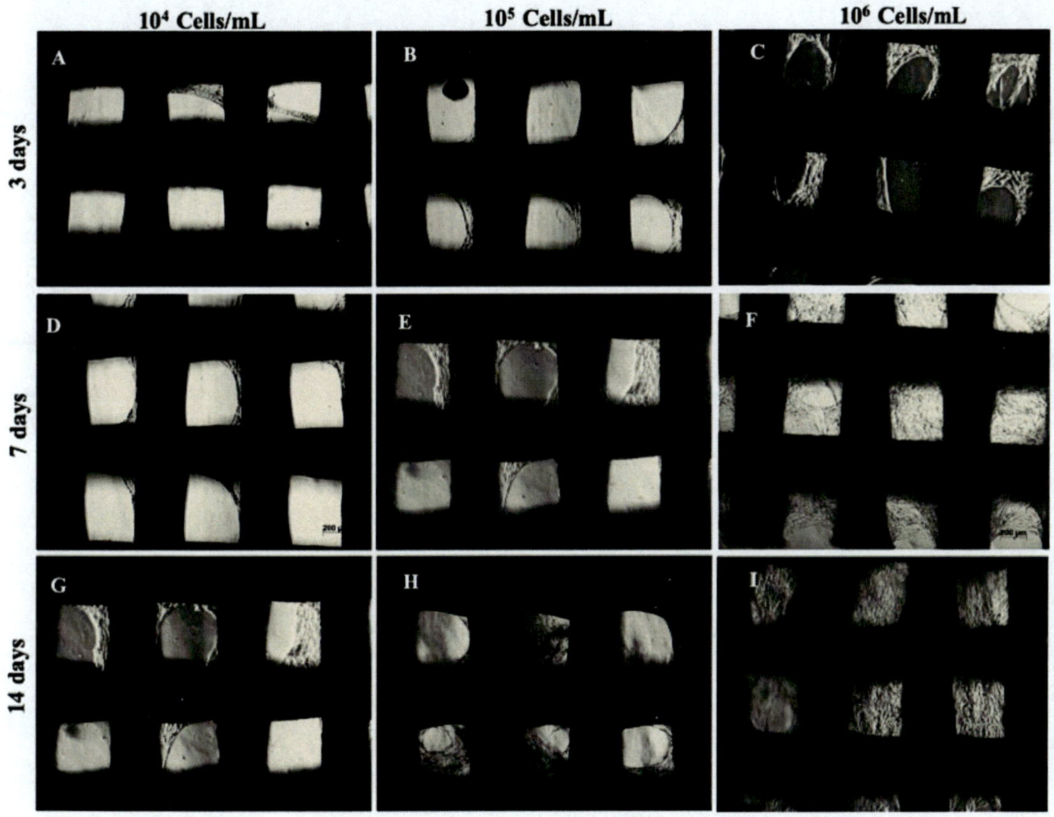

Fig. 3 Adhesion and proliferation of ADSCs on PCL scaffolds. ADSCs were cultured on PCL scaffolds at 3, 7, and 14 days with the cell seeding density of 10^4 cells/ml (**a**-**d**-**g**), 10^5 cells/ml (B, E, H), and 10^6 cells/ml (**c**, **f**, **i**)

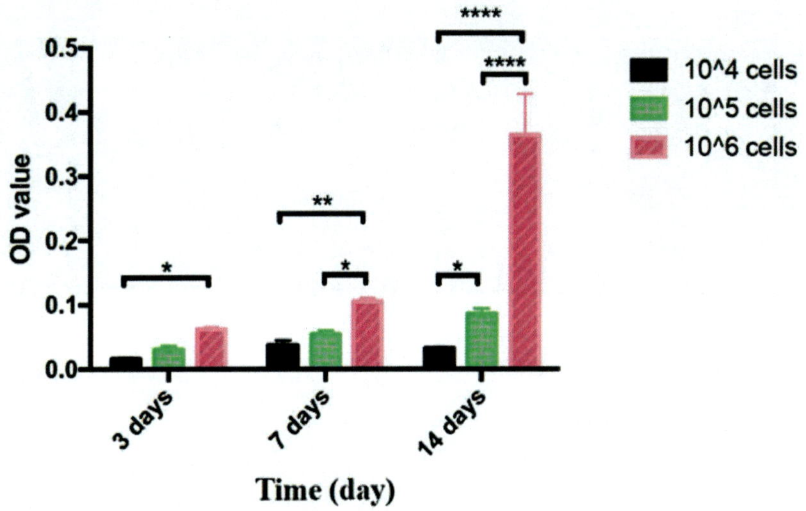

Fig. 4 ADSC proliferation on PCL scaffolds. ADSCs were cultured on PCL scaffolds at a loading density of 10^4 cells/ml, 10^5 cells/ml, and 10^6 cells/ml; proliferation rates were determined based on MTT assays. (* $P < 0.05$)

associated with carboxylate or sulfates group in aggrecan's structure. The positive of Alcian blue dye is indicative of proteoglycan (Hall, 2015). These results showed that ADSCs on PCL were successfully differentiated into chondroblasts on the PCL scaffolds.

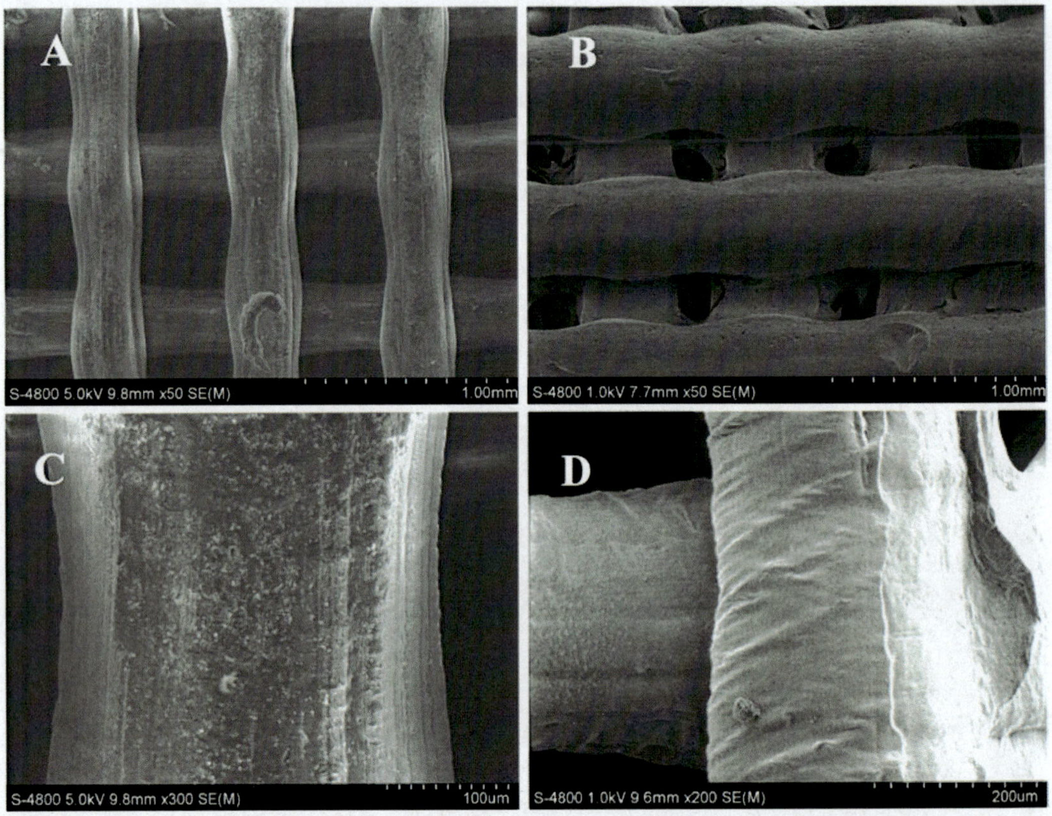

Fig. 5 The scanning electron microscopy (SEM) images PCL morphologies before and after cell seeding at 10^6 cells/mL. Morphologies of PCL at low magnification (**a**) and high magnification (**c**) before cell seeding and after culture ADSC on PCL under low magnification (**b**) and at high magnification (**d**)

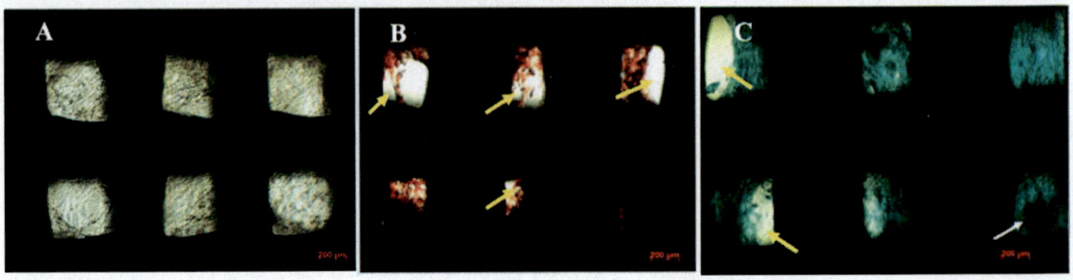

Fig. 6 ADSCs accumulated proteoglycan and aggrecan accumulation on PCL scaffolds after chondrogenesis induction. ADSC-PCL before chondrogenic differentiation (**a**) and after 31 days of differentiation (**b**, **c**). Chondrogenic differentiated cells become red (**b**) and blue (**c**) after stained with Safranin O and Alcian blue, respectively

On being induced to chondroblasts, ADSCs expressed HLA-DR in their surface. Although ADSCs are negative with HLA-DR, differentiated cells from ADSCs can be positive with HLA-DR. Indeed, the HLA-DR expression is essential for cell functions (Machado et al., 2013). After 6 months, the grafts were collected and the expression of HLA-DR was checked. The

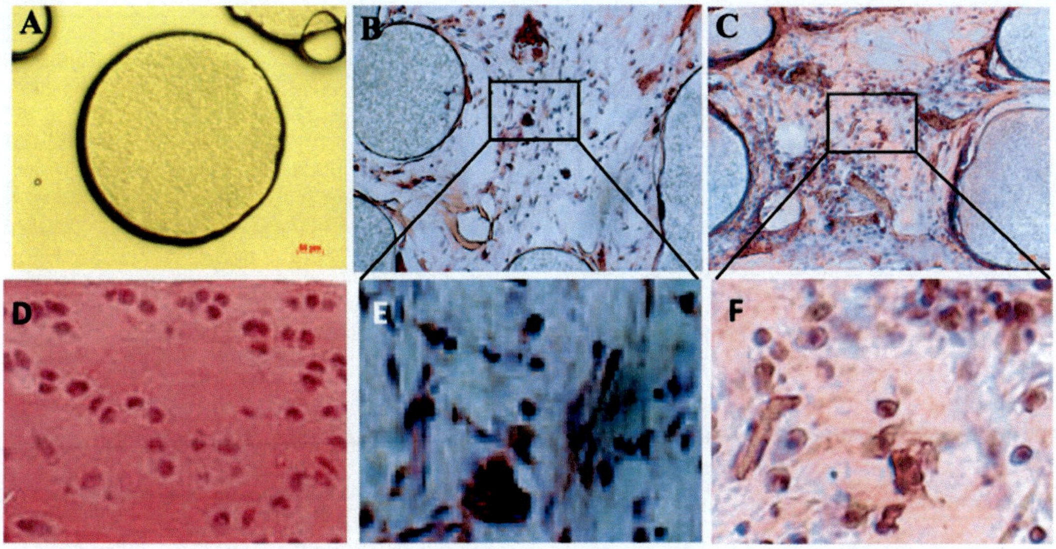

Fig. 7 Histological structure of cartilage tissues formed from ADSC-PCL complex implantation in mice after 6 months. The cross section of PCL's fibers before (**a**) and after subcutaneous implantation of PCL scaffold (without ADSCs) (**b**, **e**) and of a complex of ADSC-PCL (**c**, **f**) after 6 months. The structure of engineered cartilage from the ADSC-PCL scaffold after 6 months of subcutaneous implantation (**f**) was similar to natural cartilage (**d**)

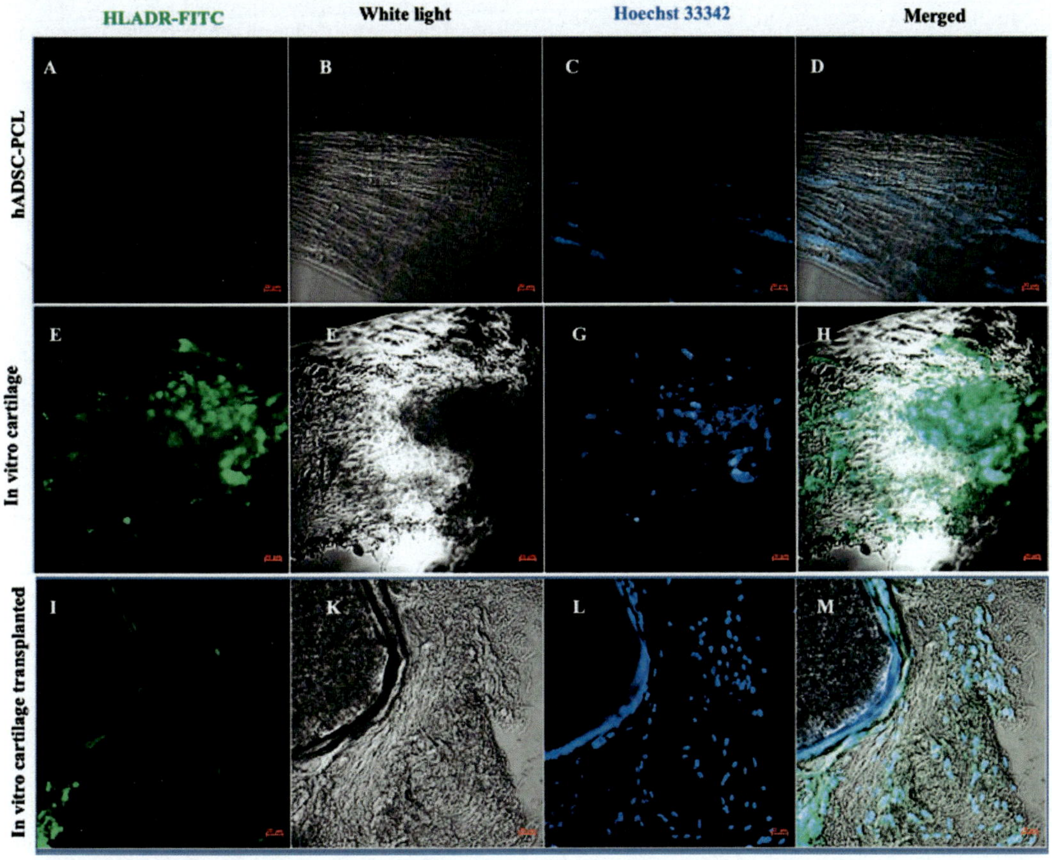

Fig. 8 HLA-DR expression of ADSCs on PCL before and after chondrogenesis induction. HLA-DR was not expressed at ADSC-PCL before chondrogenesis induction (**a–d**) but expressed after induced in vitro (**e–h**) and implantation in mice for 6 months (**i–m**)

result showed that these cells inside grafts were positive with HLA-DR. This result confirmed that chondrocytes were derived from ADSCs existed and integrated into the recipients during the 6 months of experimentation. However, the size of PCL fibers was not significantly reduced in the mice after 6 months of implantation. Although PCL is a biodegradable material, some studies have shown that it can be fragmented until complete decomposition to CO_2 and H_2O under the action of enzymes from living microorganisms but that they are not biodegradable in humans or animals (Woodruff & Hutmacher, 2010; Aleksandra et al., 2017). Other studies showed that PCL could be biodegradable in animal bodies for 2–4 years (Woodruff & Hutmacher, 2010; Middleton & Tipton, 2000).

5 Conclusion

In conclusion, we succeeded in optimizing the culture process for ADSC on PCL scaffolds. The complexes of ADSC-PCL could be induced into cartilage tissue in vitro. These engineered cartilage tissue accumulated proteoglycan and aggrecan in vitro and displayed histological structure similar to the nature cartilage after implantation into mice for 6 months. These results suggested that engineered cartilage tissues can be produced from ADSCs and PCL scaffolds.

Acknowledgements This research is funded by Vietnam National University Ho Chi Minh City (VNU-HCM) under grant number C2017-18-24/HĐ-KHCN.

Conflict of Interest The authors declare that there is no conflict of interest regarding the publication of this paper.

References

Alaribe Franca, N., Manoto Sello, L., & Motaung Shirley, C. K. M. (2016). Scaffolds from biomaterials: advantages and limitations in bone and tissue engineering. *Biologia*, 353.

Aleksandra, H., Magda, M., & Anita, B.-J. (2017). Biodegradation of poly(ε-caprolactone) in natural water environments. *Polish Journal of Chemical Technology, 19*(1), 120–126.

Bacakova, L., et al. (2004). Cell adhesion on artificial materials for tissue engineering. *Physiological Research, 53*(Suppl 1), S35–S45.

Bayliss, L. E., et al. (2017). The effect of patient age at intervention on risk of implant revision after total replacement of the hip or knee: a population-based cohort study. *Lancet (London, England), 389*(10077), 1424–1430.

Choong, C., Triffitt, J. T., & Cui, Z. F. (2004). Polycaprolactone scaffolds for bone tissue engineering: Effects of a calcium phosphate coating layer on osteogenic cells. *Food and Bioproducts Processing, 82*(2), 117–125.

Chung, C., & Burdick, J. A. (2008). Engineering cartilage tissue. *Advanced Drug Delivery Reviews, 60*(2), 243–262.

Colnot, C. (2011). Cell sources for bone tissue engineering: Insights from basic science. *Tissue Engineering. Part B, Reviews, 17*(6), 449–457.

Danisovic, L., Varga, I., & Polak, S. (2012). Growth factors and chondrogenic differentiation of mesenchymal stem cells. *Tissue & Cell, 44*(2), 69–73.

de Isla, N., et al. (2010). Introduction to tissue engineering and application for cartilage engineering. *Bio-medical Materials and Engineering, 20*(3), 127–133.

Dominici, M., et al. (2006). Minimal criteria for defining multipotent mesenchymal stromal cells. The International Society for Cellular Therapy position statement. *Cytotherapy, 8*(4), 315–317.

Frydrych, M., et al. (2015). Biomimetic poly(glycerol sebacate)/poly(l-lactic acid) blend scaffolds for adipose tissue engineering. *Acta Biomaterialia, 18*, 40–49.

Fu, W., et al. (2014). Electrospun gelatin/PCL and collagen/PLCL scaffolds for vascular tissue engineering. *International Journal of Nanomedicine, 9*, 2335–2344.

Groth, T., et al. (1999). Altered vitronectin receptor (αv integrin) function in fibroblasts adhering on hydrophobic glass. *Journal of Biomedical Materials Research, 44*(3), 341–351.

Hall, B. K. (2015). Chapter 4: Invertebrate cartilages, notochordal cartilage and cartilage origins. In B. K. Hall (Ed.), *Bones and cartilage* (2nd ed., pp. 63–78). San Diego: Academic.

He, Y., & Lu, F. (2016). Development of synthetic and natural materials for tissue engineering applications using adipose stem cells. *Stem Cells International, 2016*, 12.

Hoemann, C. D., et al. (2012). The cartilage-bone interface. *The Journal of Knee Surgery, 25*(2), 85–97.

Hubbell, J. A. (2003). Materials as morphogenetic guides in tissue engineering. *Current Opinion in Biotechnology, 14*(5), 551–558.

Iwamoto, M., et al. (2013). Toward regeneration of articular cartilage. *Birth defects research. Part C, Embryo Today: Reviews, 99*(3), 192–202.

Kawakami, Y., Rodriguez-Leon, J., & Izpisua Belmonte, J. C. (2006). The role of TGFbetas and Sox9 during limb chondrogenesis. *Current Opinion in Cell Biology, 18*(6), 723–729.

Khademhosseini, A., et al. (2014). Chapter 32: Embryonic stem cells as a cell source for tissue engineering. In R. Lanza, R. Langer, & J. Vacanti (Eds.), *Principles of tissue engineering* (4th ed., pp. 609–638). Boston: Academic.

Machado, C. d. V, Telles, P. D. d. S, & Nascimento, I. L. O. (2013). Immunological characteristics of mesenchymal stem cells. *Revista Brasileira de Hematologia e Hemoterapia, 35*(1), 62–67.

Mano, J. F., et al. (2007). Natural origin biodegradable systems in tissue engineering and regenerative medicine: Present status and some moving trends. *Journal of the Royal Society, Interface, 4*(17), 999–1030.

Middleton, J. C., & Tipton, A. J. (2000). Synthetic biodegradable polymers as orthopedic devices. *Biomaterials, 21*(23), 2335–2346.

Miljkovic, N. D., Cooper, G. M., & Marra, K. G. (2008). Chondrogenesis, bone morphogenetic protein-4 and mesenchymal stem cells. *Osteoarthritis and Cartilage, 16*(10), 1121–1130.

Mondal, D., Griffith, M., & Venkatraman, S. S. (2016). Polycaprolactone-based biomaterials for tissue engineering and drug delivery: Current scenario and challenges. *International Journal of Polymeric Materials and Polymeric Biomaterials, 65*(5), 255–265.

Murphy, C. M., et al. (2013). Cell-scaffold interactions in the bone tissue engineering triad. *European Cells & Materials, 26*, 120–132.

Oh, S. H., et al. (2007). In vitro and in vivo characteristics of PCL scaffolds with pore size gradient fabricated by a centrifugation method. *Biomaterials, 28*(9), 1664–1671.

Oldershaw, R. A. (2012). Cell sources for the regeneration of articular cartilage: The past, the horizon and the future. *International Journal of Experimental Pathology, 93*(6), 389–400.

Pisanu, G., et al. (2018). Large osteochondral allografts of the knee: Surgical technique and indications. *Joints, 6* (1), 42–53.

Sophia Fox, A. J., Bedi, A., & Rodeo, S. A. (2009). The basic science of articular cartilage: Structure, composition, and function. *Sports Health, 1*(6), 461–468.

Tabata, Y., et al. (2000). De novo formation of adipose tissue by controlled release of basic fibroblast growth factor. *Tissue Engineering, 6*(3), 279–289.

Tang, L., Wu, Y., & Timmons, R. B. (1998). Fibrinogen adsorption and host tissue responses to plasma functionalized surfaces. *Journal of Biomedical Materials Research, 42*(1), 156–163.

Toonstra, J. L., et al. (2016). Patient experiences of recovery after autologous chondrocyte implantation: A qualitative study. *Journal of Athletic Training, 51*(12), 1028–1036.

Truong, N. C., Bui, K. H.-T., & Van Pham, P. (2018). Characterization of senescence of human adipose-derived stem cells after long-term expansion. In *Tissue engineering and regenerative medicine* (pp. 109–128). Springer.

Tuan, R. S., Chen, A. F., & Klatt, B. A. (2013a). Cartilage regeneration. *The Journal of the American Academy of Orthopaedic Surgeons, 21*(5), 303–311.

Tuan, R. S., Chen, A. F., & Klatt, B. A. (2013b). Cartilage regeneration. *The Journal of the American Academy of Orthopaedic Surgeons, 21*(5), 303–311.

Tuzlakoglu, K., et al. (2004). Production and characterization of chitosan fibers and 3-D fiber mesh scaffolds for tissue engineering applications. *Macromolecular Bioscience, 4*(8), 811–819.

Valcourt, U., et al. (1999). Different effects of bone morphogenetic proteins 2, 4, 12, and 13 on the expression of cartilage and bone markers in the MC615 chondrocyte cell line. *Experimental Cell Research, 251*(2), 264–274.

Van Pham, P., et al. (2014). Good manufacturing practice-compliant isolation and culture of human adipose-derived stem cells. *Biomedical Research and Therapy, 1*(04), 133–141.

Vinatier, C., & Guicheux, J. (2016). Cartilage tissue engineering: From biomaterials and stem cells to osteoarthritis treatments. *Annals of Physical and Rehabilitation Medicine, 59*(3), 139–144.

Vinatier, C., et al. (2009). Cartilage engineering: A crucial combination of cells, biomaterials and biofactors. *Trends in Biotechnology, 27*(5), 307–314.

Wall, A., & Board, T. (2014). Chemical basis for the histological use of Safranin O in the study of articular cartilage. In P. A. Banaszkiewicz & D. F. Kader (Eds.), *Classic papers in orthopaedics* (pp. 433–435). London: Springer.

Woodruff, M. A., & Hutmacher, D. W. (2010). The return of a forgotten polymer—Polycaprolactone in the 21st century. *Progress in Polymer Science, 35*(10), 1217–1256.

Adv Exp Med Biol - Innovations in Cancer Research and Regenerative Medicine (2023) 4: 193–208
https://doi.org/10.1007/5584_2021_670

Published online: 6 November 2021

Culture and Differentiation of Human Umbilical Cord-Derived Mesenchymal Stem Cells on Growth Factor-Rich Fibrin Scaffolds to Produce Engineered Cartilages

Phat Duc Huynh, Ngoc Bich Vu, Xuan Hoang-Viet To, and Thuan Minh Le

Abstract

Introduction After injuries, the cartilage healing capacity is limited owing to its nature as a particular connective tissue without blood vessels, lymphatics, or nerves. The creation of artificial cartilage tissue mimics the biological properties of native cartilage and can reduce the need for donated tissue. Fibrin is a type of biodegradable scaffold that has great potential in tissue engineering applications. It can become good material for cell adhesion and proliferation in vitro. Therefore, this study aimed to create a cartilage tissue in vitro using umbilical cord-derived mesenchymal stem cells (UCMSC) and growth factor-rich fibrin (GRF) scaffolds.

Methods UCMSCs were isolated and expanded, and platelet-rich plasma (PRP) preparations were performed following previously published protocols. PRP was activated (aPRP) by a 0.45-μm syringe filter to release growth factors inside the platelets. Each 2.10^5 of the UCMSCs were suspended in 2 ml of aPRP to make the mixture of MSC and PRP (MSC-PRP). Then, Ca^{2+} solution was added to this mixture to produce the fibril scaffold with UCMSCs inside. UCMSCs' adhesion and proliferation inside the scaffold were evaluated by observation under inverted microscopy, H-E staining, MTT assays, and scanning electron microscopy (SEM). The fibril structure containing UCMSCs was cultured, and chondrogenesis was induced using commercial chondrogenesis media for 21 days (iMSC-GRF). The differentiation in efficacy toward cartilage was evaluated based on the accumulation of aggrecan (acan), glycosaminoglycans (GAGs), and collagen type II (Col II).

Results The results showed that we successfully created a cartilage tissue with some characteristics that mimic the properties of natural cartilage. The engineered cartilage tissue

Phat Duc Huynh and Ngoc Bich Vu contributed equally with all other contributors.

P. D. Huynh and T. M. Le
Laboratory of Stem Cell Research and Application, University of Science Ho Chi Minh City, Ho Chi Minh City, Vietnam

Vietnam National University Ho Chi Minh City, Ho Chi Minh City, Vietnam

N. B. Vu (✉) and X. H.-V. To
Vietnam National University Ho Chi Minh City, Ho Chi Minh City, Vietnam

Stem Cell Institute, University of Science Ho Chi Minh City, Ho Chi Minh City, Vietnam
e-mail: vbngoc@hcmus.edu.vn; ngocvu@sci.edu.vn

was positive with some cartilage protein, such as acan, GAG, and Coll II. In vitro cartilage presented some natural chondrocyte-like cells. The artificial cartilage tissue was positive for CD14, CD34, CD90, CD105, and HLA-DR and negative for CD44, CD45, and CD73.

Conclusion These results showed that using UCMSCs and growth factor-rich fibril from platelet-rich plasma was feasible to produce engineered cartilage tissue for further experiments or clinical usage.

Keywords

Artificial cartilage · Biodegradable · Mesenchymal stem cell · Platelet-rich plasma · Scaffold · Tissue engineering · Fibrin · Umbilical cord

Abbreviations

CD	Cluster of differentiation
ECM	Extracellular matrix.
MSC	Mesenchymal stem cell.
PBS	Phosphate-buffered saline.
UCMSCs	Human umbilical cord-derived mesenchymal stem cells.

1 Introduction

Cartilage is a connective tissue found throughout the body. It covers the surface of joints and plays an essential role in helping your body absorb shock and your bones glide smoothly against each other without pain. It also participates in the structure of many body parts, such as the ears and the nose (Liu et al., 2014; Dai et al., 2015).

Many reasons cause cartilage damage, such as sports injuries, endogenous factors, and inflammatory diseases, leading to severe consequences, like osteoarthritis (Sophia Fox et al., 2009). However, as it lacks blood vessels and nerves, the ability of cartilage to regenerate itself when damage is minimal (Martinez-Diaz et al., 2010). Therefore, healing cartilage damage is a significant challenge in the field of medicine (Sophia Fox et al., 2009).

Several methods have been developed for the treatment of cartilage injuries. Some surgical options include lavage and debridement, arthroscopic microfracture, and osteochondral autograft for small defects. For significant defects, autologous cartilage implantation techniques or knee replacement techniques are often used (Ringe & Sittinger, 2009; Tuan et al., 2003). However, each of these methods has its disadvantages. Typically, autologous cartilage transplant patients need to undergo two surgeries in order to open the knee joint. Additionally, the treatment cost is relatively high (Minas et al., 2010). On the other hand, excessive proliferation of cartilage grafts after transplantation leads to stick knee joint and blocks the movement of the knee. Further, cartilage cell isolation and in vitro proliferation are major problems in autologous chondrocyte implantation (Jagielski et al., 2017). Cell transplantation also has a significant drawback: the lack of mechanical and morphological features of normal cartilage tissue (Xu et al., 2016).

The combination of stem cells with scaffold in tissue engineering is becoming a promising new method for supplying artificial cartilage tissue and adding options for doctors in treating cartilage defects (Vinatier & Guicheux, 2016; Dzobo et al., 2018).

Mesenchymal stem cells (MSCs) are isolated from various sources, such as the bone marrow (Pontikoglou et al., 2011), adipose tissue (Lin, 2012), synovial fluid and synovial membrane (de Sousa et al., 2014), cord blood (Hussain et al., 2012), periosteum (Mara et al., 2011), and adult skeletal muscle (Jackson et al., 2010). MSCs possess advantages such as potential differentiation in functional cell types (Turksen, 2013), immunomodulation properties (Mattar & Bieback, 2015), and anti-inflammatory effects (Ogata et al., 2015). In particular, the MSCs from the umbilical cord is one of the new options being increasingly used in cell therapies due to the full characteristics of MSCs, including its ease of collection and non-violation of ethics (Yoshimura et al., 2007).

The scaffold is an important factor in tissue engineering. They support a 3D structure for cell proliferation and differentiation (Sundelacruz & Kaplan, 2009). The scaffold material used in cartilage tissue engineering may be collagen (Calabrese et al., 2017), gelatin (Ponticiello et al., 2000), agarose (Yamagata et al., 2018a), or poly(lactic-co-glycolic acids) (PLGA) (Yamagata et al., 2018b). MSCs can adhere, proliferate, and differentiate into chondrocytes well to form cartilage-like tissue on these scaffolds (Calabrese et al., 2017; Ponticiello et al., 2000; Yamagata et al., 2018a; Yamagata et al., 2018b; Sonomoto et al., 2016). Fibrin is a natural material with high potential in tissue engineering (Sheykhhasan et al., 2015; Ahmed et al., 2008). It can be isolated from peripheral blood or umbilical cord blood (Hashemi et al., 2017). These are three product types that can be formed during and after activating fibrin, including platelet-rich fibrin (PRF), concentrated growth factor (GCF), or growth factor-rich fibrin (GRF) (Fukui et al., 2015). GRF is derived from platelet-rich plasma (PRP). It contains several growth factors, including insulin-like growth factor (IGF), vascular endothelial growth factor (VEGF), platelet-derived growth factor (PDGF), and transforming growth factor (TGF-ß1) (Fukui et al., 2015). GRF is also considered a good 3D scaffold for cell adhesion and proliferation (Barbon et al., 2019). GRF has been shown to stimulate and promote cartilage healing in vivo (Fortier et al., 2011). It, therefore, might offer promising therapy for cartilage regeneration (Barbon et al., 2019; De Andrade et al., 2018). Moreover, GRF possesses biodegradable properties.

Therefore, in this study, GRF was used for scaffolding for umbilical cord-derived mesenchymal stem cell (UCMSC) adhesion, proliferation, and differentiation toward chondroblasts to produce engineered cartilage in vitro.

2 Materials and Methods

2.1 Characterization of Human Umbilical Cord-Derived Mesenchymal Stem Cells

The UCMSCs were isolated and expanded according to published protocols. The umbilical cord tissues were collected from the hospital with a consent form. They were carefully washed with washing buffer to remove blood before the arteries and veins were removed. Then the tissues were cut into tiny fragments of $1 \times 1 \times 1$ mm^3. These fragments were plated in T-75 flasks with 5 mL of MSCCult I Primary (Regenmedlab, HCMC, VN). After 5 days, 5 mL of fresh medium was added to the flask, and the cell migration outside the fragments was observed. When the cells become 70% confluent, the tissue fragments were removed and discarded; cells were subcultured by Deattachment reagent (Regenmedlab, HCMC, VN) and cultured in MSCCult I medium (Regenmedlab, HCMC, VN). The medium would be refreshed every 3 days as well as continuously subcultured to the 3rd passage. The cells in the 3rd passage were used for further experiments.

The immunophenotype of isolated cells was characterized by flow cytometry using the BD Accuri™ C6 (BD Biosciences, Franklin Lakes, NJ). UCMSCs were stained with antibodies, including CD14-FITC, CD44-PE CD73-FITC, CD90-FITC, CD105-PE, and HLA-DR-FITC antibodies (purchased from Santa Cruz Biotechnology, Dallas, TX), and CD34-FITC (BD Biosciences, Franklin Lakes, NJ). Cells were stained with appropriate antibodies for 20 min at room temperature. Then, the stained cells were analyzed with a minimum of 10,000 events.

The in vitro differentiation potential of UCMSCs into adipocytes, osteoblasts, and chondroblasts was evaluated by in vitro differentiation assays. In the adipocyte differentiation assay, UCMSCs were cultured in DMEM/F12 medium supplemented with 10% FBS, 1% of 100X antibiotic-antimycotic, 10 mM dexamethasone, 2.79 mM indomethacin, 5 mg/ml insulin, and 0.5 M of 1-Methyl-3-isobutylxanthine (IBMX) (Sigma-Aldrich, Louis St., MO). The in vitro differentiation of UCMSCs into osteoblasts and chondroblasts were carried out using the protocols and chemicals of StemPro® Osteogenesis and Chondrogenesis Differentiation Kits (Thermo Fisher Scientific, Waltham, MA), respectively. Adipocytes, osteoblasts, and chondroblasts were confirmed by staining with Oil Red O, Alizarin Red S, and Alcian blue (all reagents were purchased from Sigma-Aldrich), respectively.

2.2 Production of Growth Factor-Rich Fibrin Scaffolds

Human umbilical cord blood was collected into bags containing the anticoagulant acid citrate dextrose (ACD) at the hospital after birth with a consent form. The PRP was produced using the commercial PRP preparation kits (5PRP kit, version U, Regenmedlab, HCMC, VN). Briefly, the blood was collected in the PRP separation tubes (50 mL/tube); tubes were then centrifuged at 2500 rpm for 10 min to separate the platelets in the phase upper gel. Then, PRP was activated to make activated PRP (aPRP) using a 0.45-μm syringe filter. The samples were stored at −20 °C, which can be rapidly thawed by a thermostatic bath at 37 °C when in use. To prepare the GRF scaffold, aPRP was transferred into a shaped container, followed by the addition of 20 mM of $CaCl_2$ solution, after which it was incubated at room temperature for seconds.

Further, UCMSCs embedded in the GRF scaffolds (MSC-GRFs) were prepared by suspending each 2×10^5 of UCMSCs in 2 ml aPRP. Then, 50 μl $CaCl_2$ of 0.8 M was added to form MSC-GRFs.

2.3 Evaluation of Fibrin Degradation

A part of the back hair of mice was shaved off to expose skin. The skin was cleansed with povidone-iodine antiseptic solutions (Pharmedic, VN). A 1-cm skin incision was made to reveal the muscle. Fibrin scaffold was fixed onto the muscle and marked with a stitch; then, the wound was stitched. Povidone-iodine was soaked into and around the cut as an antiseptic. The degradation of the GRF scaffold was assessed by mouse subdermal grafting on days 7 and 21. The physiological characteristics of the mice were monitored and recorded for up to 3 months. All manipulations on mice and the usage of mice for the study were approved by the institutional ethical committee (Stem Cell Institute, University of Science, HCMC, VN).

2.4 The Surface Morphology of the Fibrin Scaffolds

The scaffold was fixed in absolute ethanol (Sigma-Aldrich, USA) overnight. Fibrin scaffold surface was exhibited by scanning electron microscopy (SEM).

2.5 MTT Assay

MSC-GRF complexes were produced as described above in 96-well plates. The MSCCult I medium was used to culture UCMSCs inside the scaffolds in the plates and refreshed every 3 days. On days 3, 6, 9, 13, 18, and 23, fresh medium was replaced, and then 10 μl of MTT solution (0.45 mg/ml) was added. These samples were incubated at 37 °C with 5% CO_2. After 40 h, the MTT solution was removed, and 100 μL DMSO (Merck, Germany) was added to dissolve the formazan crystal. The optical density (OD) values were recorded at 595 nm by a DTX-880 system (Beckman Coulter, Brea, CA).

2.6 In Vitro Cartilage Differentiation of MSC-GRF Complexes

MSC-GRF complexes were cultured in MSCCult I medium (Regenmedlab, Viet Nam) for 7 days. The cell growth medium was then removed and replaced with a chondrogenic medium (Chondrogenesis Differentiation Kits – Thermo Fisher Scientific, Waltham, MA). The fresh differentiation induction medium was renewed every 7 days for 21 days. The formation of cartilage tissue was assessed based on the accumulation of aggrecan and glycosaminoglycans.

2.7 Hematoxylin and Eosin Staining Assays

H-E staining was used to evaluate the structure of fibrin, the MSC-GRF, and the in vitro engineered cartilage samples. Firstly, they were slid into 7 μm slices. Slices were deparaffinized twice in xylene for 5 min. Tissue samples were dehydrated by dipping into a series of alcohol (100%, 90%, 70%, 50%) for 5 min each. The samples were then washed with distilled water for 2 min and air-dried. Hematoxylin (Sigma-Aldrich, USA) was added into a sample for 15 min and then washed with distilled water. The samples were differentiated in 1% acid alcohol (70% alcohol with 1% HCl) for 5 min. Eosin (Sigma-Aldrich, USA) was then added for 30 s and washed with distilled water for 3 min. Finally, the sample was washed one more time with distilled water for 3 min and mounted with a mounting medium. The structure was observed under a light microscope (Carl Zeiss, Germany).

2.8 Aggrecan and Glycosaminoglycans Expression Assay

Engineered cartilage tissue samples were fixed in 4% PFA and then slid into 7-μm-thick slices with paraffin embedding process. After being deparaffinized with xylene and alcohol, the slices were rinsed with PBS and soaked in staining solution. For Alcian blue staining, the samples were stained for 30 min, washed with destaining solution (3C2H5OH: 2CH3COOH), and then washed with PBS again to remove all acidic solutions. The samples were finally observed under a microscope. For Safranin O staining, the samples were stained for 15 min and then washed with 90–95% alcohol. The samples were rewashed with PBS and observed under a microscope.

2.9 Immunohistochemistry

Deparaffinized sections were used for immunohistochemistry. Samples were washed twice in TBST (Tris-buffered saline, 0.1% Tween 20) for 5 min each time. Blocking goat serum was then added for 2 h at room temperature. Primary antibody was added to samples, and the samples were incubated overnight at 2–8 °C. Residual primary antibodies were removed by washing the sample twice in TBST for 5 min each time. Subsequently, secondary antibodies were added and stored in the dark for 1 h. Rinsing of the sample was continued to remove excess secondary antibodies. Finally, the samples were mounted in a mounting medium containing DAPI and analyzed by LMS800 Confocal microscopy (Carl Zeiss, Germany).

3 Results

3.1 Characterization of UCMSCs

UCMSCs exhibited fibroblast-like morphology and adhered to a plastic surface (Fig. 1a). They could successfully accumulate calcium after being induced in the osteogenesis medium for 21 days. Therefore, the cells were positive with Alizarin red and formed red complexes (Fig. 1b). UCMSCs formed intracellular lipid droplets positive with Oil Red dye (Fig. 1c). In the chondrogenesis medium, the aggrecan expression positive of UCMSCs was observed after Alcian blue staining (Fig. 1d). The UCMSCs also displayed four positive MSC surface markers, including CD44, CD73, CD90, and CD105, and three negative MSC surface markers, including CD14, CD34, and HLA-DR (Fig. 1e–m).

3.2 GRF Scaffold Formation

The results show that, immediately after being formed, the GRF block has a circular structure like the shape of its container (Fig. 2a) with a diameter of 3.5 cm and a thickness of 0.5 cm (Fig. 2c). After separating from the dish surface, the GRF scaffold was shrunk. However, it still retained its round shape (Fig. 2b) with a diameter of 2.147 ± 0.04 cm and a thickness of 1 mm (Fig. 2c). Thus, GRF was an opalescent block with a certain elasticity and toughness.

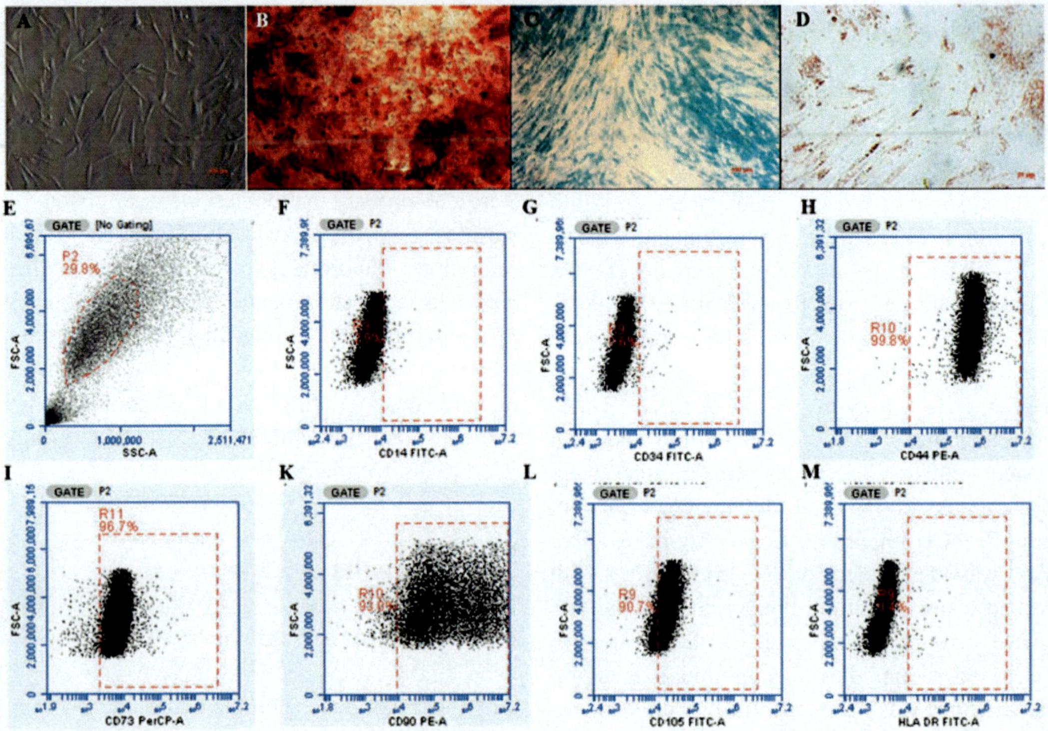

Fig. 1 **UCMSC characterization**. UCMSCs exhibited the fibroblast-like shape (**a**); differentiated into osteoblasts (**b**), chondrocytes (**c**), and adipocytes (**d**); cell population (**e**) was expressed positive for CD44 (**h**), CD73 (**i**), CD90 (**k**), and CD105 (**l**) but absent for CD14 (**f**), CD34 (**g**), and HLA-DR (**m**)

3.3 Proliferation of MSC on Fibrin Scaffolds

Immediately after the MSC-GRF formed, single spherical cells were evenly distributed in the scaffold (Fig. 3a). After 3 days, UCMSCs were adhered to and expanded on the surface of the scaffold (Fig. 3b). The cells continued to proliferate in the following days with a dense coverage of the scaffold. Some cell clusters were formed that could be observed under a microscope. The overlapping multilayer of cells was observed (Fig. 3c–d).

The cell adhesion and proliferation in GRF scaffold were evaluated by H&E staining. The results showed that the fibers of the GRF scaffold interleaved with each other to form a reasonably homogeneous network (Fig. 4a). It was easy to observe the presence of cells on MSC-GRF at day 7 (Fig. 4b). The cells scattered inside the scaffold with many different shapes, such as circles or spindle cells, showed the three-dimensional growth of cells. On day 14 (Fig. 4c) and day 28 (Fig. 4d), cell density was significantly increased compared to the day before. These results were confirmed by an MTT assay. The MTT assay-derived results also showed that, on the third day after culture, formazan crystals were presented at low concentrations in MSC-GRF (Fig. 5a), but the formazan crystal density had continuously increased on day 6 (Fig. 5b) to day 9 (Fig. 5c) to day 13 (Fig. 5d).

3.4 MSC-GRF Structure Under SEM

Immediately after loading the cells, the presence of cells on the GRF scaffold was rare. The surface of the scaffold was slightly rough with tightly interwoven fibers (Fig. 6). On day 7 of culture,

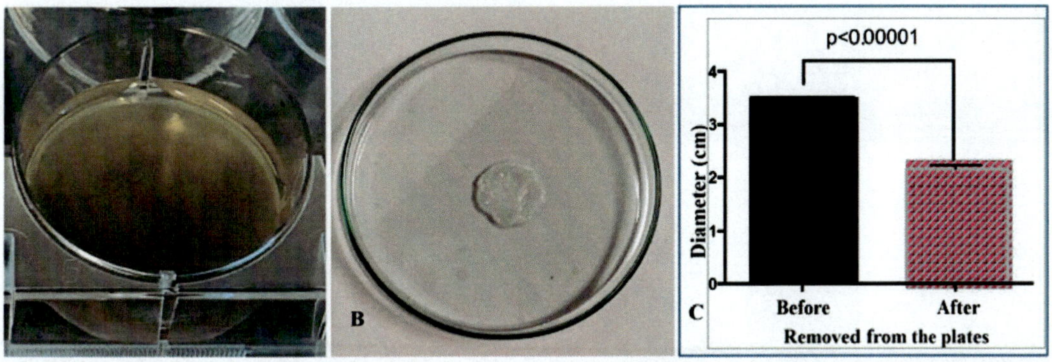

Fig. 2 The shape and size of fibrin scaffold. The fibrin scaffold shape immediately after formed (**a**); after separation from the disk surface (**b**); size of fibrin scaffold before and after being removed from the surface of the culture plates (n = 3)

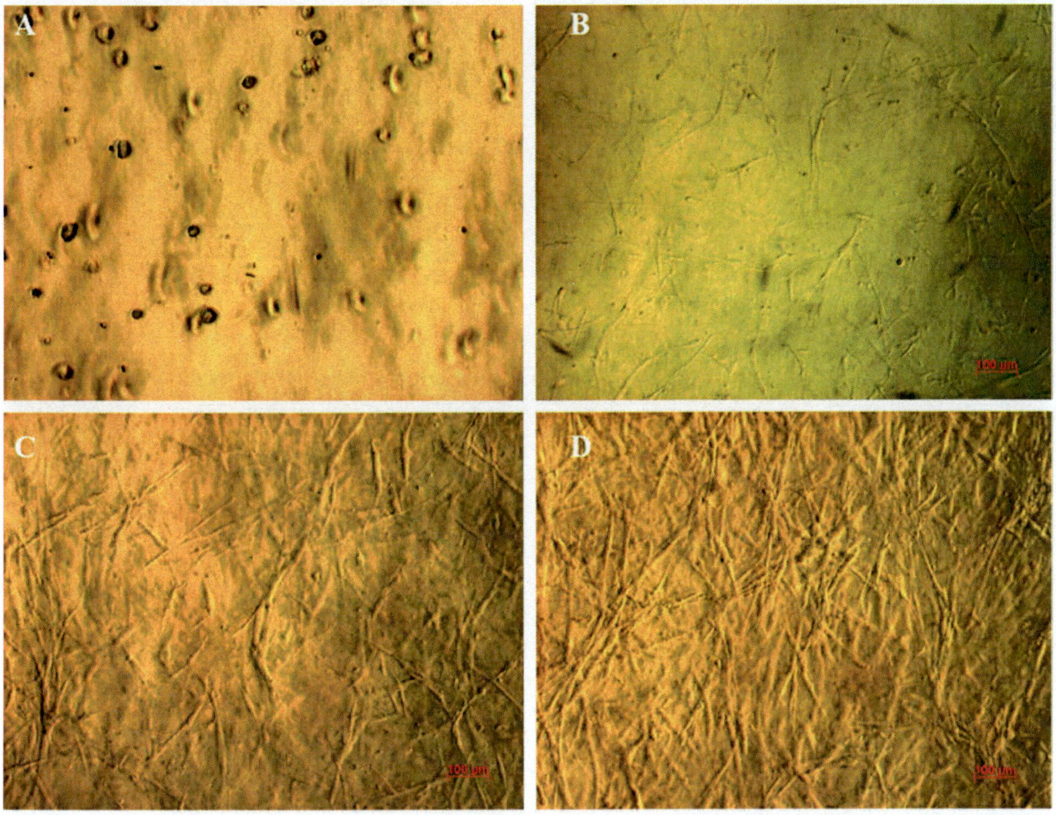

Fig. 3 The UCMSC adhesion and proliferation on the surface of the fibrin scaffolds. The scaffold immediately after cell loading was presented circular cells (**a**); cells adhered to the culture flask surface after 3 days (**b**) and continued to proliferate on day 9 (**c**) and day 13 (**d**)

the pore size between fibrin fibers of MSC-GRF was more significant than the previous days (Fig. 6). This made the MSC-GRF surface structure rougher (Fig. 6). From day 14, the cells were presented on the MSC-GRF surface with high density (Fig. 6c). The surface structure of

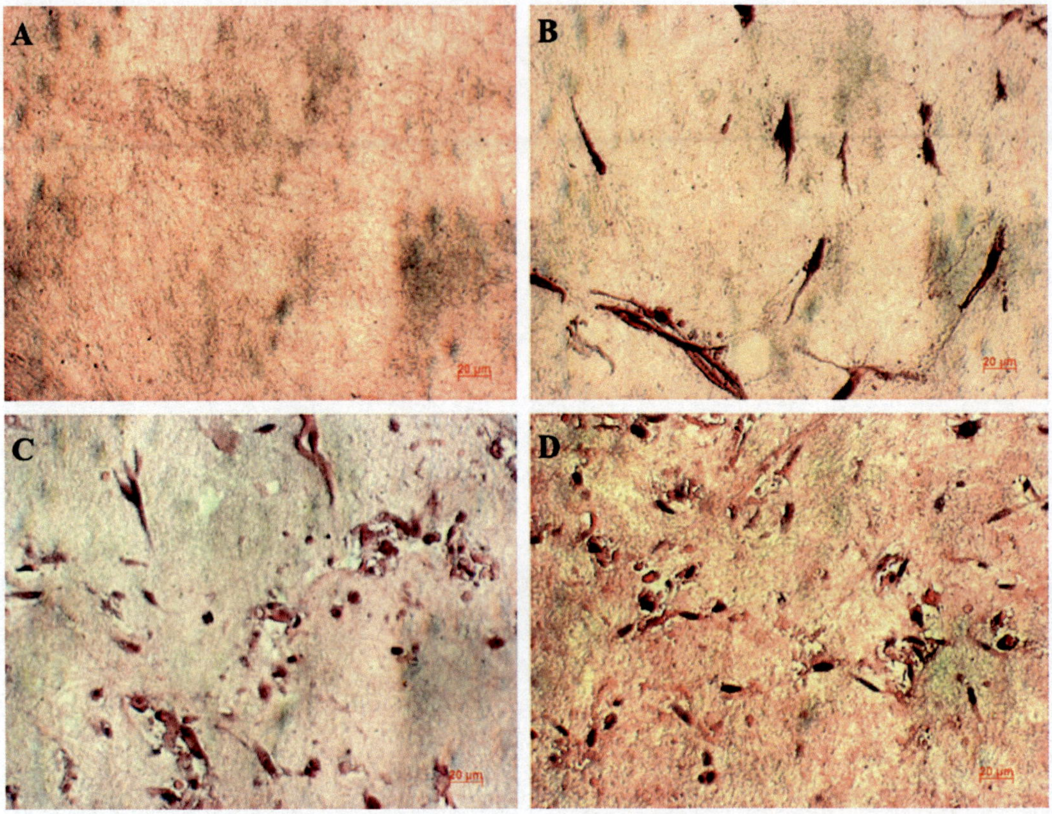

Fig. 4 H&E staining for GRF scaffolds and MSC-GRF. The GRF structure (**a**) and with MSC-GRF after 7 days (**b**), 14 days (**c**), and 28 days (**d**) of culture showed pronounced cell proliferation inside the scaffold

the scaffold had a considerable variation compared to the original. The presence of fibrin fibers was challenging to observe (Fig. 6f). This proved that there was a change in the structure of the fibrin scaffold after culturing cells on it.

Seven days after subdermal transplantation, the GRF scaffolds were disappeared (Fig. 7b) that could not be observed on the first day (Fig. 7a). No swelling or inflammation around the grafted area was observed from days 7 to 21 after transplantation (Fig. 7c). The physiological activities of the mice were still normal. This suggests that the GRF scaffold became a biodegradable scaffold within 7 days.

3.5 UCMSC Differentiation into Chondrocytes on GRF Scaffolds

The morphology of UCMSCs inside GRF scaffold was similar to the flask culture with spindle-shaped cells and large, flattened cells (Fig. 8a). After 21 days of chondrogenic differentiation induction, UCMSCs changed their shape and became rounded and smaller (Fig. 8b). This result is similar to that in the study by Tang X et al. (2012) about the feature of human cartilage cells in vivo (Fig. 8c).

3.6 The Expression of Markers of UCMSC Before and After Chondrogenic Differentiation

The results showed that UCMSCs were positive for CD44, CD73, CD90, and CD105 and negative for CD14, CD34, and CD45; HLA-DR when cultured on GRF scaffolds for 7 days (Fig. 9a). However, after chondrogenic-induced UCMSC, the expression of cell markers had changed. Differentiated cells were positive for CD14, CD34, and HLA-DR and negative for CD44 and

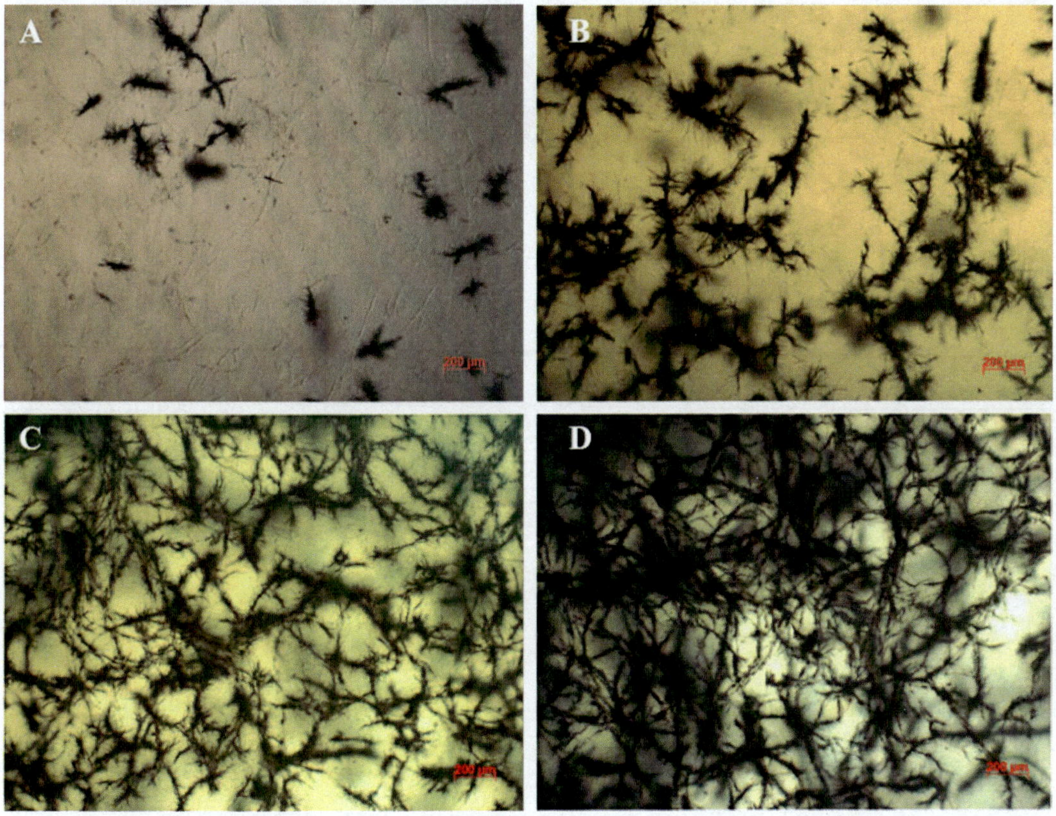

Fig. 5 Cell proliferation in MSC-GRF scaffold through formazan crystals formation assessment. Formazan crystals increase over day 3 (**a**), day 6 (**b**), day 9 (**c**), and day 13 (**d**)

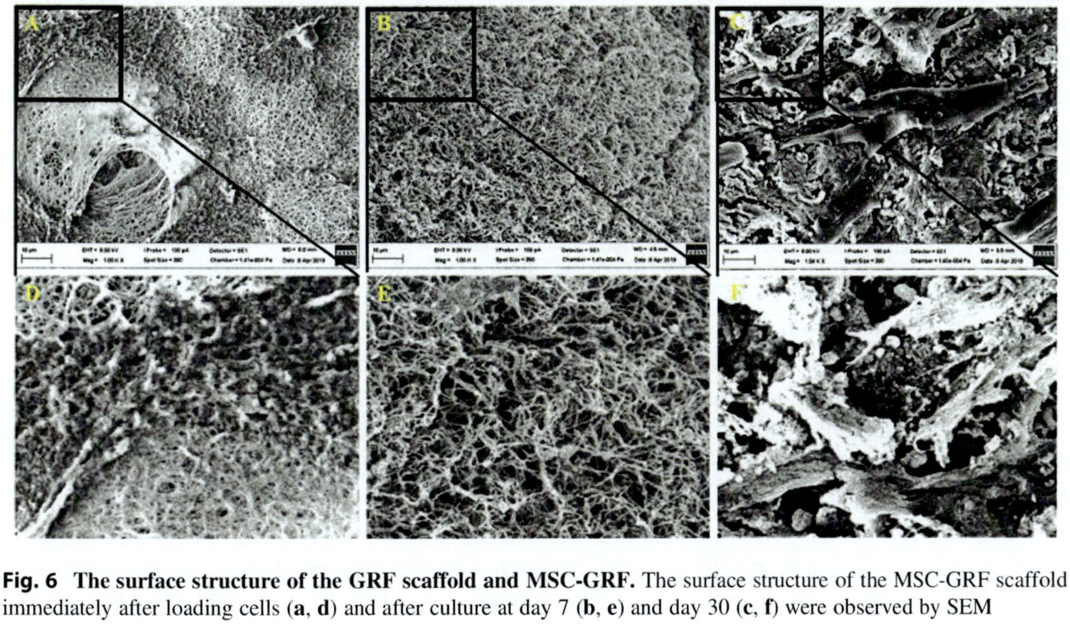

Fig. 6 The surface structure of the GRF scaffold and MSC-GRF. The surface structure of the MSC-GRF scaffold immediately after loading cells (**a**, **d**) and after culture at day 7 (**b**, **e**) and day 30 (**c**, **f**) were observed by SEM

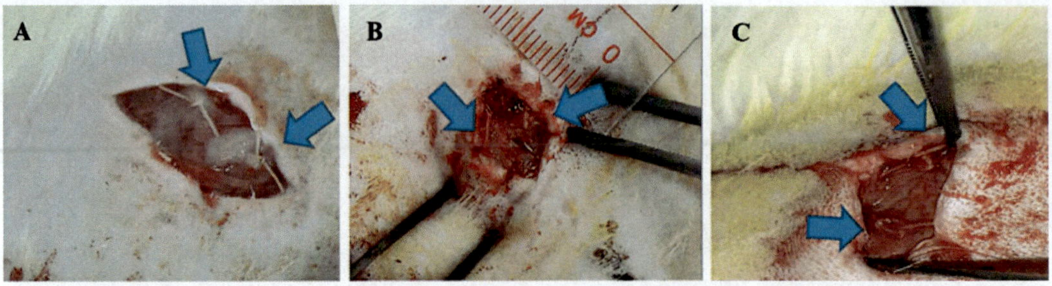

Fig. 7 **The existence of GRF scaffold in mice.** The GRF scaffold was immediately transplanted under mouse skin (**a**) 7 days (**b**) and 21 days (C) after the transplantation

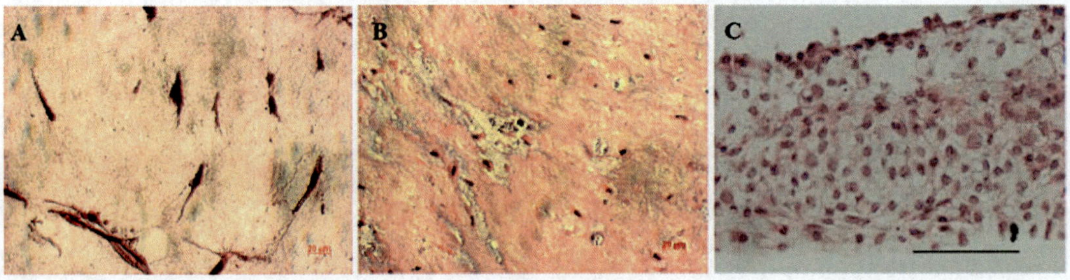

Fig. 8 **The morphology change after and before chondrogenic differentiation of MSC-GRF**. The cell morphology before (**a**) and after (**b**) differentiation and the in vivo chondrocyte following the result of Tang X et al. (**c**)

CD73. The expression of CD90 and CD105 markers was weaker than before differentiation (Fig. 9b).

Cartilage-induced MSC-GRF complexes were positive with Safranin O (Fig. 10a). They were also stained blue with Alcian blue staining assay (Fig. 10c). This meant that there were formation and accumulation of GAGs and aggrecan proteins in the extracellular matrix. In contrast, MSC-GRF without induction was negative for these assays (Fig. 10b, d).

The expression of collagen type II (Col II) in the induced UCMSC-GRF also was detected by immunohistochemistry (Fig. 11). Non-induced UCMSC-GRF complexes were negative with Col II (Fig. 11a–d), while chondrogenesis-induced UCMSC-GRF complexes significantly expressed this marker (Fig. 11e–h).

4 Discussion

Cartilage engineering is a promising strategy for producing artificial cartilage tissue to treat chondrocyte-related diseases. This study aims to make artificial cartilage tissue in vitro. In this study, we succeeded in creating engineered cartilage tissue after culturing UCMSCs on a GRF scaffold and culturing in the chondrogenic-induced medium.

In the first experiment, we successfully isolated and expanded UCMSCs with MSC phenotype that satisfied the minimal criteria for MSCs as suggested by the International Society for Cellular Therapy (Dominici et al., 2006). Indeed, UCMSCs have the ability to adhere to the plastic surface. They also can differentiate into chondrocytes, osteoblasts, and adipocytes in vitro; they also expressed the MSC-specific marker profile including testing positive for CD44, CD73, CD90, and CD105 and negative for CD14, CD34, and HLA-DR.

In the next experiment, these UCMSCs were loaded to a fibrin scaffold produced from platelet-rich plasma. The fibrin scaffold provides a three-dimensional network for cell adhesion. Furthermore, growth factors released during fibrin formation can provide safe and effective natural that cell growth took up the space of fibrin fibers. This result is similar to previous results published by W. Liu et al. (2010).

UCMSCs in UCMSC-GRF complexes produced some chondrogenic proteins including GAGs, aggrecan, and Col II and after cultured chondrogenic differentiation conditions for 21 days. Indeed, Safranin O is a dye that can strongly bind to carboxyl and sulfate groups (SO_4^{2-}) of GAGs (Kiviranta et al., 1985; Kiraly et al., 1996; Sánchez, 2012); therefore, it is usually used to determine the presence of GAGs in the cartilage tissues. Alcian blue is a dye stain aggrecans (Detrich III et al., 2016). These results showed that UCMSCs in UCMSC-GRF successfully differentiated into chondroblasts (Nguyen et al., 2018), and the complexes of UCMSC-GRF expressed the cartilage-like histological characteristics. Furthermore, the expression of Col II in cartilage-induced UCMSC-GRF complexes also confirmed this observation.

Moreover, the successful differentiation of UCMSCs toward chondroblasts in the scaffold was also observed by changes in the surface marker expression of UCMSCs before and after induction in chondrogenesis medium. We found that in contrast to hUC-MSCs, osteogenic induced cells expressed HLA-DR, CD14, CD34, and Col II, whereas absented for CD44, and CD73 marker. More interestingly, in this study, chondrocytes derived from UCMSCs still expressed some MSC-like markers, including CD90 and CD105. HLA-DR is a human leukocyte antigen class II cell surface receptor, which is found on many cells, such as a subpopulation of epidermal Langerhans cells (Wolf et al., 2001), activated T-cells (Starska et al., 2011), and peripheral blood mononuclear cells (Bone and Mills, 2013). It is also present on human cartilage chondrocytes (Pereira et al., 2016). During the chondrogenic differentiation, the expression of TGF-β family factors known to help increase the production of cartilage ECM contributed to regulating the expression of HLA-DR (Technau et al., 2011).

CD14 is a marker found on monocytes and most tissue macrophages (Naeim et al., 2013). It is also expressed at lower levels on chondrocytes in arthritic joints (Summers et al., 1995). Meanwhile, CD34 was expressed on vascular endothelial cells (Fina et al., 1990), hematopoietic stem or progenitor cells (Krause et al., 1994), and some types of fibroblasts (Naeim et al., 2018). These results are also mentioned in other publications (Pereira et al., 2016; Technau et al., 2011; Lee et al., 2013). CD44 is considered an important marker in modulating cell migration and inflammation (Gee et al., 2002). They are present in cancer cells (Senbanjo & Chellaiah, 2017), are chondrogenic subpopulation from nasal septal cartilage zone (do Amaral et al., 2012), and are also indicative of chondrocyte marker (Campbell & Pei, 2012). Therefore, the absence of CD44 has shown an increase in the potential for chondrocyte formation of stem cells.

Some published studies showed the role of CD90 in chondrogenesis. In these studies, the CD90 presence at high levels has been shown to play an essential role in the differentiation of mesenchymal stem cells into cartilage cells. It is also an indicator for the chondrogenic differentiation potential of synovium-derived stem cells (Campbell & Pei, 2012; Diaz-Romero et al., 2005; Nagase et al., 2008).

Col II, also known as COL IIA1, is the main component of collagen that constitutes articular cartilage. It makes up about 85–90% of the collagen in articular cartilage (Yamashita et al., 2009; Pouran et al., 2018). COL IIA1 plays a vital role in withstanding mechanical forces, maintaining physiological homeostasis, and providing anchor locations for chondrocytes (Aigner & Stove, 2003). On the other hand, similar to other dissolved signals, COL IIA1 is a crucial extracellular signal molecule that regulates chondrocyte proliferation, metabolism, and differentiation (Klatt et al., 2009), (Xin et al., 2015). Col II and aggrecan are used as chondrocyte markers in cell culture in vitro (Frazer et al., 1994). Our result showed that COL IIA1 was strongly expressed in 21-day differentiated samples, while it was not expressed in undifferentiated samples. This result is similar to other publications (Lee et al., 2013),(López-Ruiz et al., 2013). In a study published by Pereira et al., human cartilage chondrocytes expressed the surface markers HLA-DR, CD44, CD73, CD105, and

CD90, while lacking expression for CD45, CD14, and CD34 markers (Pereira et al., 2016). Another publication by López-Ruiz et al. showed that cultured chondrocytes were positive for CD73, CD90, and CD105 but negative for CD45 and CD34 (López-Ruiz et al., 2013).

5 Conclusion

This study successfully produced an engineered cartilage tissue by culturing and differentiating human umbilical cord-derived mesenchymal stem cells on growth factor-rich fibrin scaffolds in a suitable chondrogenesis medium. Engineered cartilage tissues expressed some native cartilage-specific properties, such as testing positive for aggrecan, glycosaminoglycans, and Col II similar to natural cartilage. These engineered cartilage tissues expressed CD14, CD34, CD90, CD105, and HLA-DR, but not CD44, CD45, and CD73. These results, therefore, suggested a promising method to produce engineered cartilage tissues for further studies or clinical usages.

Acknowledgments This research is funded by Vietnam National University Ho Chi Minh City (VNU-HCM) under grant number C2017-18-24/HĐ-KHCN.

Conflict of Interest The authors declare that there is no conflict of interest regarding the publication of this paper.

References

Ahmed, T. A., Dare, E. V., & Hincke, M. (2008). Fibrin: A versatile scaffold for tissue engineering applications. *Tissue Engineering. Part B, Reviews, 14*(2), 199–215.

Aigner, T., & Stove, J. (2003). Collagens–major component of the physiological cartilage matrix, major target of cartilage degeneration, major tool in cartilage repair. *Advanced Drug Delivery Reviews, 55*(12), 1569–1593.

Anitua, E., et al. (2019). Autologous fibrin scaffolds: When platelet- and plasma-derived biomolecules meet fibrin. *Biomaterials, 192*, 440–460.

Barbon, S., et al. (2019). Platelet-rich fibrin scaffolds for cartilage and tendon regenerative medicine: From bench to bedside. *International Journal of Molecular Sciences, 20*(7), 1701.

Bensaid, W., et al. (2003). A biodegradable fibrin scaffold for mesenchymal stem cell transplantation. *Biomaterials, 24*(14), 2497–2502.

Bone, K., & Mills, S. (Eds.). (2013). *10 – How to use the monographs, in Principles and Practice of Phytotherapy* (2nd ed., pp. 353–961). Saint Louis: Churchill Livingstone.

Calabrese, G., et al. (2017). Combination of collagen-based scaffold and bioactive factors induces adipose-derived mesenchymal stem cells chondrogenic differentiation in vitro. *Frontiers in Physiology, 8*(50).

Campbell, D. D., & Pei, M. (2012). Surface markers for chondrogenic determination: A highlight of synovium-derived stem cells. *Cell, 1*(4), 1107–1120.

Dai, L., et al. (2015). Silencing of miR-101 prevents cartilage degradation by regulating extracellular matrix-related genes in a rat model of osteoarthritis. *Molecular Therapy, 23*(8), 1331–1340.

De Andrade, L., et al. (2018). The use of platelet-rich fibrin concentrate in tissue healing and regeneration in dentistry. *International Journal of Growth Factors and Stem Cells in Dentistry, 1*(1), 23–26.

de Sousa, E. B., et al. (2014). Synovial fluid and synovial membrane mesenchymal stem cells: Latest discoveries and therapeutic perspectives. *Stem Cell Research & Therapy, 5*(5), 112–112.

Detrich, H. W., III, Westerfield, M., & Zon, L. (2016). *The Zebrafish: Cellular and developmental biology, part B developmental biology* (Vol. 134). Academic.

Diaz-Romero, J., et al. (2005). Immunophenotypic analysis of human articular chondrocytes: Changes in surface markers associated with cell expansion in monolayer culture. *Journal of Cellular Physiology, 202*(3), 731–742.

do Amaral, R. J., et al. (2012). Isolation of human nasoseptal chondrogenic cells: A promise for cartilage engineering. *Stem Cell Research, 8*(2), 292–299.

Dominici, M., et al. (2006). Minimal criteria for defining multipotent mesenchymal stromal cells. The International Society for Cellular Therapy position statement. *Cytotherapy, 8*(4), 315–317.

Dzobo, K., et al. (2018). Advances in regenerative medicine and tissue engineering: Innovation and transformation of medicine. *Stem Cells International, 2018*, 24.

Fina, L., et al. (1990). Expression of the CD34 gene in vascular endothelial cells. *Blood, 75*(12), 2417–2426.

Fortier, L. A., et al. (2011). The role of growth factors in cartilage repair. *Clinical Orthopaedics and Related Research, 469*(10), 2706–2715.

Frazer, A., et al. (1994). Studies on type II collagen and aggrecan production in human articular chondrocytes in vitro and effects of transforming growth factor-β and interleukin-1β. *Osteoarthritis and Cartilage, 2*(4), 235–245.

Fukui, N., et al. (2015). Quantification of growth factors in platelet-rich fibrin: A preliminary study. *Journal of Hard Tissue Biology, 24*(3), 231–234.

Gee, K., et al. (2002). Differential regulation of CD44 expression by lipopolysaccharide (LPS) and TNF-α in human monocytic cells: Distinct involvement of c-Jun N-terminal kinase in LPS-induced CD44 expression. *The Journal of Immunology, 169*(10), 5660–5672.

Hashemi, S.-S., et al. (2017). The role of human adult peripheral and umbilical cord blood platelet-rich plasma on proliferation and migration of human skin fibroblasts. *World Journal of Plastic Surgery, 6*(2), 198–205.

Hussain, I., et al. (2012). New approach to isolate mesenchymal stem cell (MSC) from human umbilical cord blood. *Cell Biology International, 36*(7), 595–600.

Jackson, W. M., Nesti, L. J., & Tuan, R. S. (2010). Potential therapeutic applications of muscle-derived mesenchymal stem and progenitor cells. *Expert Opinion on Biological Therapy, 10*(4), 505–517.

Jagielski, M., et al. (2017). The influence of IL-10 and TNFalpha on chondrogenesis of human mesenchymal stromal cells in three-dimensional cultures. *International Journal of Molecular Sciences, 15*(9), 15821–15844.

Kiraly, K., et al. (1996). Application of selected cationic dyes for the semiquantitative estimation of glycosaminoglycans in histological sections of articular cartilage by microspectrophotometry. *The Histochemical Journal, 28*(8), 577–590.

Kiviranta, I., et al. (1985). Microspectrophotometric quantitation of glycosaminoglycans in articular cartilage sections stained with Safranin O. *Histochemistry, 82* (3), 249–255.

Klatt, A. R., et al. (2009). Discoidin domain receptor 2 mediates the collagen II-dependent release of interleukin-6 in primary human chondrocytes. *The Journal of Pathology, 218*(2), 241–247.

Krause, D., et al. (1994). Characterization of murine CD34, a marker for hematopoietic progenitor and stem cells. *Blood, 84*(3), 691–701.

Lee, Y. S., et al. (2013). Isolation of mesenchymal stromal cells (MSCs) from human adenoid tissue. *Cellular Physiology and Biochemistry, 31*(4–5), 513–524.

Lin, F. (2012). Adipose tissue-derived mesenchymal stem cells: A fat chance of curing kidney disease? *Kidney International, 82*(7), 731–733.

Liu, W., et al. (2010). The mechanical properties of single fibrin fibers. *Journal of Thrombosis and Haemostasis, 8*(5), 1030–1036.

Liu, Q., et al. (2014). Long noncoding RNA related to cartilage injury promotes chondrocyte extracellular matrix degradation in osteoarthritis. *Arthritis & Rhematology, 66*(4), 969–978.

López-Ruiz, E., et al. (2013). Chondrocytes extract from patients with osteoarthritis induces chondrogenesis in infrapatellar fat pad-derived stem cells. *Osteoarthritis and Cartilage, 21*(1), 246–258.

Mara, C. S., et al. (2011). Periosteum as a source of mesenchymal stem cells: The effects of TGF-beta3 on chondrogenesis. *Clinics (São Paulo, Brazil), 66*(3), 487–492.

Martinez-Diaz, S., et al. (2010). In vivo evaluation of 3-dimensional polycaprolactone scaffolds for cartilage repair in rabbits. *American Journal of Sports Medicine, 38*(3), 509–519.

Mattar, P., & Bieback, K. (2015). Comparing the immunomodulatory properties of bone marrow, adipose tissue, and birth-associated tissue mesenchymal stromal cells. *Frontiers in Immunology, 6*(560).

Minas, T., et al. (2010). Autologous chondrocyte implantation for joint preservation in patients with early osteoarthritis. *Clinical Orthopaedics and Related Research, 468*(1), 147–157.

Mollapour Sisakht, M., et al. (2019). Fibrin-collagen hydrogel as a scaffold for dermoepidermal skin substitute, preparation and characterization. *Journal of Contemporary Medical Sciences, 5*(1).

Naeim, F., et al. (2013). 2 – Principles of immunophenotyping. In F. Naeim et al. (Eds.), *Atlas of hematopathology* (pp. 25–46). Academic.

Naeim, F., et al. (2018). Chapter 2: Principles of immunophenotyping. In F. Naeim et al. (Eds.), *Atlas of hematopathology* (2nd ed., pp. 29–56). Academic.

Nagase, T., et al. (2008). Analysis of the chondrogenic potential of human synovial stem cells according to harvest site and culture parameters in knees with medial compartment osteoarthritis. *Arthritis and Rheumatism, 58*(5), 1389–1398.

Nguyen, P. D.-N., et al. (2018). Engineered cartilage tissue from biodegradable Poly(ε-caprolactone) scaffold and human umbilical cord derived mesenchymal stem cells, *5*(2).

Ogata, Y., et al. (2015). Purified human synovium mesenchymal stem cells as a good resource for cartilage regeneration. *PLoS One, 10*(6), e0129096.

Pereira, R. C., et al. (2016). Human articular chondrocytes regulate immune response by affecting directly T cell proliferation and indirectly inhibiting monocyte differentiation to professional antigen-presenting cells. *Frontiers in Immunology*, **7**(415).

Ponticiello, M. S., et al. (2000). Gelatin-based resorbable sponge as a carrier matrix for human mesenchymal stem cells in cartilage regeneration therapy. *Journal of Biomedical Materials Research, 52*(2), 246–255.

Pontikoglou, C., et al. (2011). Bone marrow mesenchymal stem cells: Biological properties and their role in hematopoiesis and hematopoietic stem cell transplantation. *Stem Cell Reviews and Reports, 7*(3), 569–589.

Pouran, B., et al. (2018). Non-enzymatic cross-linking of collagen type II fibrils is tuned via osmolality switch. *Journal of Orthopaedic Research, 36*(7), 1929–1936.

Ringe, J., & Sittinger, M. (2009). Tissue engineering in the rheumatic diseases. *Arthritis Research & Therapy, 11* (1), 211.

Sánchez, H. I. A. (2012). Safranin O Dye. *Investigación en Discapacidad, 1*(2), 83–85.

Senbanjo, L. T., & Chellaiah, M. A. (2017). CD44: A multifunctional cell surface adhesion receptor is a regulator of progression and metastasis of cancer cells. *Frontiers in Cell and Developmental Biology, 5*, 18–18.

Sheykhhasan, M., Qomi, R. T., & Ghiasi, M. (2015). Fibrin scaffolds designing in order to human adipose-derived mesenchymal stem cells differentiation to chondrocytes in the presence of TGF-β3. *International Journal of Stem Cells, 8*(2), 219–227.

Sonomoto, K., et al. (2016). Spontaneous differentiation of human mesenchymal stem cells on poly-lactic-co-glycolic acid Nano-fiber scaffold. *PLoS One, 11*(4), e0153231.

Sophia Fox, A. J., Bedi, A., & Rodeo, S. A. (2009). The basic science of articular cartilage: structure, composition, and function. *Sports Health, 1*(6), 461–468.

Starska, K., et al. (2011). The role of tumor cells in the modification of T lymphocytes activity--the expression of the early CD69+, CD71+ and the late CD25+, CD26+, HLA/DR+ activation markers on T CD4+ and CD8+ cells in squamous cell laryngeal carcinoma. *Part I. Folia Histochem Cytobiol, 49*(4), 579–592.

Summers, K. L., et al. (1995). Monocyte-macrophage antigen expression on chondrocytes. *The Journal of Rheumatology, 22*(7), 1326–1334.

Sundelacruz, S., & Kaplan, D. L. (2009). Stem cell- and scaffold-based tissue engineering approaches to osteochondral regenerative medicine. *Seminars in Cell & Developmental Biology, 20*(6), 646–655.

Technau, A. F. K., Hagen, R., & Kleinsasser, N. (2011). Adipose tissue-derived stem cells show both immunogenic and immunosuppressive properties after chondrogenic differentiation. *Cytotherapy, 13*(3), 310–317.

Tuan, R. S., Boland, G., & Tuli, R. (2003). Adult mesenchymal stem cells and cell-based tissue engineering. *Arthritis Research & Therapy, 5*(1), 32–45.

Turksen, K. (2013). Adult stem cells and cardiac regeneration. *Stem Cell Reviews and Reports, 9*(5), 537–540.

Vinatier, C., & Guicheux, J. (2016). Cartilage tissue engineering: From biomaterials and stem cells to osteoarthritis treatments. *Annals of Physical and Rehabilitation Medicine, 59*(3), 139–144.

Wolf, H. M., et al. (2001). Residual expression of functional MHC class II molecules in twin brothers with MHC class II deficiency is cell type specific. *British Journal of Haematology, 115*(2), 460–471.

Xin, W., et al. (2015). Collagen II regulates chondrocyte integrin expression profile and differentiation. *Connective Tissue Research, 56*(4), 307–314.

Xu, J., et al. (2016). Chondrogenic differentiation of human mesenchymal stem cells in three-dimensional alginate gels. *Tissue Eng Part A, 14*(5), 667–680.

Yamagata, K., Nakayamada, S., & Tanaka, Y. (2018a). Use of mesenchymal stem cells seeded on the scaffold in articular cartilage repair. *Inflammation and Regeneration, 38*, 4–4.

Yamagata, K., Nakayamada, S., & Tanaka, Y. (2018b). Use of mesenchymal stem cells seeded on the scaffold in articular cartilage repair. *Inflammation and Regeneration, 38*(1), 4.

Yamashita, S., et al. (2009). Sox9 directly promotes Bapx1 gene expression to repress Runx2 in chondrocytes. *Experimental Cell Research, 315*(13), 2231–2240.

Yoshimura, H., et al. (2007). Comparison of rat mesenchymal stem cells derived from bone marrow, synovium, periosteum, adipose tissue, and muscle. *Cell and Tissue Research, 327*(3), 449–462.

Adv Exp Med Biol - Innovations in Cancer Research and Regenerative Medicine (2023) 4: 209–223
https://doi.org/10.1007/5584_2021_671

Published online: 6 November 2021

Check for updates

Treatment of Osteochondral Femoral Head Defect by Human Umbilical Cord Mesenchymal Stem Cell Sheet Transplantation: An Experimental Study in Rats

Thuan Minh Le, Ngoc Bich Vu, Phat Duc Huynh, and Phuc Van Pham

Abstract

Introduction Articular cartilage is limited in self-repair following injuries due to avascular, lymphatic, and nerve absence. Recent treatments for cartilage injuries, such as physical therapy, anti-inflammatory medication, chondrocyte implantation, and joint replacement, still have limitations. This study aimed to evaluate the treatment efficacy of human umbilical cord-derived mesenchymal stem cell sheet (UCMSCS) transplantation in rat models of the osteochondral femoral head defect.

Methods Models of osteochondral femoral head defect were produced in rats by drilling in order to reach the femoral bone tissue through the cartilage layer. Then, UCMSCS was implanted in the created cartilage lesion. The treatment efficacy was monitored by X-ray imaging. The cartilage regeneration was evaluated based on the hematoxylin and eosin staining, and proteoglycan accumulation was detected by staining Safranin O and Fast Green. The physiological, weight, or movement activity of rats were recorded during the treatment period.

Results UCMSCS transplantation showed positive effects on the cartilage regeneration in osteochondral femoral head defect grade 4 (according to ICRS score/grade). Particularly, after 12 weeks of implantation of UCMSCS, the defect was filled with hyaline cartilage-like cells and accumulated a large density of proteoglycan. The osteochondral

T. M. Le and P. D. Huynh
Laboratory of Stem Cell Research and Application, University of Science Ho Chi Minh City, Ho Chi Minh City, Vietnam

Vietnam National University Ho Chi Minh City, Ho Chi Minh City, Vietnam

N. B. Vu
Vietnam National University Ho Chi Minh City, Ho Chi Minh City, Vietnam

Stem Cell Institute, University of Science Ho Chi Minh City, Ho Chi Minh City, Vietnam

P. Van Pham (✉)
Laboratory of Stem Cell Research and Application, University of Science Ho Chi Minh City, Ho Chi Minh City, Vietnam

Vietnam National University Ho Chi Minh City, Ho Chi Minh City, Vietnam

Stem Cell Institute, University of Science Ho Chi Minh City, Ho Chi Minh City, Vietnam

Department of Animal Physiology and Biotechnology, Biology Faculty, University of Science Ho Chi Minh City, Ho Chi Minh City, Vietnam
e-mail: pvphuc@hcmus.edu.vn; phucpham@sci.edu.vn

defect score significantly increased in the treated rats compared to the untreated rats (11.67 ± 0.6 and 9.67 ± 0.6, respectively) ($p < 0.05$). The histological score also increased in treated rats compared to untreated rats (21.33 ± 1.53 vs. 18.00 ± 1.00) ($p < 0.0001$). The accumulation of proteoglycan was higher in treated rats (20.50 ± 2.23) than untreated rats (5.38 ± 0.36) ($p < 0.05$). There was no change in the physiological activities between treated and untreated rats recorded during the study.

Conclusion MSCS transplantation could promote regeneration in advanced cartilage injury.

Keywords

Mesenchymal stem cell sheet · Osteochondral femoral head defect · Platelet-rich plasma · Stem cell transplantation · Umbilical cord-derived mesenchymal stem cell

Abbreviations

FGF	Fibroblast growth factor
IGF	Insulin-like growth factor.
PDGF	Platelet-derived growth factor.
PRP	Platelet-rich plasma
TGF-b	Transforming growth factor beta.
UCMSCs	Umbilical cord-derived mesenchymal stem cells.
VEGF	Vascular endothelial growth factor.

1 Introduction

Articular cartilage lacks blood vessels, lymphatic, and nerve, as well as a harsh biomechanical environment. Therefore, it can hardly be repaired and regenerated after injury, especially for large lesions (Sophia Fox et al., 2009). The defect cartilage of traumatic and degenerative often leads to osteoarthritis (OA). These lesions are predicted as painful and attenuate for the patient (Widuchowski et al., 2007). OA affects about 3.3–3.6% of the global population and causes disability in approximately 43 million people. About 80% of the US population aged above 65 years have OA. In 2011, nearly one million patients with arthritis spent about USD 15 billion, hence making it the second most expensive disease in the USA (Bortoluzzi et al., 2018; Berenbaum et al., 2018).

To date, there are many used or developing methods for the treatment of cartilage damage, such as physical therapy, anti-inflammatory drugs, cartilage cell transplantation, or joint replacement surgery (Rönn et al., 2011; Perera et al., 2012; Gille et al., 2013). Chemotherapies seemly reduced some symptoms, while cartilage transplantation lacked suitable donation sources. Therefore, the demand for available cartilage tissues for treatment is high. Stem cell therapies are developed to treat this defect currently. These efforts included injections of mesenchymal stem cells into the lesion (Jo et al., 2014; Koh et al., 2015; Vega et al., 2015; Chimutengwende-Gordon et al., 2021), injection of MSCs in solution with hyaluronic acid hydrogel (Park et al., 2017; Buda et al., 2010), and the transplantation of aggregates of mesenchymal stem cells (Kondo et al., 2019). This study aimed to suggest and evaluate a new strategy to treat this disease using sheets of mesenchymal stem cells.

2 Materials and Methods

2.1 UCMSCS Preparation

Mesenchymal stem cells from the human umbilical cord (UCMSCs) are isolated, characterized, and expanded as the published protocols (Van Pham et al., 2016a; Van Pham et al., 2016b).

Platelet-rich plasma (PRP) was prepared according to the manufacturer's instruction of the 5PRP kit (Regenmedlab, HCMC, VN) (Fig. 1). Umbilical cord blood samples were collected from donors after birth with consent forms. The institutional ethical committee approved the blood collection manipulations and usage of the blood samples for study. These samples were rechecked with HIV-1/HIV-2, HBV, HCV, CMV, and HTLV contamination by real-time PCR in the laboratory. Only the samples that were negative

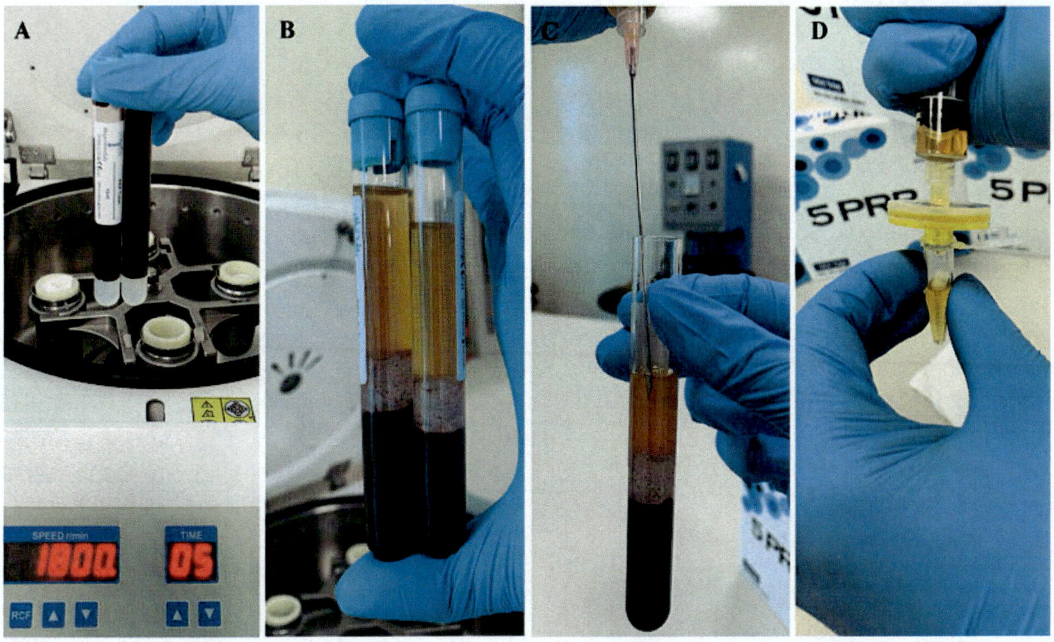

Fig. 1 Platelet-rich plasma preparation procedure from human umbilical cord blood. Blood is transferred into 5PRP tubes and centrifuged (**a**), blood after centrifugation (**b**), suspend platelets with half volume of plasma (**c**), filter platelet-rich plasma (**d**)

with these viruses were used for further experiments. A total of 24 mL of anticoagulated blood were collected into two tubes of 5PRP tubes (12 mL/tube) and centrifuged at 1.800 g for 5 min. Then, half of the upper plasma layer in the 5PRP tubes (about 3 mL) was removed. One half of plasma with a high concentration of platelets (about 3 mL) near to the separation gel of the tubes was collected and filtered through a 0.45 um filter to sterile the samples. The PRP solution was stored at −20 °C until use.

One million UCMSCs were suspended in 200 μL of MSCCult I medium (Regenmedlab, HCMC, VN) and 400 μL filtered PRP. Then, this suspension was poured to a well of 4-well plate. Immediately, a volume of 20 μL CaCl2 0.8 M was directly added to this suspension and kept standing until gelation to form sheets (about 15 min). Finally, these plates containing sheets were incubated for 5 days at 37 °C in a CO_2 incubator before the sheets (UCMSCSs) were collected for further experiments.

2.2 Animals

Healthy 6-month-old male rats of 350–450 g body weight were used to produce models of osteochondral defects. All manipulations on rats were approved by the Institutional Ethics Committee (Stem Cell Institute, University of Science, VNU-HCM, VN). An osteochondral defect at the crest of the rat femur with a diameter of 2 mm and depth of 1 mm was produced by drilling through the cartilage layer to the bone using a small drill. In each rat, only an osteochondral defect was caused in a leg.

All rat models of osteochondral defects were randomly gathered into the treated group and untreated group with 12 rats per group. All rats in the treated group were implanted with a UCMSCS, while rats in the untreated group did not receive any therapy. All mice of both groups were kept in the same care conditions of ventilated air cages for 3 months after implantation.

2.3 Evaluation of the Motion of Rats

The motion of rats was an indication of injury healing of the osteochondral defect. In this study, the motion of the hindlimb with the osteochondral defect was compared to the normal hindlimb in the same rat. The rats were placed on a flat surface to move freely in the area within 1 min. The number of steps that normal limb and injured limb could make per minute was counted. The physical performance efficiency (PE value) of the injured limb was calculated as a percentage ratio of the number of steps of the injured limb and the normal limb in 1 min.

2.4 Body Weight Change

After surgery, all rats in both groups were kept in clean conditions with 12-h lighting time per day. A total of 100 g of food was provided to each rat per day. The weight of rats will be recorded at days 0 and 3 and weeks 1, 2, 4, 8, and 12 after surgery.

2.5 X-Ray Examination

Rats are continuously anesthetized with isoflurane. After the anesthesia process, the rat was placed in an X-ray chamber. X-ray images were recorded at focal length 7.2, aperture 2.8, exposure time of 3 s, and 1 × 1-pixel resolution. Optical images of the femoral head were taken using the In-Vivo Xtreme II system (Bruker, USA).

2.6 Macroscopic Evaluation of Cartilage Damage

The femoral heads of the rats were collected and soaked in 0.9% saline solution. The femoral head structure images were recorded using the microscope at 2x magnification. Recovery of damaged femoral head cartilage was evaluated according to the International Cartilage Repair Society (ICRS) score (Hoemann et al., 2011) (Table 1) (Kazemi et al., 2017).

2.7 Histology Analysis

The femoral head was fixed in 10% paraformaldehyde (PFA) (Sigma-Aldrich, Louis St, MO) for 72 h, decalcified around 15 days in 10% disodium ethylenediaminetetraacetate (EDTA 2Na) (Sigma-Aldrich, St. Louis, MO) at 4 °C, and then immersed in 30% sucrose solution until the tissues sink. The tissue was embedded in paraffin and cut into 3 μm sections. The tissue samples were then stained with hematoxylin and eosin (H&E) and Safranin O/Fast Green dye. Sections were examined under a light microscope. Histological results were assessed according to the O'Driscoll histological grading scale (Kazemi et al., 2017) (Table 2).

2.8 Quantify GAGs Content

Tissue samples after staining by Safranin O and Fast Green were quantified GAG content using ImageJ software.

2.9 Statistical Analysis

In this study, data were analyzed by the software GraphPad Prism 8. Statistical analysis was performed by t-test. P-value <0.05 was considered significant. All data were presented as mean ± SD.

3 Results

3.1 UCMSC Sheets

Immediately after fibrin has been seeded, UCMSCs were round and distributed throughout the fibrin scaffold (Fig. 2a). After a day of culture,

Table 1 The International Cartilage Repair Society (ICRS) score

Categories	*Score*
Degree of defect repair	
In level with surrounding cartilage	4
75% repair of defect depth	3
50% repair of defect depth	2
25% repair of defect depth	1
0% repair of defect depth	0
Integration to border zone	
Complete integration with surrounding cartilage	4
Demarcating border <1 mm	3
¾ of graft integrated, ¼ with a notable border >1 mm width	2
½ of graft integrated with surrounding cartilage, ½ with a notable border >1 mm	1
From no contact to ¼ of graft integrated with surrounding cartilage	0
Macroscopic appearance	
Intact smooth surface	4
Fibrillated surface	3
Small, scattered fissures or cracks	2
Several, small or few but large fissures	1
Total degeneration of grafted area	0
Overall repair assessment	
Grade I: Normal	12
Grade II: Nearly normal	11–8
Grade III: Abnormal	7–4
Grade IV: Severely abnormal	3–1

UCMSCs were elongated and had a spindle shape (Fig. 2b). Cells continued to proliferate with higher density on the fibrin scaffold in the following days, which could be detected by observations under an inverted microscope (Fig. 2c). The UCMSCS had a flattened cylinder shape when cultured in a 4-well disk (Fig. 2d). Following its separation from the disk, the stem cell sheet retained its morphology (Fig. 2e).

3.2 Rat Osteochondral Defect Model

An osteochondral defect was caused by a drilling method with a diameter of 2 mm and depth of 1 mm. The defect area displayed the rough surface (Fig. 3d). X-ray evaluation showed that the defect site was an area of unusually low density (Fig. 3f). Indeed, the drilling method removed the surface cartilage as well as the bone under the cartilage layer (Fig. 3e).

3.3 Effects of UCMSCS Transplantation on Models

3.3.1 Survival, Weight, and Movement of Rats

For 12 weeks, all rats in both groups (untreated and treated) survived. One week after surgery, the weight of the rats in both groups decreased but then gradually increased after 2 weeks. There were no significant differences in the reduction of body weight between the treated and untreated groups ($1.0 \pm 0.3\%$ g vs. $1.5 \pm 1.03\%$ for untreated rats vs. treated rats, respectively) ($p > 0.05$) (Fig. 4a).

In the 12th week, the body weight of rats in the untreated group increased $2.4 \pm 1.2\%$ compared to that of the first week, while this value was $9.3 \pm 4.3\%$ in the treated rats. However, the statistical analysis showed no significant difference in the increasing rate between treated and untreated rats ($p > 0.05$).

Table 2 O'Driscoll histological grading scale

Characteristics	*Score*
Nature of predominant tissue	
Cellular morphology	
Hyaline articular cartilage	4
Incompletely differentiated mesenchyme	2
Fibrous tissue or bone	0
Safranin O staining of the matrix	
Normal or nearly normal	3
Moderate	2
Slight	1
None	0
Structural characteristics	
Surface regularity	
Smooth and intact	3
Superficial horizontal lamination	2
Fissures 25% to 100% of the thickness	1
Severe disruption including fibrillation	0
Structural integrity	
Normal	2
Slight disruption including cysts	1
Severe disintegration	0
Thickness	
100% of normal adjacent cartilage	2
50% to 100% of normal cartilage	1
0% to 50% of normal cartilage	0
Bonding to the adjacent cartilage	
Bonded at both ends of graft	2
Bonded at one end or partially at both ends	1
Not bonded	0
Freedom from cellular changes of degeneration	
Hypocellularity	
Normal cellularity	3
Slight hypocellularity	2
Moderate hypocellularity	1
Severe hypocellularity	0
Chondrocyte clustering	
No clusters	2
<25% of the cells	1
25–100% of the cells	0
Freedom from degenerative changes in adjacent cartilage	
Normal cellularity, no clusters, normal staining	3
Normal cellularity, mild clusters, slight staining	2
Mild or moderate hypocellularity, slight staining	1
Severe hypocellularity, poor or no staining	0
Total	**24**

The results showed that during 12 weeks, the physical performance efficiency (PE) of treated limbs in the treated rats was higher in the untreated limbs of untreated rats. However, the statistical analysis showed that these increases were not significant ($p > 0.05$). After surgery,

Fig. 2 UCMSC sheet morphology. The distribution and morphology of UCMSCS (**a**) immediately after being seeded after 1 day (**b**) and 5 days (**c**) culture on fibrin scaffold (**d**). The shape of the UCMSCS when cultured on a 4-well plate and before being transplanted into the cartilage defect (**e**)

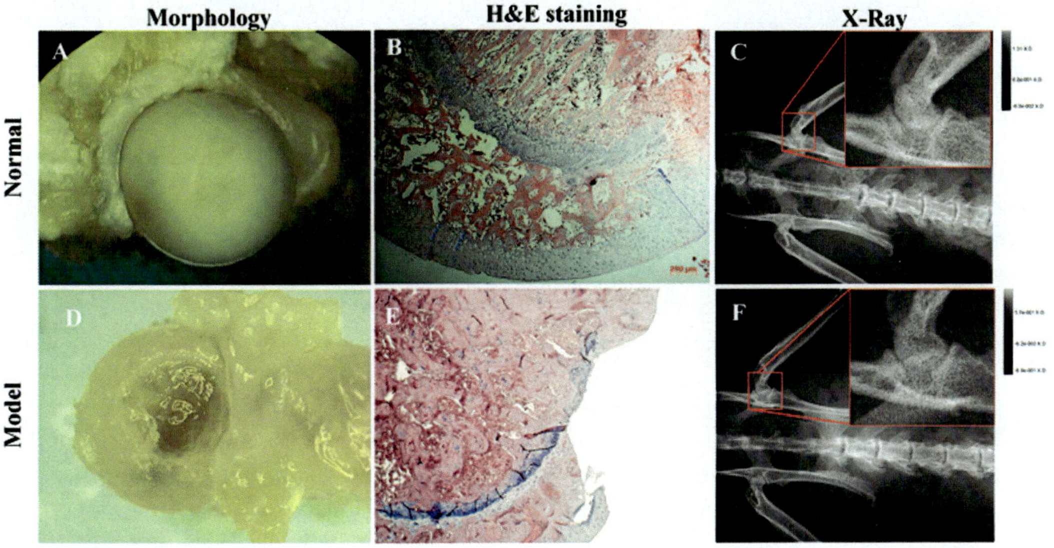

Fig. 3 Structure of femoral heads in normal rats and osteochondral defect
Characterization of normal and osteochondral heads in rats. Images of macroscopic evaluation (**a** and **d**), cartilage-bone tissue by staining HE (**b** and **e**), and X-ray (**c** and **f**)

the EP values were 18.3 ± 5.8% in the untreated rats compared to 20.0 ± 6.0% in treated rats (Fig. 4). On day 4 after treatment, this value was 34.2 ± 6.7% in the treated rats and 32.5 ± 7.5% in the untreated rats ($p > 0.05$). After 1 week, the EP values increased to 64.2 ± 9.0% in the untreated rat and 66.7 ± 8.9% in the treated rats. From the second week to the 12th week, PE values were 100% for both groups and like normal limbs.

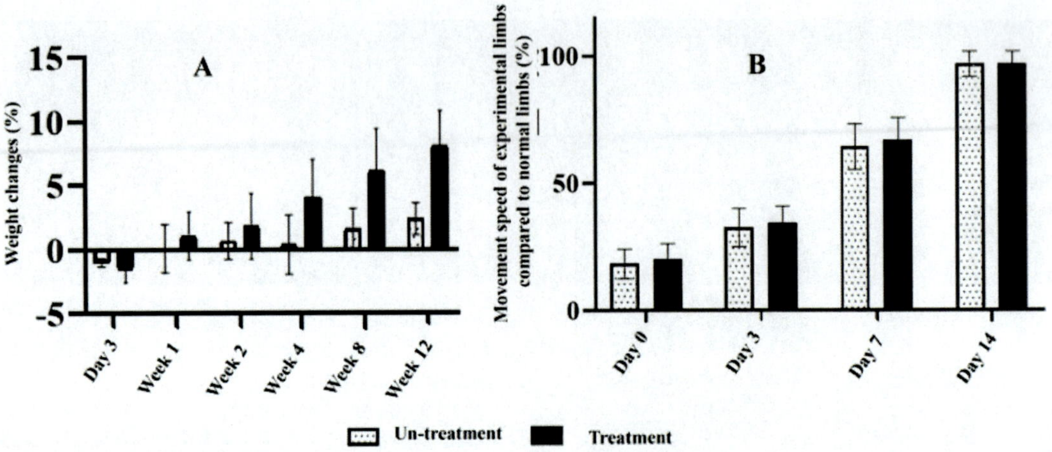

Fig. 4 The changes in body weight and physical performance efficiency of limbs in both treated and untreated rats. (**a**) Treated rats gained body weight slightly more than untreated rats; (**b**) there was no difference in the physical performance efficiency of untreated and treated limbs at day 0, 3, 7, and 14

3.3.2 Macroscopic Findings

The injury was filled with cartilage-like tissue in the treatment group 2 weeks after surgery, and this tissue was well integrated with host tissues (Fig. 5e). In the untreated rats, the osteochondral defect was also filled with a neo-tissue, but the surface of this tissue was rough and not integrated with the native tissue (Fig. 5a).

In the 4th week, in the treated rats, the new tissues at the defect site were milky in color and well integrated with the natural tissue area, and the lesion boundary was faded (Fig. 5f). The osteochondral defect was filled with thicker new tissue in the untreated rat and started to integrate into normal tissues. However, the margins between the new tissue and native cartilage were clear (Fig. 5b).

At 8 weeks after surgery, the new tissues in both groups were opaque white and integrated into the natural tissue. In the untreated group, the defect was filled with a clear white new tissue that was so different from native tissues (Fig. 5c), while in the treatment group, the size of the defect was clearly reduced with the new tissue similar to native cartilage tissue (Fig. 5g).

At 12 weeks after the surgery, the osteochondral defect in the treatment group was filled with new tissue similar to the native cartilage tissue that well-integrated the normal cartilage, significantly reduced lesion area more than the untreated rats (Fig. 5h).

The results also showed that ICRS microscopic scores of the treated rats were consistently higher in the untreated rats at all times of sampling. At the 12th week, the ICRS scores were 11.67 ± 0.6 in treated rats and 9.67 ± 0.6 in untreated rats ($p < 0.05$).

3.3.3 X-Ray Images

At 2 weeks after the surgery, the X-ray images showed that the density of the osteochondral lesion was lower than the surrounding normal tissue and the joint cavity narrowed in both untreated rats and treated rats (Fig. 6). At 4, 8, and 12 weeks, both treated and untreated groups showed a gradual increase in the intensity of the lesion area. However, at 8th and 12th weeks, the femoral head structure displayed like a normal femoral head in the treated rats (Fig. 6g, h). In contrast, in untreated rats, there was still an abnormal structure and cavity at the 12th week (Fig. 6d).

3.3.4 Histological Analysis

At 2 and 4 weeks after surgery, there was no remarkable difference between the two groups after 2 weeks. Both groups showed disarrayed and damaged cells. The predominant cellular

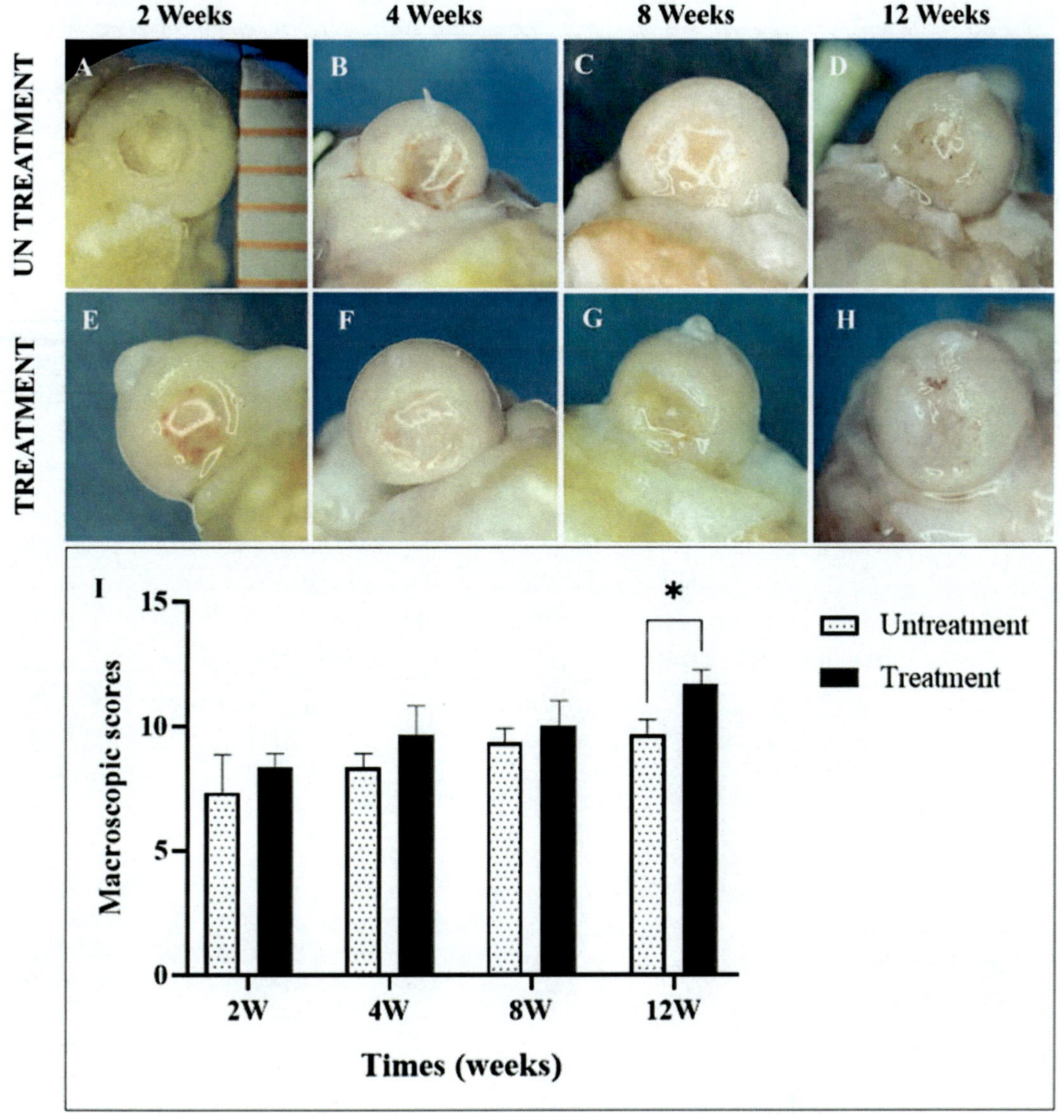

Fig. 5 Macroscopic findings of osteochondral defects in untreated rats and treated rats. Although new tissues formed at defects in untreated and treated rats, there were differences in structures between treated and untreated rats (**a–h**). ICRS scores were determined in untreated rats and treated rats (**i**). *: $p < 0.05$

composition was elongated, spindle-shaped cells (Fig. 7b, f, j, m). The repair tissue was still not stable and was poorly integrated with the native tissue in both groups 2 weeks postoperatively. The control group showed negative for Fast Green and Safranin O at 2 and 4 weeks after surgery (Fig. 7d, h).

At 8 weeks postoperatively, the treated group showed slight improvement with the appearance of some cartilage cells (Fig. 8e). The untreated group showed the clear formation of scarring fibrous tissue and collagen fiber mesh (Fig. 8b). In the untreated rats, the osteochondral defect lesion was filled with fibrous connective tissue and fibroblast cells. This new tissue contained elongated fibroblast cells and slightly positive staining Safranin O (Fig. 8c). However in the treatment group, the new tissue contained

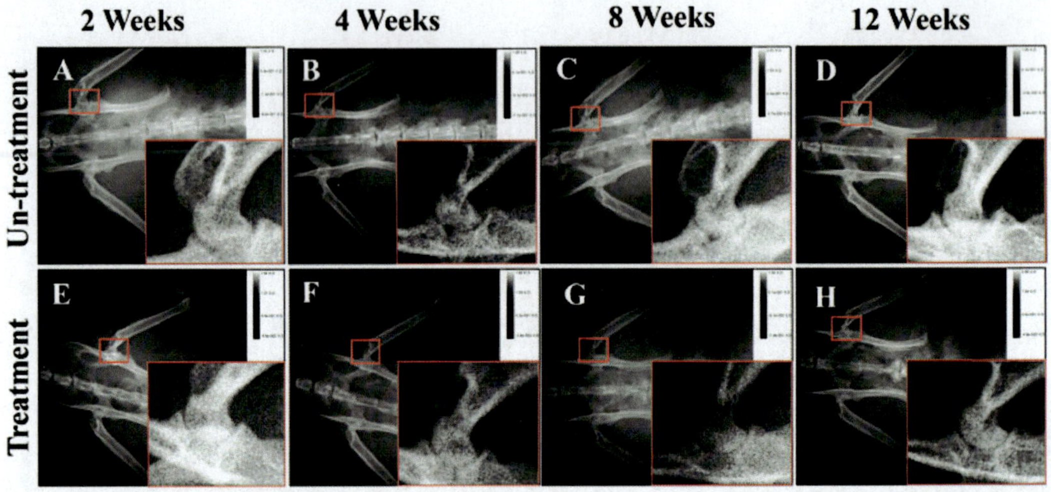

Fig. 6 X-ray images for osteochondral defects were recorded in untreated and treated rats at second, fourth, eighth, and 12th weeks

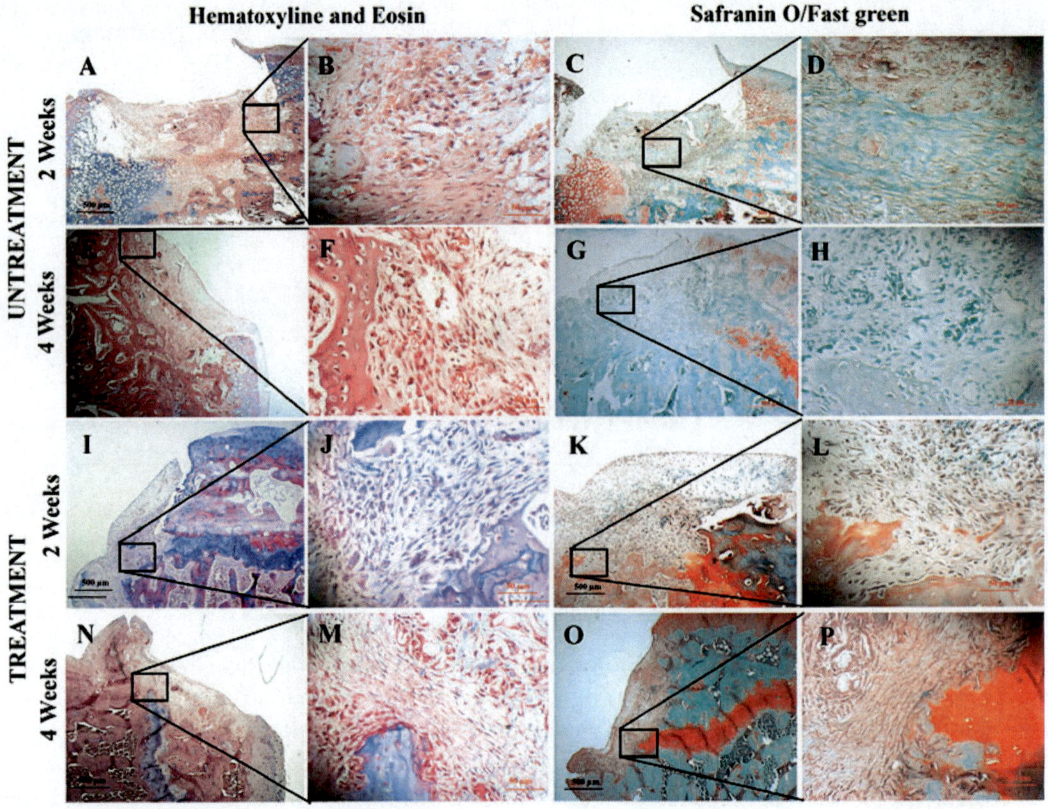

Fig. 7 Results of histological analysis of femoral heads using hematoxylin and eosin (HE), Safranin O, and Fast Green staining at the second and fourth weeks after treatment in untreated and treated rats. Magnification 10X for **a**, **e**, **i**, **n**, **c**, **g**, **k**, and **o** and 40X for **b**, **f**, **j**, **m**, **d**, **h**, **l**, and **p**

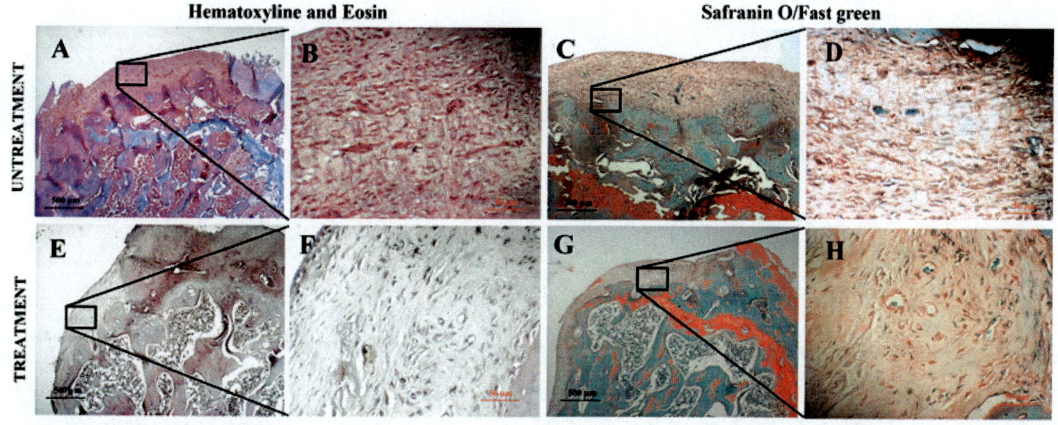

Fig. 8 Results of histological analysis of femoral heads using hematoxylin and eosin (HE), Safranin O, and Fast Green staining at eighth week after treatment in untreated and treated rats. Magnification 10X for **a**, **e**, **c**, and **g** and 40X for **b**, **f**, **d**, and **h**

fibroblasts and cell-like chondrocytes in lacunae. These cells were positive with Safranin O (Fig. 8h).

At 12 weeks after surgery, the new tissue formed in the treatment group displayed the same characteristics as immature cartilage (Fig. 9f, k, respectively). Almost all cells in this tissue were round and nested in lacunae and glassy-looking matrix. Most cells seem to be isolated and randomly distributed in the matrix (Fig. 9e, f). Few cells come together as an isogenous group. The osteochondral defect showed positive for Safranin O staining (Fig. 9g, h). In the untreated group, cellular replenishment bulges out of the curved boundary of the articular cartilage layer, as seen in the normal femoral head. The new tissue showed that cells' morphology was round and nested in lacunae as seen in mature cartilage (Fig. 9b). However, the few cells were presented in this tissue.

O'Driscoll's guidance for evaluating histological results in repairing osteochondral defects was applied in this study. Results showed a difference in scores between treatment and control groups at 8 and 12 weeks after surgery. At 8 weeks, the control group scored (8.33 ± 0.58) and the treatment group scored (10.33 ± 0.58) ($p < 0.05$). At 12 weeks, the control group had a score of (18.00 ± 1.00) and the treatment group had a score of (21.33 ± 1.523) ($p < 0.05$) (Fig. 10a).

At 2, 4, and 8 weeks, proteoglycan protein was lowly expressed; however, the expression gradually increased to 12 weeks (20.50 ± 2.23 in the treated group compared to 5.38 ± 0.36 in the untreated group) (Fig. 10b).

4 Discussion

Stem cell therapies have shown promising results for repairing cartilage defects. However, at present, not many studies apply mesenchymal stem cells combination with fibrin for cartilage repair. This study demonstrated the feasibility of using human umbilical cord-derived mesenchymal stem cells to repair severe cartilage damage.

H. Katagiri et al. investigated the seriousness of severe cartilage lesions on rats. They showed poor cartilage regeneration with lesions larger than 1.4 mm in diameter and 1 mm in depth (Katagiri et al., 2017). Therefore, this study performed a femoral head lesion with a diameter of 2 mm and depth of 1 mm. UCMSCS with a 3 mm diameter and 1 mm thickness is grafted right at the lesion site.

The results showed that 100% of rats could survive during the experiment. This result

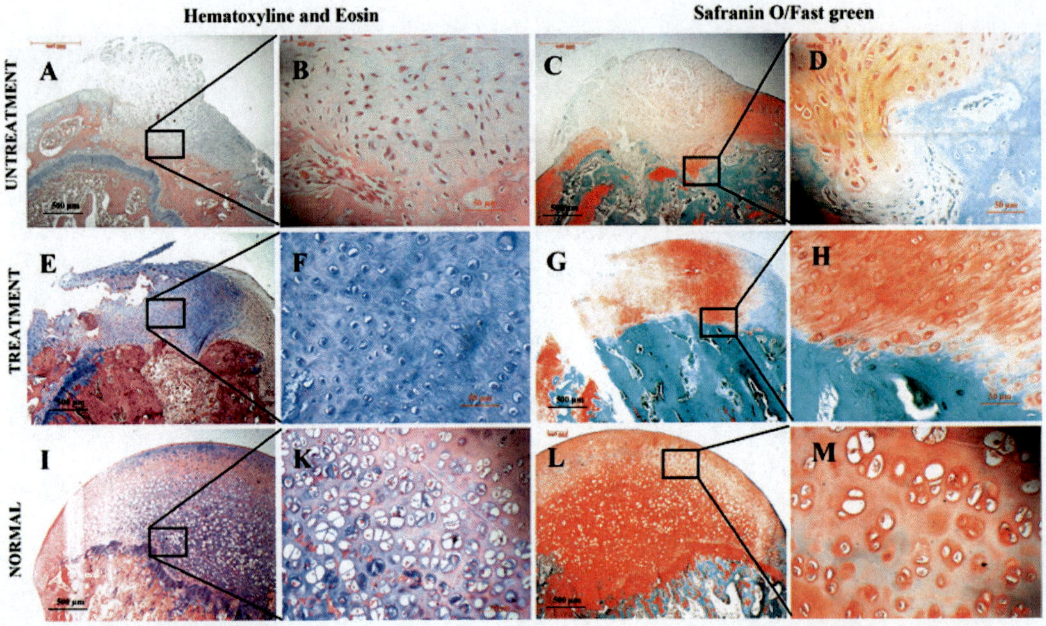

Fig. 9 Results of histological analysis of femoral heads using hematoxylin and eosin (HE), Safranin O, and Fast Green staining at 12th week after treatment in untreated and treated rats. Magnification 10X for **a**, **e**, **i**, **c**, **g**, and **l** and 40X for **b**, **f**, **k**, **d**, **h**, and **m**

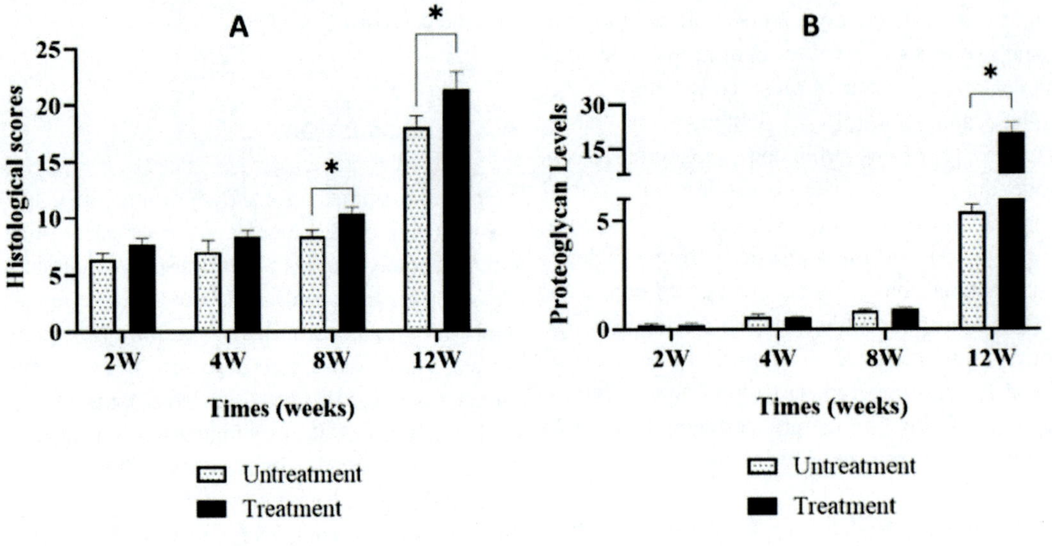

Fig. 10 Histological scores and proteoglycan accumulation levels at osteochondral defects in treated and untreated rats. (**a**) O'Driscoll histology scores at 2, 4, 8, and 12 weeks after surgery. (**b**) Proteoglycan levels at 2, 4, 8, and 12 weeks after surgery using ImageJ software. *: $p < 0.05$, ***: $p < 0.0001$

demonstrated that the therapy did not affect rats' survival. On the other hand, rats' body weight and mobility did not differ between untreated and treated groups. Hence, basic physiological indicators were not significantly changed if the UCMSCS was used.

In the next assays, microscopic, X-ray images, and histology results showed UCMSCS transplantation could provide better cartilage regeneration than non-treatment. Although the physical performance efficiency was not different between treated rats and untreated rats, the X-ray and histological analysis images clearly showed that UCMSCS transplantation drove the healing toward cartilage regeneration. In untreated rats, the healing process was scarring. Indeed, the new tissue that formed and filled the defects in rats transplanted with UCMSCS displayed that the histological structure was similar to the cartilage tissue.

UCMSCS transplantation also boosted the regeneration process faster than the untreated rats, with higher histological scores in treated rats at weeks 4 and 8. Besides, X-ray images showed that the healing process in the treated group was more stable than in the untreated group. In particular, at week 12 after surgery, the tissue recovered similar to hyaline cartilage. Kazami et al. also investigated BMMSC with fibrin to effectively repair cartilage tissue in dogs at weeks 16 and 24 after surgery (Kazemi et al., 2017). Xie et al. also showed the effectiveness of MSCs with PRP in the treatment of cartilage damage. Macroscopic and histological scores were higher in the treatment group than in the control group at week 12 after surgery (Oprea et al., 2003). Another study by JC Lee et al. that used amniotic membrane-derived mesenchymal stem cells and PRP for damaged cartilage repairing showed the results in macroscopic and histological characteristics of cartilage tissue (Oprea et al., 2003).

These results could come from the positive effects of UCMSCs and both PRP. PRP contains many growth factors such as platelet-derived growth factor (PDGF), insulin-like growth factor (IGF), basic fibroblast growth factor (bFGF), transforming growth factor (TGF-b), and vascular endothelial growth factor (VEGF). These factors stimulate chondral cell proliferation and prepared matrix (Fortier et al., 2011; Sun et al., 2010). Particularly, PRP can be enabled easily to form a fibrin biological scaffold by thrombin or calcium. Fibrin is an important factor in supporting stem cells for repairing the damage. Fibrin has created a three-dimensional environment that supports proliferation, differentiation, and secretion of cytokines of mesenchymal stem cells. In addition, fibrin promotes the migration of mesenchymal stem cells to wound sites and the proliferation of cells (Oprea et al., 2003). A number of growth factors are also released by MSCs such as FGF, PDGF, TGF-b, IGF, and VEGF (Alsousou et al., 2009). TGF-b is known to help stimulate collagen II and proteoglycan production (Shapiro et al., 1993). PDGF is a factor that allows cartilage cells to remain in the state as cartilage cells, stimulating proliferation and proteoglycan synthesis (Schmidt et al., 2006). The remaining factors play a similar role in the formation and development of cartilage (Stewart et al., 2007; Kim et al., 2011).

The study also showed that the important effects of UCMSCS transplantation led the healing toward tissue regeneration and reduced the scarring. Published studies showed that MSCs could attenuate the scarring during wound healing (Jackson et al., 2012). In a review about the employment of MSCs to treat hypertrophic and keloid scars, Bojanic et al. (2021) also showed that MSC transplantation could be used to treat scars. Xie et al. (2021) suggested that MSCs can inhibit cell proliferation and migration and suppress extracellular matrix synthesis in hypertrophic and keloid fibroblasts (Xie et al., 2021).

5 Conclusion

Transplantation of UCMSCS is a new and promising therapy for repairing osteochondral defects. Fibrin from PRP is a natural scaffold that can be easily prepared and is suitable for cell adhesion and proliferation. In this study, the transplantation of UCMSCSs from UCMSCs and PRP-derived fibrin showed good effects of repairing cartilage damage, attenuating the scarring, and producing a new tissue like hyaline cartilage 12 weeks of transplantation in rat models. The proteoglycan accumulation of the cartilage matrix is also improved in the treated group. These findings suggested a new strategy to treat osteochondral defects using UCMSCSs.

Acknowledgements This research is funded by National University Ho Chi Minh City (VNU-HCM) under grant number NV2018-18-02.

Conflict of Interest The authors declare that there is no conflict of interest regarding the publication of this paper.

References

Alsousou, J., Thompson, M., Hulley, P., Noble, A., & Willett, K. (2009). The biology of platelet-rich plasma and its application in trauma and orthopaedic surgery: A review of the literature. *The Journal of Bone and Joint Surgery British Volume, 91*(8), 987–996.

Berenbaum, F., Wallace, I. J., Lieberman, D. E., & Felson, D. T. (2018). Modern-day environmental factors in the pathogenesis of osteoarthritis. *Nature Reviews Rheumatology, 14*(11), 674–681.

Bortoluzzi, A., Furini, F., & Scirè, C. A. (2018). Osteoarthritis and its management-epidemiology, nutritional aspects and environmental factors. *Autoimmunity Reviews, 17*(11), 1097–1104.

Buda, R., Vannini, F., Cavallo, M., Grigolo, B., Cenacchi, A., & Giannini, S. (2010). Osteochondral lesions of the knee: A new one-step repair technique with bone-marrow-derived cells. *The Journal of Bone and Joint Surgery American Volume, 92*(Suppl 2), 2–11.

Chimutengwende-Gordon, M., Ahmad, M. A., Bentley, G., Brammah, J., Carrington, R., Miles, J., et al. (2021). Stem cell transplantation for the treatment of osteochondral defects of the knee: Operative technique for a single-stage transplantation procedure using bone marrow-derived mesenchymal stem cells. *The Knee, 28*, 400–409.

Fortier, L. A., Barker, J. U., Strauss, E. J., McCarrel, T. M., & Cole, B. J. (2011). The role of growth factors in cartilage repair. *Clinical Orthopaedics and Related Research®, 469*(10), 2706–2715.

Gille, J., Behrens, P., Volpi, P., de Girolamo, L., Reiss, E., Zoch, W., et al. (2013). Outcome of autologous matrix induced Chondrogenesis (AMIC) in cartilage knee surgery: Data of the AMIC registry. *Archives of Orthopaedic and Trauma Surgery, 133*(1), 87–93.

Hoemann, C., Kandel, R., Roberts, S., Saris, D. B., Creemers, L., Mainil-Varlet, P., et al. (2011). International cartilage repair society (ICRS) recommended guidelines for histological endpoints for cartilage repair studies in animal models and clinical trials. *Cartilage, 2*(2), 153–172.

Jackson, W. M., Nesti, L. J., & Tuan, R. S. (2012). Mesenchymal stem cell therapy for attenuation of scar formation during wound healing. *Stem Cell Research & Therapy, 3*(3), 20.

Jo, C. H., Lee, Y. G., Shin, W. H., Kim, H., Chai, J. W., Jeong, E. C., et al. (2014). Intra-articular injection of mesenchymal stem cells for the treatment of osteoarthritis of the knee: a proof-of-concept clinical trial. *Stem Cells (Dayton, Ohio), 32*(5), 1254–1266.

Katagiri, H., Mendes, L., & Luyten, F. (2017). Definition of a critical size osteochondral knee defect and its negative effect on the surrounding articular cartilage in the rat. *Osteoarthritis and Cartilage, 25*(9), 1531–1540.

Kazemi, D., Shams Asenjan, K., Dehdilani, N., & Parsa, H. (2017). Canine articular cartilage regeneration using mesenchymal stem cells seeded on platelet rich fibrin: Macroscopic and histological assessments. *Bone & Joint Research, 6*(2), 98–107.

Kim, J. H., Lee, M. C., Seong, S. C., Park, K. H., & Lee, S. (2011). Enhanced proliferation and chondrogenic differentiation of human synovium-derived stem cells expanded with basic fibroblast growth factor. *Tissue Engineering Part A, 17*(7–8), 991–1002.

Koh, Y. G., Choi, Y. J., Kwon, S. K., Kim, Y. S., & Yeo, J. E. (2015). Clinical results and second-look arthroscopic findings after treatment with adipose-derived stem cells for knee osteoarthritis. *Knee Surgery, Sports Traumatology, Arthroscopy: Official Journal of the ESSKA., 23*(5), 1308–1316.

Kondo, S., Nakagawa, Y., Mizuno, M., Katagiri, K., Tsuji, K., Kiuchi, S., et al. (2019). Transplantation of aggregates of autologous synovial mesenchymal stem cells for treatment of cartilage defects in the femoral condyle and the femoral groove in microminipigs. *The American Journal of Sports Medicine, 47*(10), 2338–2347.

Oprea, W. E., Karp, J. M., Hosseini, M. M., & Davies, J. E. (2003). Effect of platelet releasate on bone cell migration and recruitment in vitro. *Journal of Craniofacial Surgery, 14*(3), 292–300.

Park, Y.-B., Ha, C.-W., Lee, C.-H., & Park, Y.-G. (2017). Restoration of a large osteochondral defect of the knee using a composite of umbilical cord blood-derived mesenchymal stem cells and hyaluronic acid hydrogel: a case report with a 5-year follow-up. *BMC Musculoskelet Disord, 18*(1), 59.

Perera, J., Gikas, P., & Bentley, G. (2012). The present state of treatments for articular cartilage defects in the knee. *The Annals of The Royal College of Surgeons of England, 94*(6), 381–387.

Rönn, K., Reischl, N., Gautier, E., & Jacobi, M. (2011). Current surgical treatment of knee osteoarthritis. *Art, 2011*.

Schmidt, M., Chen, E., & Lynch, S. (2006). A review of the effects of insulin-like growth factor and platelet derived growth factor on in vivo cartilage healing and repair. *Osteoarthritis and Cartilage, 14*(5), 403–412.

Shapiro, F., Koide, S., & Glimcher, M. J. (1993). Cell origin and differentiation in the repair of full-thickness defects of articular cartilage. *JBJS, 75*(4), 532–553.

Sophia Fox, A. J., Bedi, A., & Rodeo, S. A. (2009). The basic science of articular cartilage: Structure, composition, and function. *Sports Health, 1*(6), 461–468.

Stewart, A. A., Byron, C. R., Pondenis, H., & Stewart, M. C. (2007). Effect of fibroblast growth factor-2 on equine mesenchymal stem cell monolayer expansion and chondrogenesis. *American Journal of Veterinary Research, 68*(9), 941–945.

Sun, Y., Feng, Y., Zhang, C., Chen, S., & Cheng, X. (2010). The regenerative effect of platelet-rich plasma on healing in large osteochondral defects. *International Orthopaedics, 34*(4), 589–597.

Van Pham, P., Truong, N. C., Le, P. T., Tran, T. D., Vu, N. B., Bui, K. H., et al. (2016a). Isolation and proliferation of umbilical cord tissue derived mesenchymal stem cells for clinical applications. *Cell and Tissue Banking, 17*(2), 289–302.

Van Pham, P., Bich, N. V., & Phan, N. K. (2016b). Umbilical cord-derived stem cells (Modulatist TM) show strong immunomodulation capacity compared to adipose tissue-derived or bone marrow-derived mesenchymal stem cells. *Biomedical Research and Therapy, 3*(6), 1–10.

Vega, A., Martín-Ferrero, M. A., Del Canto, F., Alberca, M., García, V., Munar, A., et al. (2015). Treatment of knee osteoarthritis with allogeneic bone marrow mesenchymal stem cells: A randomized controlled trial. *Transplantation, 99*(8), 1681–1690.

Widuchowski, W., Widuchowski, J., & Trzaska, T. (2007). Articular cartilage defects: Study of 25,124 knee arthroscopies. *The Knee, 14*(3), 177–182.

Xie, F., Teng, L., Xu, J., Lu, J., Zhang, C., Yang, L., et al. (2021). Adipose-derived mesenchymal stem cells inhibit cell proliferation and migration and suppress extracellular matrix synthesis in hypertrophic-scar and keloid fibroblasts. *Experimental and Therapeutic Medicine, 21*(2), 139.

Adv Exp Med Biol - Innovations in Cancer Research and Regenerative Medicine (2023) 4: 225–226
https://doi.org/10.1007/978-3-031-38613-8

Index